D1620444

Edited by
Jane Wu

Posttranscriptional Gene Regulation

Related Titles

Meyers, R. A. (ed.)

Epigenetic Regulation and Epigenomics

2012
ISBN: 978-3-527-32682-2

Stamm, S., Smith, C., Lührmann, R. (eds.)

Alternative pre-mRNA Splicing
Theory and Protocols

2012
ISBN: 978-3-527-32606-8

Harbers, M., Kahl, G. (eds.)

Tag-based Next Generation Sequencing

2012
ISBN: 978-3-527-32819-2

Snustad, D. P., Simmons, M. J.

Genetics International Student Version

2011
ISBN: 1-118-09242-2

Caetano-Anollés, G.

Evolutionary Genomics and Systems Biology

2010
ISBN: 978-0-470-19514-7

Edited by Jane Wu

Posttranscriptional Gene Regulation

RNA Processing in Eukaryotes

The Editor

Dr. Jane Wu
Northwestern University
Center for Genetic Medicine
303 E. Superior St
Chicago, IL 60611
USA

Cover
Illustrated on the cover is a simplified summary of our current understanding of eukaryotic gene regulation. Most of our knowledge about gene expression and regulation has come from studies of protein coding genes ("Coding"), although nonprotein coding genes ("Noncoding") may account for the vast majority of eukaryotic genomes. Following transcription from DNA (depicted as the double-stranded helix), precursor noncoding RNA gene products (diagrammed as the stem-looped structure) or messenger RNA precursors (pre-mRNAs) undergo multistep posttranscriptional RNA processing before they become functionally active in engaging their target genes or serving as messengers (mRNAs) to direct polypeptide synthesis in the protein synthesis machinery (Ribosomes). Each step of these complex gene expression processes (depicted as green triangles) is under intricate regulation to meet the cell's needs in adapting to its constantly changing environment. (Contributed by Mengxue Yang and Jane Y. Wu.)

Library of Congress Card No.: applied for

British Library Cataloguing-in-Publication Data
A catalogue record for this book is available from the British Library.

Bibliographic information published by the Deutsche Nationalbibliothek
The Deutsche Nationalbibliothek lists this publication in the Deutsche Nationalbibliografie; detailed bibliographic data are available on the Internet at <http://dnb.d-nb.de>.

Wiley-Blackwell is an imprint of John Wiley & Sons, formed by the merger of Wiley's global Scientific, Technical, and Medical business with Blackwell Publishing.

Print ISBN: 978-3-527-32202-2
ePDF ISBN: 978-3-527-66542-6
ePub ISBN: 978-3-527-66541-9
mobi ISBN: 978-3-527-66540-2
oBook ISBN: 978-3-527-66543-3

Cover Design Formgeber, Eppelheim
Typesetting Toppan Best-set Premedia Limited, Hong Kong
Printing and Binding Markono Print Media Pte Ltd, Singapore·

Printed on acid-free paper

Contents

Foreword

RNA is the "working substance" of genetics, a multipart actor playing almost every part in the expression of genetic information. The central dogma of molecular biology places RNA centrally between the cell's genetic library of DNA, and its catalytic engines the proteins, underscoring RNA's key importance in the flux of information. Yet the deceptive simplicity of the original scheme as set out by Crick greatly underrepresents the myriad functions of RNA, and its critical role in orchestrating the almost impossibly complex regulatory networks of the cell.

There are many reasons why RNA plays such a big part in the genetic life of the cell. The chemical nature of DNA has been selected to be relatively immutable to change; it has a rather staid, stay-at-home character. RNA by contrast is its rather racy cousin that is into everything. The 2′-hydroxyl group of RNA adds immeasurably to its chemical reactivity and structural repertoire. Moreover, as a formally single-stranded molecule, RNA can fold up into complex structures that bind ligands with huge selectivity (and there is a whole dimension of genetic regulation that is based on riboswitch function) and even exhibit catalytic activity. These properties were most probably the key to the early evolution of life, and the strongest indicator of this is telegraphed in the catalysis of peptidyl transfer by RNA in the modern ribosome. While most catalysis in the cell is now performed by proteins with their greater chemical flexibility, RNA has taken up most of the remaining genetic functions and run with them. The ancient role of RNA may be one of the reasons why it occupies such a central role in contemporary biology.

These additional functions have revealed themselves over a long period. I am old enough to remember the tremendous surprise we all felt when it transpired that eukaryotic genes were interrupted by introns, requiring splicing processes accurate to the nucleotide and a hugely complex apparatus to achieve that. This machine, the spliceosome, is a dynamic assembly of functional RNA and proteins, yet at its heart is probably at least in part a ribozyme. Unsurprisingly, splicing is precisely and dynamically regulated in a coordinated way in the cell, and alternative splicing events can increase genetic coding power by enabling exon skipping that is important in development.

Another unexpected development was the discovery that specific nucleotides could be chemically changed to "edit" the genetic content of the RNA. The classic example is the deamination of adenosine to generate inosine by the enzyme adenosine deaminases acting on RNA (ADAR). Of course, the high level of nucleotide modification in the transfer RNAs (tRNAs) should indicate that the chemical alteration of "plain vanilla" RNA is likely to be considerable. In archaea and eukaryotes, RNA is subject to site-specific methylation and pseudouridylation by relatively small RNA–protein machines, the box C/D and H/ACA small nucleolar ribonucleoprotein particles (snoRNPs). These exhibit an ordered assembly that begins with the binding of a protein that stabilizes a ubiquitous structural motif in RNA called the kink turn, and ends when the methylase or uridine isomerase binds to the complex.

Transcription of RNA is tightly regulated but does not have the same checks on accuracy that DNA replication requires. There is no major requirement for repair processes for RNA; it lives by a more throwaway philosophy. And that is just what happens. In nonsense-mediated decay, RNA containing a premature termination codon is selectively degraded. This process is also used to regulate the synthesis of RNA-binding proteins. At the other end of the central dogma, translation is also regulated, and remarkably this can be involved in learning and memory.

Nobody can be unaware of the revolution that has swept through RNA biology with the discovery of gene regulation mediated by small RNA species. While the fraction of the genome that encodes proteins is really quite small, it transpires that much more of the genome encodes regulatory RNAs that are transcribed from untranslated regions and introns. Presumably for decades, RNA investigators were running these species off the ends of their gels, yet these noncoding RNA species have now assumed a vast importance and have become an industry both as a regulatory mechanism and a powerful laboratory tool. They could even yet be the source of therapeutic agents, although as ever the big problem there remains delivery. Regulatory RNA networks of great complexity will ultimately require the methods of systems biology and mathematical modeling to be fully understood.

Given the central importance of RNA in the life of the cell, it is obvious that defects will occur with serious consequences for function. These can lead to important diseases, exemplified by the role of noncoding RNA species in neurodevelopment and neurodegeneration. The role of RNA in disease is a significant issue in human health that we neglect at our peril. I suspect we have only begun to scratch the surface of this topic and await exciting developments.

This book provides a guided tour through the key functions of posttranscriptional gene regulation, that is, pretty much an overview of RNA biology, written by the experts in each area. Importantly, it also covers some aspects of the pathology of RNA when things go wrong. I congratulate Jane Wu in putting this book together – it is a highly valuable resource for many of us who are interested in eukaryotic gene regulation.

Beijing, May 2012
University of Dundee

David M.J. Lilley, PhD, FRS

List of Contributors

Jennifer A. Aoki
Columbia University
Department of Biological Sciences
1117 Fairchild Center
New York, NY 11740
USA

Stephen Buratowski
Harvard Medical School
Department of Biological Chemistry
and Molecular Pharmacology
240 Longwood Avenue
Boston, MA 02115
USA

Piero Carninci
Omics Science Center
LSA Technology Development Group
Functional Genomics Technology
Team
1-7-22 Suehiro-cho
Tsurumi-ku
Yokohama
Kanagawa 230-0045
Japan

and

Omics Science Center
LSA Technology Development Group
Omics Resource Development Unit
1-7-22 Suehiro-cho
Tsurumi-ku
Yokohama
Kanagawa 230-0045
Japan

Mengmeng Chen
Chinese Academy of Sciences
Institute of Biophysics
State Laboratory of Brain and
Cognition
Beijing 100101
China

Ying Chen
The University of Chicago
Department of Ecology and Evolution
Committee on Genetics
1101 E 57 Street
Chicago, IL 60637
USA

Hongzheng Dai
The University of Chicago
Department of Ecology and Evolution
Committee on Genetics
1101 E 57 Street
Chicago, IL 60637
USA

Jill A. Dembowski
University of Pittsburgh
Department of Biological Sciences
4249 Fifth Avenue
Pittsburgh, PA 15260
USA

Jianwen Deng
Chinese Academy of Sciences
Institute of Biophysics
State Laboratory of Brain and Cognition
Beijing 100101
China

Sebastian M. Fica
University of Chicago
Department of Molecular Genetics & Cell Biology
CLSC 821A
920 East 58th St
Chicago, IL 60637
USA

Katherine S. Godin
Medical Research Council Laboratory of Molecular Biology
Structural Studies Division
Hills Road
Cambridge CB1 2QW
UK

and

University of Washington
Department of Chemistry
Box 351700
Seattle, WA 98195
USA

Graydon B. Gonsalvez
Georgia Health Sciences University
Department of Cellular Biology and Anatomy
1459 Laney Walker Blvd
Augusta, GA 30912
USA

Paula J. Grabowski
University of Pittsburgh
Department of Biological Sciences
4249 Fifth Avenue
Pittsburgh, PA 15260
USA

Matthias Harbers
DNAFORM Inc.
Leading Venture Plaza 2
75-1 Ono-cho
Tsurumi-ku
Yokohama
Kanagawa 230-0046
Japan

Hisashi Iizasa
The Wistar Institute
Department of Gene Expression & Regulation
3601 Spruce Street
Philadelphia, PA 19104
USA

Yukio Kawahara
The Wistar Institute
Department of Gene Expression & Regulation
3601 Spruce Street
Philadelphia, PA 19104
USA

Minkyu Kim
Seoul National University
WCU Department of Biophysics and Chemical Biology
1 Gwanak-Ro, Gwanka-Gu, Seoul 151-747
Korea

Jianghong Liu
Chinese Academy of Sciences
Institute of Biophysics
State Laboratory of Brain and Cognition
Beijing 100101
China

Manyuan Long
The University of Chicago
Department of Ecology and Evolution
Committee on Genetics
1101 E 57 Street
Chicago, IL 60637
USA

James L. Manley
Columbia University
Department of Biological Sciences
1117 Fairchild Center
New York, NY 11740
USA

A. Gregory Matera
University of North Carolina
Department of Biology
Program in Molecular Biology &
Biotechnology and Lineberger
Comprehensive Cancer Center
3352 Genome Science Building
Chapel Hill, NC 27599-3280
USA

Melissa Mefford
University of Chicago
Department of Molecular Genetics &
Cell Biology
CLSC 821A
920 East 58th St
Chicago, IL 60637
USA

Kazuko Nishikura
The Wistar Institute
Department of Gene Expression &
Regulation
3601 Spruce Street
Philadelphia, PA 19104
USA

Eliza C. Small
University of Chicago
Department of Molecular Genetics &
Cell Biology
CLSC 821A
920 East 58th St
Chicago, IL 60637
USA

Jonathan P. Staley
University of Chicago
Department of Molecular Genetics &
Cell Biology
CLSC 821A
920 East 58th St
Chicago, IL 60637
USA

Louis Valente
The Wistar Institute
Department of Gene Expression &
Regulation
3601 Spruce Street
Philadelphia, PA 19104
USA

Gabriele Varani
University of Washington
Department of Chemistry
Box 351700
Seattle, WA 98195
USA

and

University of Washington
Department of Biochemistry
Seattle, WA 98195
USA

Jane Y. Wu
Northwestern University Feinberg School of Medicine
Center for Genetic Medicine
Lurie Comprehensive Cancer Center
Department of Neurology
303 E. Superior
Chicago, IL 60611
USA

Mengxue Yang
Chinese Academy of Sciences
Institute of Biophysics
State Laboratory of Brain and Cognition
Beijing 100101
China

and

Northwestern University Feinberg School of Medicine
Center for Genetic Medicine
Lurie Comprehensive Cancer Center
Department of Neurology
303 E. Superior
Chicago, IL 60611
USA

Kun Zhu
Chinese Academy of Sciences
Institute of Biophysics
State Laboratory of Brain and Cognition
Beijing 100101
China

Li Zhu
Chinese Academy of Sciences
Institute of Biophysics
State Laboratory of Brain and Cognition
Beijing 100101
China

Boris Zinshteyn
The Wistar Institute
Department of Gene Expression & Regulation
3601 Spruce Street
Philadelphia, PA 19104
USA

1
The Role of Cotranscriptional Recruitment of RNA-Binding Proteins in the Maintenance of Genomic Stability

Jennifer A. Aoki and James L. Manley

1.1
Introduction

All steps in transcription and pre-mRNA processing are extensively coordinated. The carboxy-terminal domain (CTD) of the large subunit of RNA polymerase (RNAP) II plays an important role in cotranscriptionally recruiting factors necessary for capping, splicing, polyadenylation, and other mRNA processing events [1–4]. The CTD acts as a platform for these factors to bind, and this process is coordinated by phosphorylation changes that occur during transcription [5]. Although transcription and RNA processing steps are not obligatorily coupled, as seen by the fact that these processes have been studied for many years as individual steps, some posttranscriptional modifications have been shown to be functionally coupled *in vitro* such as transcription capping [6] and transcription 3′ end processing [7]. There has also been recent evidence for "reverse coupling," where a proximal 5′ splice site enhances the recruitment of basal transcription initiation factors to the promoter [8]. While it is still unclear if transcription and splicing are functionally coupled in the cell [9, 10], there is evidence that cotranscriptional recruitment of serine-arginine (SR) proteins onto pre-mRNA is vital in maintaining genomic stability [11, 12]. Indeed, recent work shows that ASF/SF2, an SR protein first discovered for its role in constitutive and alternative splicing [13, 14], is a component of a high-molecular-weight (HMW) complex formed on pre-mRNA during cotranscriptional splicing assays [15], reflecting the early recruitment of ASF/SF2 and other SR proteins to nascent RNA during transcription.

All the steps in transcription and RNA processing appear to function together to produce an export-competent and translatable mRNP. In yeast, where splicing is less frequent, transcription is coupled to loading of export factors and mRNP formation through the THO/TREX complex [16]. Mutation of factors in the THO/TREX complex also results in genomic instability [16]. In metazoans, transcription is linked to mRNP formation through splicing [17], formation of the exon junction complex (EJC) [18, 19], and THO/TREX recruitment [16].

Below we will discuss the coordination of transcription and pre-mRNA processes that inherently protects the genome from invasion of nascent RNA into DNA of

Posttranscriptional Gene Regulation: RNA Processing in Eukaryotes, First Edition. Edited by Jane Wu.

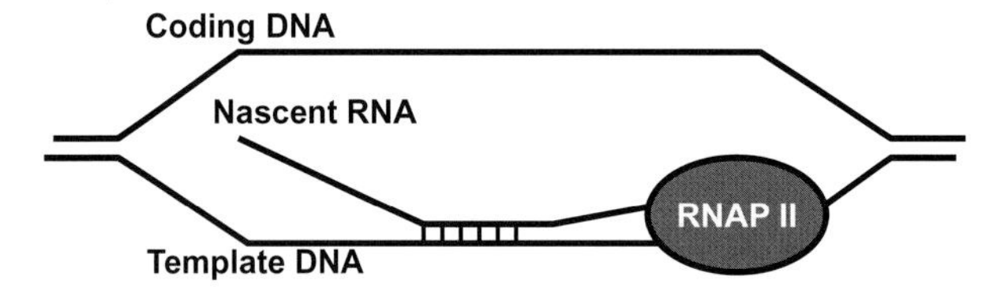

Figure 1.1 Schematic of a cotranscriptionally formed R-loop structure. Nascent RNA hybridizes with template DNA, leaving coding DNA single stranded.

the transcribing locus. The invading RNA can then hybridize to the template DNA, producing an aberrant R-loop structure, leaving the coding strand of the DNA single stranded and subject to DNA damage and strand breakage (Figure 1.1). We will describe other examples of cotranscriptionally formed R-loops and speculate on mechanisms that cause such structures to lead to genomic instability.

1.2 THO/TREX

1.2.1 THO/TREX in *Saccharomyces cerevisiae*

The THO complex proteins were first discovered in genetic screens for their role in transcription elongation of GC-rich genes in *S. cerevisiae* [20]. The complex consists of Hpr1, Tho2, Mft2, and Thp2, which are recruited to elongating RNAP II complexes. In addition to impairment of transcriptional elongation, THO mutants cause reduced efficiency of gene expression and an increase in hyper-recombination between direct repeats [21]. Mutations in *hpr1, tho2,* and *mft1* can also produce export defects and retention of transcripts at sites of transcription [16]. This reflects the association of THO with export factors Yra1 and Sub1 to form the TREX complex (transcription/export). TREX is recruited early to actively transcribing genes and travels entire length of genes with RNAP II [16]. Interestingly, mutants of the export machinery, Sub2, Yra1, Mex67, and Mtr2 also have THO-like phenotypes of defective transcription and hyper-recombination [16]. Further investigation of 40 selected mutants representing various steps in biogenesis and export of mRNP showed a weak but significant effect on recombination and transcript accumulation [22]. In particular, mutants of the nuclear exosome and 3′ end processing machinery showed inefficient transcription elongation and genetic interactions with THO. The TREX complex exemplifies the importance of the link between transcription and export-competent mRNP formation in yeast with the maintenance of the genomic integrity.

Further investigation of the association of the yeast TREX complex with actively transcribed DNA showed that the THO components play a critical role in the loading of the export machinery onto newly synthesized RNA [23]. Hpr1 was shown to associate with DNA templates through its association with the CTD. While Sub2 was only bound to nascent RNA, Yra1 was associated with both DNA and RNA on intronless genes. Yra1 is recruited to THO and helps to load Sub2 onto the nascent RNA. While Hpr1 was able to associate with both intronless and intron-containing templates similarly, there was a large decrease in the ability of Yra1p and Sub2 to be deposited onto the intron-containing RNA. Data suggested that spliceosome assembly interfered with the transfer of TREX components onto the RNA in these *in vitro* transcription assays.

1.2.2 THO/TREX in Higher Eukaryotes

Recruitment of TREX may not be transcription coupled in mammals but coupled to splicing instead [24]. When Tho2 immunodepleted HeLa nuclear extracts were used for *in vitro* transcription, there was no elongation defect detected, as was seen in yeast. There was also no effect on spliceosome assembly, splicing, or RNA stability even though all components of the THO complex have previously been detected in purified spliceosomes [25]. Immunoprecipitation assays showed that human Tho2 only associated with *in vitro* spliced mRNA but not unspliced pre-mRNA. Further *in vitro* experiments showed that TREX bound to the 5′ cap-binding complex (CBC) in a splicing-dependent manner [26]. Immunoprecipitation assays showed that TREX preferentially associated with *in vitro* spliced and capped mRNA compared with uncapped or unspliced. The interaction of TREX with the 5′ cap is mediated by protein–protein interactions between REF/Aly and CBP80. Microinjection of these preassembled mRNP into *Xenopus* oocytes showed then to be export competent.

The above *in vitro* data seem to conflict with the *in vivo* data produced by Hrp1 depletion in HeLa cells. While it is evident that export-component REF/Aly directly interacts with CBP80 in a splicing-dependent manner, Hpr1 associates with DNA not RNA in yeast [23], so it is possible that recruitment of hHpr1 and hTho2 to the CBC in the immunoprecipitation assay may be due to their affinity for REF/Aly. Also, in *Drosophila melanogaster*, only the depletion of both *THO2* and *HPR1* by siRNA shows significant nuclear accumulation of poly(A)+ RNA [27], which could signify their divergent roles in transcription elongation and formation of export-component mRNP. THO is essential for heat-shock mRNA export in *D. melanogaster*, which may perhaps reflect its role in stress conditions. If the recruitment of REF/Aly is only dependent on cap formation and splicing, this might also explain the THO-independent recruitment of UAP56 in *D. melanogaster*. In any event, these data together indicate that the recruitment of the export machinery in higher eukaryotes is not linked to THO/TREX in a manner similar to *S. cerevisiae*.

1.2.3 THO/TREX and R-loop Formation

How do defects in THO/TREX cause hyper-recombination? Huertas and Aguilera proposed that mutations affecting THO/TREX components cause cotranscriptional production of aberrant R-loop structures [28]. They provided evidence of R-loop formation utilizing a hammerhead ribozyme to release hybridized nascent RNA.

This ribozyme was able to suppress transcription elongation impairment and hyper-recombination phenotype in THO mutants. Further evidence was provided by overexpression of RNase H to degrade RNA moiety of RNA:DNA hybrids, which also suppressed the THO phenotypes [28]. A more recent study of the point mutant *hpr1-101*, which has a transcriptional defect but does not cause R-loop formation, shows no hyper-recombination phenotype. This indicates that while the transcriptional defect by THO mutants may be further aggravated by R-loops, they are not mediated by them. The RNA:DNA hybrids do appear to lead to the hyper-recombination phenotype associated with THO mutants [29].

Therefore, in yeast, early recruitment of THO/TREX plays a key role in protecting against or preventing the formation of R-loop structures. In mammals though, the late recruitment of THO/TREX suggests that it plays a less important role, or perhaps no role, in protecting against genomic instability. This points to a possible role for earlier cotranscriptional processes in protecting the genome from DNA damage.

1.3 Linking Transcription to Export of mRNP

Early in transcription, the cell is already preparing to package nascent pre-mRNA into mRNP. As mentioned earlier, the RNAP II CTD coordinates the recruitment of RNA processing factors to transcribing genes. Spt5, a subunit of the DRB sensitivity-inducing factor (DSIF) transcriptional elongation factor, plays an early role in integrating the various steps of pre-mRNA processing to guarantee an export-competent mRNP. Immediately after transcription is induced, Spt5 helps to recruit a capping enzyme (CE) [30, 31], which is then activated together with the phosphorylated CTD to cap the 5′ end of the growing pre-mRNA (Figure 1.2a). *In vitro*, the recruitment of CE has been shown to cause the formation of R-loops [32]. However, these may be prevented by the recruitment of ASF/SF2 to the RNA, which prevents the formation of these aberrant RNA:DNA structures [11, 32]. Spt5 has been implicated not only in transcription elongation [33], CE recruitment, and splicing [34, 35], but also in transcription-coupled repair [36] and the recruitment of the exosome [37], which plays a key role in mRNP quality control [38, 39].

1.3.1 The Thp1-Sac3-Sus1-Cdc31 (THSC) Complex

In *S. cerevisiae*, in addition to THO/TREX, another complex has been shown to play a role in the export of properly formed mRNPs. The Thp1-Sac3-Sus1-Cdc31 (THSC) complex was shown to be recruited to transcribing genes to localize active genes to the nuclear periphery [40–42]. Sus1 is also a component of the SAGA histone acetylase complex, which is involved in facilitating transcription initiation [43]. Sus1 acts to link transcription to localization of actively transcribing genes to the nuclear pore. Deletion of *sus1* but not other genes coding SAGA components

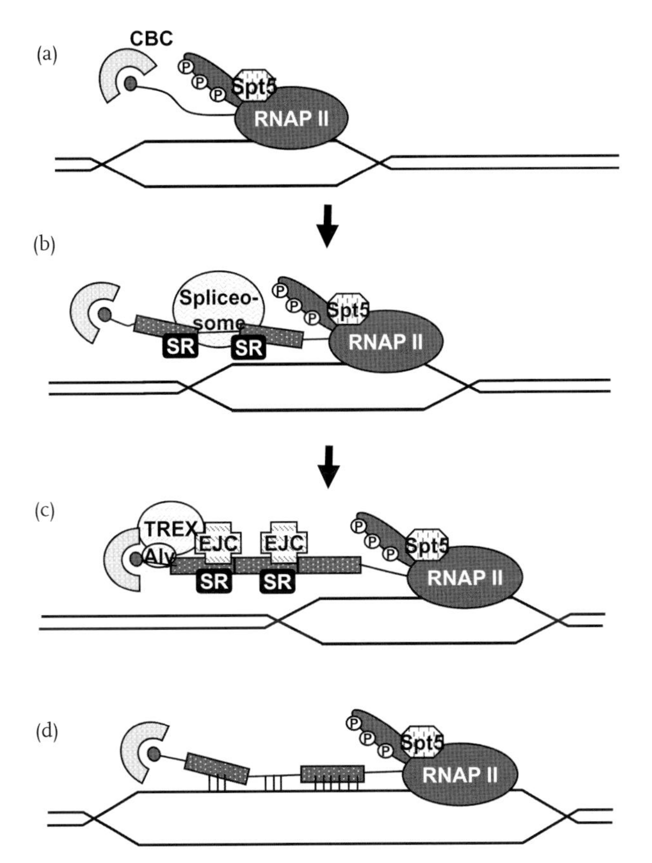

Figure 1.2 Early steps in cotranscriptional mRNA processing protect against genomic instability. (a) The phosphorylated CTD of RNAP II, along with Spt5, recruits a capping enzyme to cap the 5′ end of a growing pre-mRNA. These steps work to (b) cotranscriptionally load the spliceosome and SR proteins onto pre-mRNA to remove the introns. (c) The EJC is then deposited onto the mRNA in a splicing-dependent manner. REF/Aly and other components of the TREX complex are stabilized onto the 5′ end of the mRNA. (d) Disruption of any of these early steps of mRNA processing can leave nascent RNA open to bind to template DNA forming an R-loop structure.

confers a transcription-dependent hyper-recombination phenotype, similar to phenotypes seen in THO/TREX [40]. It is interesting to note that mutations in other nucleoporins lead to an accumulation of DNA breaks, visualized by Rad52 foci formation, but not in conjunction with transcription [44, 45].

1.3.2
SR Proteins

In metazoan cells, where most genes are spliced, SR proteins play an important role in linking transcription to spliceosome assembly in order to form an export-competent mRNP. SR proteins can be recruited and cotranscriptionally loaded onto the pre-mRNA, indicated, in part, by interactions with the CTD of RNAP II [46] (Figure 1.2b). SR proteins can be recruited onto pre-mRNA *in vitro* by RNAP II but not by T7 RNAP during transcription assays [15]. Also, Li and Manley showed that ASF/SF2, SC35, and SRp20 can prevent the formation of cotranscriptional R-loop structures *in vitro*, most likely participating in mRNP formation on nascent transcripts and again in a CTD-specific manner [11]. It is not clear whether the function of SR proteins in the mRNP formation is directly related to their roles in splicing.

1.3.3
The Exon Junction Complex

The exon junction complex (EJC) is a complex of proteins that are deposited onto mRNA 20–24 nucleotides upstream of the exon–exon junction in a splicing-dependent fashion [47–51] (Figure 1.2c). The most recent biochemical and structural studies show that the EJC core complex consists of eIF4A3, the Y14:Magoh heterodimer, and MLN51 [47, 52–56]. All the core components have been shown to associate with the spliceosome [25]. eIF4A3 binds onto spliced RNA in a sequence-independent manner with an 8- to 10-nt footprint upstream of the splice junction [57–60] and acts as a platform for the other EJC core proteins. Various other proteins involved in nuclear export and non-sense-mediated decay have been shown to interact with this minimal core of the EJC. Early experiments showing the recruitment of the export factors UAP56, REF/Aly, and TAP/p15 to the EJC led to the belief that the EJC played a key role in export [18, 19], but later biochemical studies showed that RNPS1, REF/Aly, and UAP56 can load onto spliced RNA in the absence of the key component of the core EJC, eIF4A3 [17].

While the EJC factor is not a key factor in the recruitment of the export machinery, splicing has been shown to be mandatory for the loading of export factors [17, 26]. The core components of eIF4A3, Magoh, and Y14 also require splicing to take place in order to be deposited onto mRNA [17]. In addition, there is an enrichment of spliced mRNA in complexes containing REF/Aly and Upf3b/3a, indicating that splicing stabilizes the interactions of these proteins with the mRNA [17]. Although REF/Aly has been found to associate with the cap binding complex (CBC) on the 5′ end of the mRNA, association of the export machinery with mRNPs is splicing dependent [17] (Figure 1.2c). Therefore, splicing, not THO/TREX recruitment, plays an important role in loading of mRNP remodeling proteins to produce an export-competent mRNP.

1.3.4
The Exosome

The exosome plays a vital role in the recognition of mRNA that has not been properly capped, spliced, or 3′ end processed [38, 39, 61]. Specifically, mutations in genes encoding THO components and components of the pre-mRNA 3′ processing complex, such as Rna14 and Rna15, can cause accumulation of improperly processed mRNA retained at sites of transcription [62]. This retention can be relieved by deletion of the gene encoding the 3′–5′ exonuclease Rrp6. Inserting a self-cleaving ribozyme can also relieve retention. Interestingly, these aberrant mRNPs can then be exported to the cytoplasm following cleavage, despite lacking properly processed 5′ ends. In higher eukaryotes, splice site mutations can also cause Rrp6-mediated nuclear retention of unspliced RNA [63, 64]. In cells expressing RNAP II with CTD truncations that affect splicing, a self-cleaving ribozyme can release an improperly processed mRNA from the site of transcription. The released transcript can then be properly spliced, while the small portion of mRNA 3′ of the cleavage site remains associated with the DNA [64]. This shows that pre-mRNA association with RNAP II is not required for splicing, but proper splicing is required for the release of mRNPs from sites of transcription in order to be exported. Interesting questions arise from these observations. Could R-loops form due to the lack of splicing? And might R-loops be a mechanism to recognize improperly processed transcripts?

1.4 Cotranscriptional R-loop Formation

It is noteworthy that defects in early steps in mRNA production appear to have a higher impact on genomic stability than steps later in the process. This is most likely due to the ability to cause cotranscriptional R-loops before the pre-mRNA is stably packaged into mRNP. As described earlier, transcription and splicing play important roles in the recruitment of the THSC, THO/TREX, and EJC, which are all loaded onto the pre-mRNA, necessary for proper export, and protect against deleterious RNA:DNA hybrid formations (Figure 1.2d). Deletion mutants of later pre-mRNA processing steps, such as 3′ end processing or quality control components of the 5′–3′ exosome, cause lower levels of hyper-recombination in deletion mutants [22]. Most likely this reflects the reduced opportunity to form R-loops since transcription has progressed to the far 3′ end of the gene.

To better understand the role of R-loops in the impairment of transcription elongation, R-loop formation was studied in an *in vitro* system [65]. RNA was hybridized to plasmid DNA to form a 300-bp R-loop structure, which was subsequently purified and studied using yeast whole cell extract. The artificial R-loops reduced transcription elongation efficiencies compared with plasmid DNA without these aberrant structures. Removal of the RNA moiety by RNase H prior to the transcription reaction improved efficiency directly proving that R-loops can cause deficiencies in transcription elongation *in vitro* [65]. Early pulse labeling experiments employing DT40 cells depleted of ASF/SF2, however, did not detect a transcriptional defect *in vivo*, suggesting that R-loops may not cause a significant elongation effect in vertebrates [66].

1.4.1
R-loops in *Escherichia coli*

Cotranscriptional R-loop formation was first suggested to occur in *E. coli* and has been extensively studied [67, 68]. RNase H was recovered in a genetic complementation screen of growth defects of topoisomerase mutants Δ*topA* in *E. coli* [69]. The ability of RNase H to digest the RNA moiety of RNA:DNA hybrids led to the belief that R-loops are involved in the observed growth defects. The negative supercoiling caused by deletion of *topA* leaves DNA more open for hybridization to RNA, most likely by nascent transcripts. Later, RNase H overproduction was shown to correct defects in transcription elongation [70] and reinstate full-length RNA synthesis in Δ*topA E. coli* [71]. Topoisomerase I is upregulated in the heat-shock response [72] likely to protect the accumulation of hypernegative supercoiling and R-loop formation at induced stress genes loci [73]. Interestingly, this is reminiscent of the need for THO in heat-shock poly(A)+ export in *D. melanogaster* [27]. The uncoupling of transcription and translation in *E. coli* by drugs that inhibit translation can also cause an increase in R-loop formation, suggesting that ribosomes in prokaryotes, like RNP structures in eukaryotes, play a role in inhibiting the re-annealing of nascent RNA to template DNA [74].

1.4.2
Naturally Occurring R-loops

Stable R-loops can also form during transcription through G-rich templates such as those used by Aguilera and colleagues in the THO/TREX mutant studies in yeast [28] and in the immunoglobulin (Ig) class switch region of mammalian B cells [75]. This is probably because rG:dC hybrids are exceptionally stable [76]. In addition, natural R-loops have also been shown to form both *in vitro* and in bacteria by transcription of the Friedreich ataxia triplet repeat, GAA-TTC, by T7 RNAP polymerase [77]. These repeats were able to form these natural R-loops in bacterial cells with normal levels of RNase H, showing that these hybrids formed more rapidly than they could be removed. We must keep in mind that T7 RNAP polymerizes at a rate of 200–400 nt s^{-1} [78], while human RNAP II only transcribes at 15–20 nt s^{-1} [79], which could explain the inability of RNase HI to digest the R-loops. Nonetheless, these results indicate that very stable RNA:DNA hybrids can form *in vivo*.

1.4.3
TREX Protects against R-loop Formation

When the recruitment of TREX was investigated in yeast, it was shown that Hpr1 is recruited to both intronless and intron-containing genes by RNAP II, but Yra1 and Sub2, while recruited efficiently to intronless genes, were poorly recruited to intron-containing genes [23]. Suppression by siRNA of human Hrp1 in HeLa cells, like its yeast homolog, caused gene expression and transcription elongation defects

[80]. These defects could be suppressed by expression of RNase H. Other studies showed that hTho2 immunodepleted HeLa nuclear extract did not show any transcription elongation defects [24]. These studies later concluded that the components of the human TREX complex could only be recruited via a splicing-dependent mechanism [26]. It was thus concluded that the EJC might play a role in the stabilization of the export machinery after splicing either by protein–protein interaction or remodeling of mRNP structure. This is consistent with the previously discussed enhancement of export machinery recruitment by EJC. While these studies show some discrepancies in the role of THO in transcription and R-loop formation in mammals, the data support the underlying need of cotranscriptional processes, either direct THO/TREX recruitment or recruitment of other factors, for example, SR proteins, to protect against deleterious R-loop formation and subsequent genomic instability.

1.4.4 SR Proteins Protect against R-loop Formation

As mentioned above, genetic inactivation of ASF/SF2, in the chicken B-cell line DT40, provided evidence of genomic instability, including the production of high molecular weight (HMW) DNA fragmentation, hyper-recombination, and G2 cell cycle arrest [11]. *In vitro* transcription experiments showed that cotranscriptional R-loops could be suppressed by a dose-dependent addition of ASF/SF2 in the presence of phosphorylated CTD. Later experiments showed that overexpression of RNPS1 can rescue phenotypes of HMW DNA fragmentation, hypermutation, and G2 cell cycle arrest of ASF/SF2-depleted DT40 cells, and can cause HMW DNA fragmentation when depleted from HeLa cells [12]. In contrast, SRp20 and SC35 were not able to or only partially able to rescue the HMW DNA fragmentation phenotype.

In ASF/SF2-depleted cells, it was shown that expression of RNPS1 alleviates genomic instability [12]. It is argued that since RNPS1 cannot compensate for ASF/SF2 function in splicing, RNPS1 has a role in forming RNP complexes on nascent transcripts to prevent R-loop formation. It was not investigated whether RNPS1 could prevent cotranscriptional R-loop formation *in vitro*. It is interesting to note that SRp20 and SC35 were able to partially rescue from R-loop formation *in vitro* [11] but incapable of suppression of HMW DNA fragmentation *in vivo* [12]. This could be a reflection of the role of recruitment of these factors onto the nascent transcript, for it has been shown that RNPS1 can be loaded onto RNA before the splicing reaction takes place.

RNPS1 has many functions beyond its role as a general splicing activator, especially in its diverse roles in the EJC. The cellular concentration of RNPS1 in HeLa cell strains appears to play a role in eliciting a strong non-sense-mediated decay response [81]. Another function of RNPS1 is the recruitment of Acinus and SAP18 to the EJC [54]. Acinus, SAP18, and RNPS1 are part of the apoptosis- and splicing-associated protein (ASAP) complex [82]. Microinjection of excess ASAP complex into cells causes an increased rate of apoptosis. It would be interesting if RNPS1 could function to sequester Acinus and SAP18 as a mechanism to protect against apoptosis during normal splicing and mRNP formation. If this was true, improper splicing or mRNP formation signaled by improper loading of RNPS1 and ASAP to RNA could increase the concentration of free ASAP in the cell to cause acceleration of apoptosis [82]. However, this is most likely be downstream of HMW DNA fragmentation, since ASAP only accelerates cells that are already stimulated for apoptosis. With the various functions of ASF/SF2 and RNPS1, it is difficult to distinguish the exact mechanism by which RNPS1 suppresses ASF/SF2 depletion-induced R-loop formation.

1.5 R-loop-induced Double-Stranded (ds) DNA Breaks

1.5.1 Class Switch Recombination

It is still unknown how R-loop formation can lead to dsDNA breaks in ASF/SF2-depleted cells and indeed by R-loops more generally. Investigation of naturally occurring R-loops in class switch recombination (CSR) in stimulated B cells may provide a hint to a mechanism. CSR is a DNA recombination event that switches DNA segments located upstream of each heavy chain constant region. CSR results in the switch of Ig isotype from IgM to either IgG, IgE, or IgA (reviewed in [83]). It has been shown that class switching is a transcription-dependent event that has R-loop structural intermediates [75]. Activation-induced deaminase (AID), a protein essential for CSR and somatic hypermutation, is expressed in activated B cells and specifically binds to G-quartets, tertiary structures on single-stranded (ss) DNA composed of four stacked guanidines [84]. Interestingly, it has been shown that the RNA exosome targets AID to these DNA strands [85]. Cytidines on ssDNA can then be deaminated by AID [86, 87]. Uracil, the product of cytidine deamination, is then removed by uracil-DNA glycosylase, which results in a nick or ssDNA break [88].

1.5.2 Formation of Double-Strand Breaks

How are such ssDNA breaks converted to dsDNA breaks, which then lead to DNA rearrangements, hyper-recombination, and genetic instability? There are several possibilities. For example, if two ssDNA breaks on opposing sides are in close proximity, it will form a dsDNA break. In bacteria, if an ssDNA is not properly repaired, it can be converted into dsDNA breaks [89]. This has been proposed to be by collision of the DNA replication machinery with barriers to its procession, leading to fork collapse, which will result in dsDNA breaks. Barriers can include protein bound to dsDNA or aberrant DNA structures, such as nicked DNA or possibly R-loops.

The HMW DNA fragmentation observed in the ASF/SF2-depleted DT40 [11] may be caused in conjunction with AID, whose expression is limited to activated B cells. Indeed, in *S. cerevisiae*, exogenously expressed AID activity strongly stimulated hyper-recombination by R-loop forming THO mutants [40]. However, depletion of ASF/SF2 by siRNA in HeLa cells also produced DNA breaks [11], and it is believed that dsDNA breaks occur generally when R-loops form. It is possible that related proteins in the AID family could function similarly to AID itself in these and other cell types (reviewed in [90, 91]). AID family member APOBEC1, an RNA-editing enzyme expressed in colorectal tissue, can also deaminate cytidine of ssDNA *in vitro* [91]. Overexpression of APOBEC1 induces cancer in these cells [92], reflecting its ability to cause damage to DNA *in vivo*. Thus, it is possible that related members of APOBEC family can cause dsDNA breaks observed in response to R-loop formation in other cells types. This is consistent with the fact that APOBEC family proteins are highly mutagenic [93], and this may in part reflect an ability to cause ssDNA in response to R-loops.

DNA breaks observed in cells producing R-loops can occur by mechanisms not involving APOBEC proteins. In yeast, it has been shown that *hpr1* and *tho2* mutants have increased sensitivity to DNA damage induced by UV irradiation [94]. In general, ssDNA is more susceptible to DNA damage and breaks than dsDNA. For example, spontaneous deamination of cytidine on ssDNA is 140-fold more efficient than on dsDNA [95]. Increased genetic instability was linked to impaired replication in *in vitro* experiments utilizing *HPR1* and *THO2* mutants [96]. Further evidence for this was revealed by the discovery of point mutants in *hpr1* that uncoupled hyper-recombination phenotypes from transcription elongation and transcript retention [97]. In these mutants, hyper-recombination was correlated with an impairment of replication. Impaired replication could reflect the role of fork collapse due to aberrant DNA structures and subsequent dsDNA breaks as mentioned above. This theory is supported by the fact that replication fork progression could be partially restored by hammerhead ribozyme cleavage of nascent RNA in these experiments [97].

1.5.3
Rrm3 and Pif1 DNA Helicases

Rrm3 is a 5′ to 3′ DNA helicase that facilitates replication past non-nucleosomal protein–DNA complexes, and which is conserved from yeast to humans [98–100]. This helicase is necessary for normal fork progression through an estimated 1400 discrete sites. The inability to replicate past protein–DNA complex causes fork breakage, which could lead to DNA damage and genomic instability. Indeed, *RRM3* was first identified in a genetic study because its absence increased recombination in tandemly repeated ribosomal DNA genes [101]. *In vitro* studies have shown that both Pif1 and hPif1, related proteins in the same family as Rrm3, are able to unwind RNA:DNA hybrids [101, 102]. Pif1 has been shown to inhibit the lengthening of telomeres, a structure that protects chromosome ends, by its interaction and removal of telomerase [101, 102]. Telomerase contains a component called telomerase RNA component (TERC), a G-rich RNA that can form G-quadraplexes, which are reminiscent of the G-quartets found in CSR. It is an intriguing but interesting possibility that Rrm3p or Pif1p might play a role in protecting against R-loop formation in eukaryotic cells.

1.6 Concluding Remarks

Research for many years has been focused on functional coupling of transcription and splicing. It was surprising that a role of cotransciptional splicing, or at least recruitment of splicing factors, is to protect chromosomes from genomic instability. While a considerable amount of work has been done to understand how cotranscriptional events function to preserve genomic stability, the exact mechanism by which R-loop formation leads to dsDNA breaks is still a mystery. Understanding this process is especially interesting in light of the fact that genome instability is a key cause of cancer [103]. Dissection of the steps by which R-loop structures can form DNA breaks and consequently genomic rearrangements may indeed provide insight into cancer and degeneration.

References

1 de Almeida, S.F. and Carmo-Fonseca, M. (2008) The CTD role in cotranscriptional RNA processing and surveillance. *FEBS Lett.*, **582**, 1971–1976.

2 Hirose, Y. and Manley, J.L. (2000) RNA polymerase II and the integration of nuclear events. *Genes Dev.*, **14**, 1415–1429.

3 Maniatis, T. and Reed, R. (2002) An extensive network of coupling among gene expression machines. *Nature*, **416**, 499–506.

4 Proudfoot, N.J., Furger, A., and Dye, M.J. (2002) Integrating mRNA processing with transcription. *Cell*, **108**, 501–512.

5 Phatnani, H.P. and Greenleaf, A.L. (2006) Phosphorylation and functions of the RNA polymerase II CTD. *Genes Dev.*, **20**, 2922–2936.

6 Kim, H.J., Jeong, S.H., Heo, J.H., Jeong, S.J., Kim, S.T., Youn, H.D., Han, J.W., Lee, H.W., and Cho, E.J. (2004) mRNA capping enzyme activity is coupled to an early transcription elongation. *Mol. Cell. Biol.*, **24**, 6184–6193.

7 Adamson, T.E., Shutt, D.C., and Price, D.H. (2005) Functional coupling of cleavage and polyadenylation with transcription of mRNA. *J. Biol. Chem.*, **280**, 32262–32271.

8 Damgaard, C.K., Kahns, S., Lykke-Andersen, S., Nielsen, A.L., Jensen, T.H., and Kjems, J. (2008) A 5' splice site enhances the recruitment of basal transcription initiation factors in vivo. *Mol. Cell*, **29**, 271–278.

9 Das, R., Dufu, K., Romney, B., Feldt, M., Elenko, M., and Reed, R. (2006) Functional coupling of RNAP II transcription to spliceosome assembly. *Genes Dev.*, **20**, 1100–1109.

10 Lazarev, D. and Manley, J.L. (2007) Concurrent splicing and transcription are not sufficient to enhance splicing efficiency. *RNA*, **13**, 1546–1557.

11 Li, X.L. and Manley, J.L. (2005) Inactivation of the SR protein splicing factor ASF/SF2 results in genomic instability. *Cell*, **122**, 365–378.

12 Li, X.L., Niu, T.H., and Manley, J.L. (2007) The RNA binding protein RNPS1

alleviates ASF/SF2 depletion-induced genomic instability. *RNA*, **13**, 2108–2115.

13 Ge, H. and Manley, J.L. (1990) A protein factor, ASF, controls cell-specific alternative splicing of SV40 early pre-mRNA in vitro. *Cell*, **62**, 25–34.

14 Krainer, A.R., Conway, G.C., and Kozak, D. (1990) The essential pre-mRNA splicing factor SF2 influences 5' splice site selection by activating proximal sites. *Cell*, **62**, 35–42.

15 Das, R., Yu, J., Zhang, Z., Gygi, M.P., Krainer, A.R., Gygi, S.P., and Reed, R. (2007) SR proteins function in coupling RNAP II transcription to pre-mRNA splicing. *Mol. Cell*, **26**, 867–881.

16 Strasser, K., Masuda, S., Mason, P., Pfannstiel, J., Oppizzi, M., Rodriguez-Navarro, S., Rondon, A.G., Andres, A.K., Struhl, K., Reed, R., and Hurt, E. (2002) TREX is a conserved complex coupling transcription with messenger RNA export. *Nature*, **417**, 304–308.

17 Zhang, Z. and Krainer, A.R. (2007) Splicing remodels messenger ribonucleoprotein architecture via elF4A3-dependent and -independent recruitment of exon junction complex components. *Proc. Natl. Acad. Sci. U.S.A.*, **104**, 11574–11579.

18 Gatfield, D., Le Hir, H., Schmitt, C., Braun, I.C., Kocher, T., Wilm, M., and Izaurralde, E. (2001) The DExH/D box protein HEL/UAP56 is essential for mRNA nuclear export in *Drosophila*. *Curr. Biol.*, **11**, 1716–1721.

19 Le Hir, H., Gatfield, D., Izaurralde, E., and Moore, M.J. (2001) The exon-exon junction complex provides a binding platform for factors involved in mRNA export and nonsense-mediated mRNA decay. *EMBO J.*, **20**, 4987–4997.

20 Chavez, S., Garcia-Rubio, M., Prado, F., and Aguilera, A. (2001) Hpr1 is preferentially required for transcription of either long or G+C-rich DNA sequences in *Saccharomyces cerevisiae*. *Mol. Cell. Biol.*, **21**, 7054–7064.

21 Rondon, A.G., Jimeno, S., Garcia-Rubio, M., and Aguilera, A. (2003) Molecular evidence that the eukaryotic THO/TREX complex is required for efficient transcription elongation. *J. Biol. Chem.*, **278**, 39037–39043.

22 Luna, R., Jimeno, S., Marin, M., Huertas, P., Garcia-Rubio, M., and Aguilera, A. (2005) Interdependence between transcription and mRNP processing and export, and its impact on genetic stability. *Mol. Cell*, **18**, 711–722.

23 Abruzzi, K.C., Lacadie, S., and Rosbash, M. (2004) Biochemical analysis of TREX complex recruitment to intronless and intron-containing yeast genes. *EMBO J.*, **23**, 2620–2631.

24 Masuda, S., Das, R., Cheng, H., Hurt, E., Dorman, N., and Reed, R. (2005) Recruitment of the human TREX complex to mRNA during splicing. *Genes Dev.*, **19**, 1512–1517.

25 Zhou, Z.L., Licklider, L.J., Gygi, S.P., and Reed, R. (2002) Comprehensive proteomic analysis of the human spliceosome. *Nature*, **419**, 182–185.

26 Cheng, H., Dufu, K., Lee, C.S., Hsu, J.L., Dias, A., and Reed, R. (2006) Human mRNA export machinery recruited to the 5' end of mRNA. *Cell*, **127**, 1389–1400.

27 Rehwinkel, J., Herold, A., Gari, K., Kocher, T., Rode, M., Ciccarelli, F.L., Wilm, M., and Izaurralde, E. (2004) Genome-wide analysis of mRNAs regulated by the THO complex in *Drosophila melanogaster*. *Nat. Struct. Mol. Biol.*, **11**, 558–566.

28 Huertas, P. and Aguilera, A. (2003) Cotranscriptionally formed DNA:RNA hybrids mediate transcription elongation impairment and transcription-associated recombination. *Mol. Cell*, **12**, 711–721.

29 Gomez-Gonzalez, B. and Aguilera, A. (2009) R-loops do not accumulate in transcription-defective hpr1-101 mutants: implications for the functional role of THO/TREX. *Nucleic Acids Res.*, **37**, 4315–4321.

30 Pei, Y. and Shuman, S. (2002) Interactions between fission yeast mRNA capping enzymes and elongation factor Spt5. *J. Biol. Chem.*, **277**, 19639–19648.

31 Wen, Y. and Shatkin, A.J. (1999) Transcription elongation factor hSPT5 stimulates mRNA capping. *Genes Dev.*, **13**, 1774–1779.

32 Kaneko, S., Chu, C., Shatkin, A.J., and Manley, J.L. (2007) Human capping enzyme promotes formation of transcriptional R loops in vitro. *Proc. Natl. Acad. Sci. U.S.A.*, **104**, 17620–17625.

33 Yamaguchi, Y., Narita, T., Inukai, N., Wada, T., and Handa, H. (2001) SPT genes: key players in the regulation of transcription, chromatin structure and other cellular processes. *J. Biochem.*, **129**, 185–191.

34 Lindstrom, D.L., Squazzo, S.L., Muster, N., Burckin, T.A., Wachter, K.C., Emigh, C.A., McCleery, J.A., Yates, J.R., 3rd, and Hartzog, G.A. (2003) Dual roles for Spt5 in pre-mRNA processing and transcription elongation revealed by identification of Spt5-associated proteins. *Mol. Cell. Biol.*, **23**, 1368–1378.

35 Xiao, Y., Yang, Y.H., Burckin, T.A., Shiue, L., Hartzog, G.A., and Segal, M.R. (2005) Analysis of a splice array experiment elucidates roles of chromatin elongation factor Spt4-5 in splicing. *PLoS Comput. Biol.*, **1**, e39.

36 Ding, B., LeJeune, D., and Li, S. (2010) The C-terminal repeat domain of Spt5 plays an important role in suppression of Rad26-independent transcription coupled repair. *J. Biol. Chem.*, **285**, 5317–5326.

37 Andrulis, E.D., Werner, J., Nazarian, A., Erdjument-Bromage, H., Tempst, P., and Lis, J.T. (2002) The RNA processing exosome is linked to elongating RNA polymerase II in *Drosophila*. *Nature*, **420**, 837–841.

38 Vanacova, S. and Stefl, R. (2007) The exosome and RNA quality control in the nucleus. *EMBO Rep.*, **8**, 651–657.

39 Houseley, J., LaCava, J., and Tollervey, D. (2006) RNA-quality control by the exosome. *Nat. Rev. Mol. Cell Biol.*, **7**, 529–539.

40 Gonzalez-Aguilera, C., Tous, C., Gomez-Gonzalez, B., Huertas, P., Luna, R., and Aguilera, A. (2008) The THP1-SAC3-SUS1-CDC31 complex works in transcription elongation-mRNA export preventing RNA-mediated genome instability. *Mol. Biol. Cell*, **19**, 4310–4318.

41 Chekanova, J.A., Abruzzi, K.C., Rosbash, M., and Belostotsky, D.A. (2008) Sus1, Sac3, and Thp1 mediate post-transcriptional tethering of active genes to the nuclear rim as well as to non-nascent mRNP. *RNA*, **14**, 66–77.

42 Jani, D., Lutz, S., Marshall, N.J., Fischer, T., Kohler, A., Ellisdon, A.M., Hurt, E., and Stewart, M. (2009) Sus1, Cdc31, and the Sac3 CID region form a conserved interaction platform that promotes nuclear pore association and mRNA export. *Mol. Cell*, **33**, 727–737.

43 Rodriguez-Navarro, S., Fischer, T., Luo, M.J., Antunez, O., Brettschneider, S., Lechner, J., Perez-Ortin, J.E., Reed, R., and Hurt, E. (2004) Sus1, a functional component of the SAGA histone acetylase complex and the nuclear pore-associated mRNA export machinery. *Cell*, **116**, 75–86.

44 Loeillet, S., Palancade, B., Cartron, M., Thierry, A., Richard, G.F., Dujon, B., Doye, V., and Nicolas, A. (2005) Genetic network interactions among replication, repair and nuclear pore deficiencies in yeast. *DNA Repair (Amst)*, **4**, 459–468.

45 Palancade, B., Liu, X., Garcia-Rubio, M., Aguilera, A., Zhao, X., and Doye, V. (2007) Nucleoporins prevent DNA damage accumulation by modulating Ulp1-dependent sumoylation processes. *Mol. Biol. Cell*, **18**, 2912–2923.

46 Misteli, T., Caceres, J.F., Clement, J.Q., Krainer, A.R., Wilkinson, M.F., and Spector, D.L. (1998) Serine phosphorylation of SR proteins is required for their recruitment to sites of transcription in vivo. *J. Cell Biol.*, **143**, 297–307.

47 Mishler, D.M., Christ, A.B., and Steitz, J.A. (2008) Flexibility in the site of exon junction complex deposition revealed by functional group and RNA secondary structure alterations in the splicing substrate. *RNA*, **14**, 2657–2670.

48 Gehring, N.H., Lamprinaki, S., Hentze, M.W., and Kulozik, A.E. (2009) The hierarchy of exon-junction complex assembly by the spliceosome explains key features of mammalian nonsense-mediated mRNA decay. *PLoS Biol.*, **7**, e1000120. doi: 10.1371/journal.pbio.1000120.

49 Kataoka, N., Yong, J., Kim, V.N., Velazquez, F., Perkinson, R.A., Wang, F., and Dreyfuss, G. (2000) Pre-mRNA splicing imprints mRNA in the nucleus with a novel RNA-binding protein that persists in the cytoplasm. *Mol. Cell*, **6**, 673–682.

50 Le Hir, H., Izaurralde, E., Maquat, L.E., and Moore, M.J. (2000) The spliceosome deposits multiple proteins 20–24 nucleotides upstream of mRNA exon-exon junctions. *EMBO J.*, **19**, 6860–6869.

51 Le Hir, H., Moore, M.J., and Maquat, L.E. (2000) Pre-mRNA splicing alters mRNP composition: evidence for stable association of proteins at exon-exon junctions. *Genes Dev.*, **14**, 1098–1108.

52 Stroupe, M.E., Tange, T.O., Thomas, D.R., Moore, M.J., and Grigorieff, N. (2006) The three-dimensional architecture of the EJC core. *J. Mol. Biol.*, **360**, 743–749.

53 Andersen, C.B.F., Ballut, L., Johansen, J.S., Chamieh, H., Nielsen, K.H., Oliveira, C.L.P., Pedersen, J.S., Seraphin, B., Le Hir, H., and Andersen, G.R. (2006) Structure of the exon junction core complex with a trapped DEAD-box ATPase bound to RNA. *Science*, **313**, 1968–1972.

54 Tange, T.O., Shibuya, T., Jurica, M.S., and Moore, M.J. (2005) Biochemical analysis of the EJC reveals two new factors and a stable tetrameric protein core. *RNA*, **11**, 1869–1883.

55 Ballut, L., Marchadier, B., Baguet, A., Tomasetto, C., Seraphin, B., and Le Hir, H. (2005) The exon junction core complex is locked onto RNA by inhibition of eIF4AIII ATPase activity. *Nat. Struct. Mol. Biol.*, **12**, 861–869.

56 Le Hir, H. and Andersen, G.R. (2008) Structural insights into the exon junction complex. *Curr. Opin. Struct. Biol.*, **18**, 112–119.

57 Bono, F., Ebert, J., Lorentzen, E., and Conti, E. (2006) The crystal structure of the exon junction complex reveals how it maintains a stable grip on mRNA. *Cell*, **126**, 713–725.

58 Lau, C.K., Diem, M.D., Dreyfuss, G., and Van Duyne, G.D. (2003) Structure of the Y14-Magoh core of the exon junction complex. *Curr. Biol.*, **13**, 933–941.

59 Shibuya, T., Tange, T.O., Sonenberg, N., and Moore, M.J. (2004) eIF4AIII binds spliced mRNA in the exon junction complex and is essential for nonsense-mediated decay. *Nat. Struct. Mol. Biol.*, **11**, 346–351.

60 Shibuya, T., Tange, T.O., Stroupe, M.E., and Moore, M.J. (2006) Mutational analysis of human eIF4AIII identifies regions necessary for exon junction complex formation and nonsense-mediated mRNA decay. *RNA*, **12**, 360–374.

61 Assenholt, J., Mouaikel, J., Andersen, K.R., Brodersen, D.E., Libri, D., and Jensen, T.H. (2008) Exonucleolysis is required for nuclear mRNA quality control in yeast THO mutants. *RNA*, **14**, 2305–2313.

62 Libri, D., Dower, K., Boulay, J., Thomsen, R., Rosbash, M., and Jensen, T.H. (2002) Interactions between mRNA export commitment, 3'-end quality control, and nuclear degradation. *Mol. Cell. Biol.*, **22**, 8254–8266.

63 de Almeida, S.F., Garcia-Sacristan, A., Custodio, N., and Carmo-Fonseca, M. (2010) A link between nuclear RNA surveillance, the human exosome and RNA polymerase II transcriptional termination. *Nucleic Acids Res.*, **38**, 8015–8026.

64 Custodio, N., Vivo, M., Antoniou, M., and Carmo-Fonseca, M. (2007) Splicing- and cleavage-independent requirement of RNA polymerase II CTD for mRNA release from the transcription site. *J. Cell Biol.*, **179**, 199–207.

65 Tous, C. and Aguilera, A. (2007) Impairment of transcription elongation by R-loops in vitro. *Biochem. Biophys. Res. Commun.*, **360**, 428–432.

66 Wang, J., Takagaki, Y., and Manley, J.L. (1996) Targeted disruption of an essential vertebrate gene: ASF/SF2 is required for cell viability. *Genes Dev.*, **10**, 2588–2599.

67 Drolet, M. (2006) Growth inhibition mediated by excess negative supercoiling: the interplay between

transcription elongation, R-loop formation and DNA topology. *Mol. Microbiol.*, **59**, 723–730.

68 Drolet, M., Broccoli, S., Rallu, F., Hraiky, C., Fortin, C., Masse, E., and Baaklini, I. (2003) The problem of hypernegative supercoiling and R-loop formation in transcription. *Front. Biosci.*, **8**, D210–D221.

69 Drolet, M., Phoenix, P., Menzel, R., Masse, E., Liu, L.F., and Crouch, R.J. (1995) Overexpression of RNase H partially complements the growth defect of an *Escherichia coli* delta topA mutant–R-loop formation is a major problem in the absence of DNA topoisomerase I. *Proc. Natl. Acad. Sci. U.S.A.*, **92**, 3526–3530.

70 Hraiky, C., Raymond, M.A., and Drolet, M. (2000) RNase H overproduction corrects a defect at the level of transcription elongation during rRNA synthesis in the absence of DNA topoisomerase I in *Escherichia coli*. *J. Biol. Chem.*, **275**, 11257–11263.

71 Baaklini, I., Hraiky, C., Rallu, F., Tse-Dinh, Y.C., and Drolet, M. (2004) RNase HI overproduction is required for efficient full-length RNA synthesis in the absence of topoisomerase I in *Escherichia coli*. *Mol. Microbiol.*, **54**, 198–211.

72 Rui, S. and Tse-Dinh, Y.C. (2003) Topoisomerase function during bacterial responses to environmental challenge. *Front. Biosci.*, **8**, D256–D263.

73 Cheng, B., Rui, S., Ji, C., Gong, V.W., Van Dyk, T.K., Drolet, M., and Tse-Dinh, Y.C. (2003) RNase H overproduction allows the expression of stress-induced genes in the absence of topoisomerase I. *FEMS Microbiol. Lett.*, **221**, 237–242.

74 Broccoli, S., Rallu, F., Sanscartier, P., Cerritelli, S.M., Crouch, R.J., and Drolet, M. (2004) Effects of RNA polymerase modifications on transcription-induced negative supercoiling and associated R-loop formation. *Mol. Microbiol.*, **52**, 1769–1779.

75 Yu, K.F., Chedin, F., Hsieh, C.L., Wilson, T.E., and Lieber, M.R. (2003) R-loops at immunoglobulin class switch regions in the chromosomes of stimulated B cells. *Nature Immunol.*, **4**, 442–451.

76 Sugimoto, N., Nakano, S., Katoh, M., Matsumura, A., Nakamuta, H., Ohmichi, T., Yoneyama, M., and Sasaki, M. (1995) Thermodynamic parameters to predict stability of RNA/DNA hybrid duplexes. *Biochemistry*, **34**, 11211–11216.

77 Grabczyk, E. and Usdin, K. (2000) The GAA*TTC triplet repeat expanded in Friedreich's ataxia impedes transcription elongation by T7 RNA polymerase in a length and supercoil dependent manner. *Nucleic Acids Res.*, **28**, 2815–2822.

78 Steitz, T.A. (2004) The structural basis of the transition from initiation to elongation phases of transcription, as well as translocation and strand separation, by T7 RNA polymerase. *Curr. Opin. Struct. Biol.*, **14**, 4–9.

79 Uptain, S.M., Kane, C.M., and Chamberlin, M.J. (1997) Basic mechanisms of transcript elongation and its regulation. *Annu. Rev. Biochem.*, **66**, 117–172.

80 Li, Y.P., Wang, X.L., Zhang, X.J., and Goodrich, D.W. (2005) Human hHpr1/p84/Thoc1 regulates transcriptional elongation and physically links RNA polymerase II and RNA processing factors. *Mol. Cell. Biol.*, **25**, 4023–4033.

81 Viegas, M.H., Gehring, N.H., Breit, S., Hentze, M.W., and Kulozik, A.E. (2007) The abundance of RNPS1, a protein component of the exon junction complex, can determine the variability in efficiency of the Nonsense Mediated Decay pathway. *Nucleic Acids Res.*, **35**, 4542–4551.

82 Schwerk, C., Prasad, J., Degenhardt, K., Erdjument-Bromage, H., White, E., Tempst, P., Kidd, V.J., Manley, J.L., Lahti, J.M., and Reinberg, D. (2003) ASAP, a novel protein complex involved in RNA processing and apoptosis. *Mol. Cell. Biol.*, **23**, 2981–2990.

83 Selsing, E. (2006) Ig class switching: targeting the recombinational mechanism. *Curr. Opin. Immunol.*, **18**, 249–254.

84 Duquette, M.L., Pham, P., Goodman, M.F., and Maizels, N. (2005) AID binds to transcription-induced structures in c-MYC that map to regions associated with translocation and hypermutation. *Oncogene*, **24**, 5791–5798.

85 Basu, U., Meng, F.L., Keim, C., Grinstein, V., Pefanis, E., Eccleston, J., Zhang, T., Myers, D., Wasserman, C.R., Wesemann, D.R., Januszyk, K., Gregory, R.I., Deng, H., Lima, C.D., and Alt, F.W. (2011) The RNA exosome targets the AID cytidine deaminase to both strands of transcribed duplex DNA substrates. *Cell*, **144**, 353–363.

86 Yu, K.F., Roy, D., Bayramyan, M., Haworth, I.S., and Lieber, M.R. (2005) Fine-structure analysis of activation-induced deaminase accessibility to class switch region R-loops. *Mol. Cell. Biol.*, **25**, 1730–1736.

87 Muramatsu, M., Kinoshita, K., Fagarasan, S., Yamada, S., Shinkai, Y., and Honjo, T. (2000) Class switch recombination and hypermutation require activation-induced cytidine deaminase (AID), a potential RNA editing enzyme. *Cell*, **102**, 553–563.

88 Di Noia, J. and Neuberger, M.S. (2002) Altering the pathway of immunoglobulin hypermutation by inhibiting uracil-DNA glycosylase. *Nature*, **419**, 43–48.

89 Kuzminov, A. (2001) Single-strand interruptions in replicating chromosomes cause double-strand breaks. *Proc. Natl. Acad. Sci. U.S.A.*, **98**, 8241–8246.

90 Beale, R.C.L., Petersen-Mahrt, S.K., Watt, I.N., Harris, R.S., Rada, C., and Neuberger, M.S. (2004) Comparison of the differential context-dependence of DNA deamination by APOBEC enzymes: correlation with mutation spectra *in vivo*. *J. Mol. Biol.*, **337**, 585–596.

91 Conticello, S.G., Thomas, C.J.F., Petersen-Mahrt, S.K., and Neuberger, M.S. (2005) Evolution of the AID/APOBEC family of polynucleotide (deoxy)cytidine deaminases. *Mol. Biol. Evol.*, **22**, 367–377.

92 Yamanaka, S., Balestra, M.E., Ferrell, L.D., Fan, J.L., Arnold, K.S., Taylor, S., Taylor, J.M., and Innerarity, T.L. (1995) Apolipoprotein-B messenger-RNA-editing protein induces hepatocellular-carcinoma and dysplasia in transgenic animals. *Proc. Natl. Acad. Sci. U.S.A.*, **92**, 8483–8487.

93 Conticello, S.G. (2008) The AID/APOBEC family of nucleic acid mutators. *Genome Biol.*, **9**, 229.

94 Gonzalez-Barrera, S., Prado, F., Verhage, R., Brouwer, J., and Aguilera, A. (2002) Defective nucleotide excision repair in yeast hpr1 and tho2 mutants. *Nucleic Acids Res.*, **30**, 2193–2201.

95 Frederico, L.A., Kunkel, T.A., and Shaw, B.R. (1990) A sensitive genetic assay for the detection of cytosine deamination: determination of rate constants and the activation energy. *Biochemistry*, **29**, 2532–2537.

96 Wellinger, R.E., Prado, F., and Aguilera, A. (2006) Replication fork progression is impaired by transcription in hyperrecombinant yeast cells lacking a functional THO complex. *Mol. Cell. Biol.*, **26**, 3327–3334.

97 Huertas, P., Garcia-Rubio, M.L., Wellinger, R.E., Luna, R., and Aguilera, A. (2006) An hpr1 point mutation that impairs transcription and mRNP biogenesis without increasing recombination. *Mol. Cell. Biol.*, **26**, 7451–7465.

98 Azvolinsky, A., Dunaway, S., Torres, J.Z., Bessler, J.B., and Zakian, V.A. (2006) The *S. cerevisiae* Rrm3p DNA helicase moves with the replication fork and affects replication of all yeast chromosomes. *Genes Dev.*, **20**, 3104–3116.

99 Torres, J.Z., Schnakenberg, S.L., and Zakian, V.A. (2004) *Saccharomyces cerevisiae* Rrm3p DNA helicase promotes genome integrity by preventing replication fork stalling: viability of rrm3 cells requires the intra-S-phase checkpoint and fork restart activities. *Mol. Cell. Biol.*, **24**, 3198–3212.

100 Ivessa, A.S., Lenzmeier, B.A., Bessler, J.B., Goudsouzian, L.K., Schnakenberg, S.L., and Zakian, V.A. (2003) The *Saccharomyces cerevisiae* helicase Rrm3p facilitates replication past nonhistone protein-DNA complexes. *Mol. Cell*, **12**, 1525–1536.

101 Keil, R.L. and McWilliams, A.D. (1993) A gene with specific and global effects on recombination of sequences from

tandemly repeated genes in *Saccharomyces cerevisiae*. *Genetics*, **135**, 711–718.

102 Boule, J.B., Vega, L.R., and Zakian, V.A. (2005) The yeast Pif1p helicase removes telomerase from telomeric DNA. *Nature*, **438**, 57–61.

103 Michor, F., Iwasa, Y., and Nowak, M.A. (2004) Dynamics of cancer progression. *Nat. Rev. Cancer*, **4**, 197–205.

2
Transcription Termination by RNA Polymerase II

Minkyu Kim and Stephen Buratowski

Considerable progress has been made in understanding transcription initiation by RNA polymerase II (RNApII). In contrast, much less is known about the mechanisms that terminate transcription, although recent years have seen significant advances in this area. Accurate termination is important because stopping transcription too late or even too early can disrupt normal gene expression. Like initiation, termination is likely to be a complex, regulated process controlled by multiple mechanisms. This review summarizes recent progress in the field of RNApII termination.

2.1
Messenger RNA Gene Termination

Early studies of eukaryotic gene expression established that termination of mRNA encoding genes is tightly coupled to the cleavage/polyadenylation reaction that modifies mRNA 3′ ends. Two models were proposed to explain this coupling. The “allosteric” or “antiterminator” model hypothesizes that the poly(A) signals at 3′ ends of genes, sensed as either DNA or RNA, can change the properties of the elongating RNApII complex independently of transcript cleavage. Termination could be triggered by dissociation of positive elongation factors or recruitment of termination factors [1]. The “torpedo” model postulates that cleavage of the nascent RNA transcript at the poly(A) site signals to RNApII, leading to destabilization of the template/transcript/RNApII ternary complex [2]. The important difference between the two models is that cleavage of the nascent RNA at the poly(A) site is not required for transcription termination in the “allosteric” model, whereas it is an obligatory step for the “torpedo” model. However, it is important to note that the two models are not mutually exclusive.

Posttranscriptional Gene Regulation: RNA Processing in Eukaryotes, First Edition. Edited by Jane Wu.

2.1.1 The Allosteric Model

This model suggests that RNApII becomes less competent for elongation when it reaches the poly(A) signal at the 3′ end of genes, perhaps due to an exchange of protein factors or a conformational change within the RNApII itself. It is clear that various proteins bind to RNApII only at certain stages of the transcription cycle and many of these dynamic transitions are mediated by the carboxy-terminal domain (CTD) of RNApII. The CTD consists of multiple repeats of the amino acid sequence YSPTSPS with Ser2, Thr4, Ser5, and Ser7 being phosphorylated at various stages of transcription [3–6], comprising a "CTD code" [7, 8]. Recent genome-wide studies show that the changing patterns of these phosphorylation marks (except Thr4, which is not fully analyzed yet) are similar at all RNApII-transcribed genes and not scaled to gene length [9–11].

Ser5 is phosphorylated by basal transcription factor TFIIH (Kin28 in yeast and Cdk7 in higher eukaryotes) at the promoter, and the mRNA capping enzyme is recruited to early transcription complexes by binding specifically to this modified form of the CTD [12] (Figure 2.1). As RNApII escapes from the promoter, Ser5P is reduced and the capping enzyme dissociates. Ser2 is first phosphorylated at a low level by Bur1 (Cdk9 in higher eukaryotes) and then increasingly phosphorylated by the Ctk1 kinase (Cdk12 in higher eukaryotes) as RNApII moves further downstream [3, 4, 13, 14]. Similar to the transition in CTD phosphorylation and capping enzyme association at the 5′ ends of genes, another transition occurs at the 3′ ends. Downstream of polyadenylation sites, levels of Ser2P begin to drop, and several polymerase-

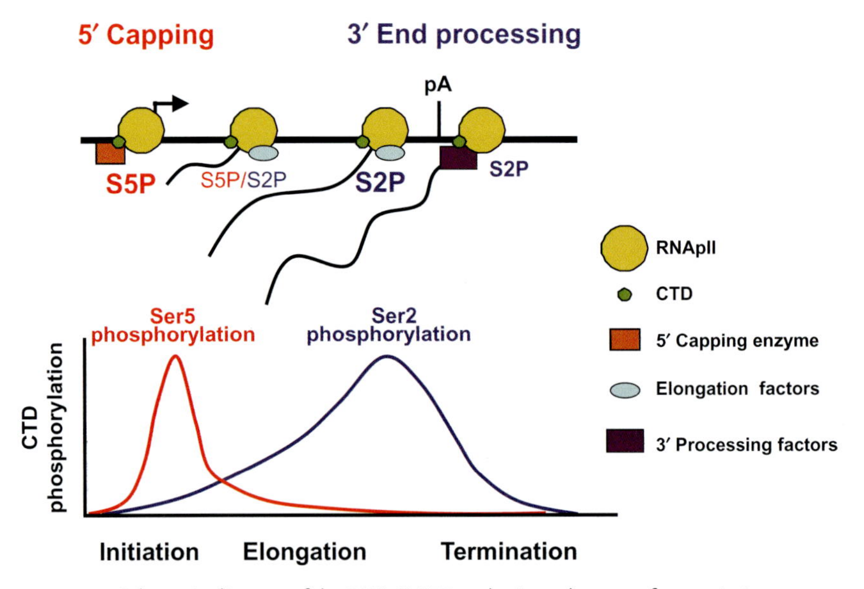

Figure 2.1 Schematic diagram of the RNApII CTD cycle. At each stage of transcription, a particular CTD phosphorylation pattern dominates and is recognized by the appropriate elongation or mRNA processing factors.

associated factors (e.g., PAF1 and TREX complexes) are apparently released. At roughly the same time, RNA 3′ cleavage and polyadenylation factors are strongly recruited to genes near the 3′ ends [15, 16]. These complementary patterns of recruitment are consistent with an exchange of factors bound to RNApII. Along these lines, the serine/arginine-rich (SR)-like protein Npl3 was recently shown to promote elongation and antagonize the binding of polyadenylation factors (e.g., Rna15) to nascent RNA transcripts until it is released from RNApII at the 3′ end of genes [17]. Similarly, Sub1, a yeast homolog of the mammalian transcriptional coactivator PC4, was suggested to function as an antiterminator to prevent premature termination during elongation phase through its interaction with Rna15 [18]. These data all provide experimental support for the "allosteric" model.

Visual support for cleavage-independent termination comes from an electron microscopy (EM) survey of transcription of more than 100 genes in *Drosophila* [19]. Although cotranscriptional cleavage of RNA transcripts was observed at the 3′ end of some genes, in many cases, the intact full-length RNA transcripts were still attached to RNApII up to a discrete termination site along the gene, indicating that prior RNA cleavage at the poly(A) site might not be a prerequisite to RNApII termination in the fly. In yeast, termination of the *CYC1* gene was not affected by a cleavage-defective *pcf11* mutant, arguing against an obligatory role of RNA cleavage in termination [20]. Instead, the CTD-interacting domain (CID) of Pcf11 was shown to be important for the termination of this gene. One possibility is that *CYC1* can terminate by a different mechanism (see the small nucleolar RNA (snoRNA) pathway below). Alternatively, the RNA cleavage/polyadenylation factors may promote the termination of RNApII independently of the actual cleavage at the poly(A) site. Interestingly, the Pcf11 protein was shown to bridge the CTD of RNApII and the nascent RNA transcript, leading to dismantling of an RNApII elongation complex *in vitro* [21].

2.1.2 The Torpedo Model

This model postulates an essential role for endonucleolytic RNA cleavage, which may in turn generate an entry site for a 5′–3′ exoribonuclease that can trigger termination [2]. Supporting this notion, RNA cleavage factors (Rna14, Rna15, Pcf11) but not polyadenylation factors (Pap1, Fip1, Yth1) are required for termination [22]. Strong support for the torpedo model emerged from the discovery that the 5′–3′ exonuclease Rat1 is necessary for mRNA termination. While studying the yeast CTD Ser2P-binding protein Rtt103, a termination complex was isolated containing Rtt103, RNApII, and the Rat1–Rai1 complex [23]. Interestingly, these proteins were preferentially recruited to the 3′ ends of genes, but mutation of Rat1 (*rat1-1*) did not affect RNA cleavage or polyadenylation. Rat1 was shown to degrade the transcript downstream of the poly(A) site. This degradation apparently triggers transcription termination via a still unknown mechanism, presumably by destabilizing the transcript/template/RNApII interaction [23] (Figure 2.2). The exonuclease activity of Rat1 is critical for this process since an exonuclease-inactive

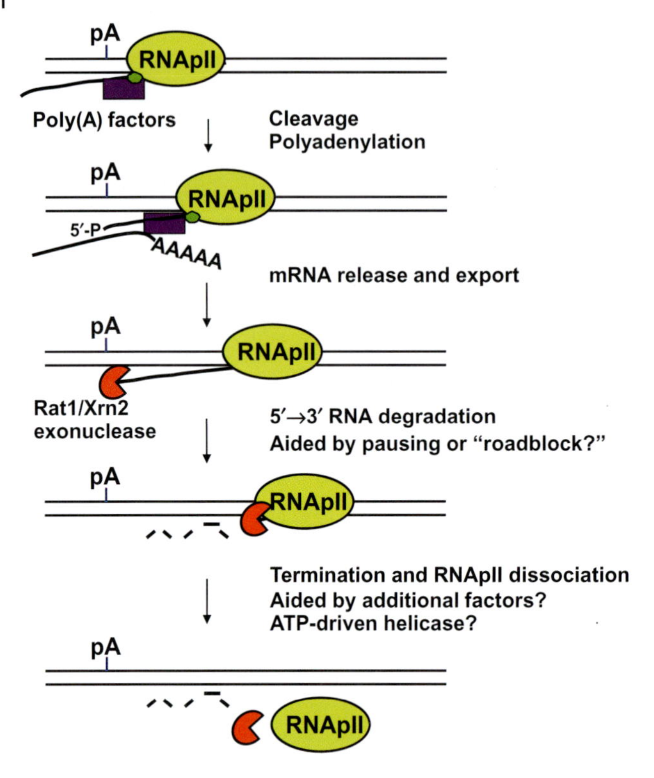

Figure 2.2 Torpedo termination model by Rat1/Xrn2. Cleavage at the polyadenylation site creates an entry site for Rat1/Xrn2 5′–3′ exoribonuclease that degrades the transcripts downstream of the poly(A) site. Upon catching up to the elongating RNApII, this nuclease somehow triggers termination by destabilizing the RNA/DNA/RNApII ternary complex.

rat1 (*D235A*) mutant could not support termination. Complementary results in mammalian cells [24] indicate an evolutionarily conserved role for Rat1 (Xrn2 in human) in termination. Interestingly, these studies also suggested that mRNA cleavage events other than that performed by the polyadenylation machinery can also trigger Rat1 entry and termination. These findings revealed an exonuclease involved in termination, and strongly support the "torpedo" model.

Several studies have implicated RNApII pausing in termination [25–27]. Such pausing can be intrinsic to certain nucleotide sequences, but it can also be induced when RNApII is blocked by a DNA-binding protein or perhaps a particular chromatin feature. The torpedo model provides a satisfying explanation for these observations. Given that the RNApII continues transcribing past the poly(A) site, the Rat1 exonuclease must "catch up" to the elongation complex to trigger termination. A pause site would be predicted to increase the probability of Rat1 reaching the elongating RNApII.

The essential role in termination of RNA degradation downstream of a 3′ cleavage site by Rat1/Xrn2 was questioned by Luo *et al.* [28]. This study found that relocating the cytoplasmic 5′–3′ exonuclease Xrn1 to the nucleus (NLS-Xrn1) could restore downstream RNA degradation in *rat1-1* cells, but did not reverse the termination defect [28]. Bentley and colleagues suggested that Rat1 could help recruit polyadenylation factors and that the termination defect in the *rat1-1* mutant might be due to a deficiency in this function [28]. However, inactivation of Rat1 or depletion of Xrn2 from cells did not affect RNA cleavage and polyadenylation [23, 24], which argues against this model. More likely, NLS-Xrn1 can degrade the downstream RNA (perhaps posttranscriptionally), but fails to rescue the termination defect in *rat1-1* cells because it cannot make the same interactions as Rat1 with Rai1, Rtt103, or other key factors necessary for RNApII termination.

It is still not clear how Rat1 promotes termination. One possibility is that when Rat1 catches up RNApII by degrading RNA transcripts, it may be able to directly displace RNApII from the DNA template by exerting force on the polymerase or the transcript. However, a Rat1-Rai1 complex was insufficient to trigger termination in an *in vitro* assay using purified RNApII [29]. Alternatively, Rat1 may simply render RNApII vulnerable to termination by halting its elongation. An *in vitro* study shows that RNApII becomes arrested if the nascent transcript is shortened to less than 50 nucleotides [30]. So, when the RNA is trimmed down to a certain length by Rat1, RNApII could arrest and become susceptible to additional factor(s) that displace it from the DNA. This last step might be carried out by an ATP-driven helicase or translocase, similar to the bacterial proteins Rho and Mfd, which displace RNA polymerase at termination or DNA damage sites, respectively [31, 32]. Pcf11, the CTD-interacting RNA 3′ processing factor is another candidate, based on its ability to dismantle elongation complexes *in vitro* [21].

2.2
Small Nucleolar RNA Gene Termination Pathway

In addition to mRNAs, RNApII transcribes noncoding RNAs such as the small nuclear RNAs (snRNAs) and small nucleolar RNAs (snoRNAs). snoRNAs are required for the maturation of pre-rRNAs (cleavage and base modification) and are divided into two classes (C/D and H/ACA) based on conserved secondary structures [33]. The 3′ ends of snoRNAs are not polyadenylated by the mRNA poly(A) polymerase (Pap1), instead being made as longer precursors that are trimmed to a final length of 100–200 nucleotides. These observations raise the question of whether they are terminated by the same mechanisms used for mRNAs. While there are some factors common to the mRNA and snoRNA termination pathways, mutation of Rat1 or depletion of the mRNA endonuclease Brr5 caused mRNA termination defects but did not lead to transcriptional read-through at many snoRNA genes tested [34]. These results suggest that snoRNA termination does not use the mRNA "torpedo" model.

Similar experiments using other mRNA cleavage/polyadenylation factors (Rna14, Rna15) revealed that none of these is required for snoRNA termination, with one interesting exception. The Pcf11 protein is a component of Cleavage Factor IA (CFIA) and has at least two distinct functions: one domain interacts with the

CTD of RNApII, while the other promotes the mRNA cleavage reaction. By introducing a set of mutations into different parts of Pcf11, these two different functions of the protein were shown to be genetically separable [20]. Intriguingly, mutations that disrupt the cleavage reaction cause read-through at mRNA genes, consistent with the torpedo model, but not at snoRNA genes. In contrast, the CTD interaction mutants cause defective termination at snoRNAs, but not mRNAs [34]. These results not only argue that Pcf11 is required for both pathways (or that the pathways partially overlap), but also indicate that any cleavage of the snoRNAs is not mediated by the same factor(s) that act at mRNAs. It is known that snoRNAs are processed at their 3′ ends by Rnt1 (RNAse III) and the exosome (3′–5′ exonuclease complex) [35–37], but so far, mutations in these factors show no effect on termination of either snoRNAs or mRNAs [34].

Experiments from several laboratories have shown that termination at snoRNA genes in yeast requires a protein complex consisting of the sequence-specific RNA-binding proteins Nrd1 and Nab3 (recognizing GUAA/G and UCUU, respectively [38]) as well as the DNA/RNA helicase Sen1 [39–43]. Nrd1 also has a CTD-interacting domain (CID) that helps its targeting to RNApII [42, 44]. Once targeted to the RNA and the transcription complex, the Nrd1–Nab3–Sen1 complex terminates the transcription by a mechanism that is not yet understood. Sen1 may be functionally analogous to the bacterial termination factor Rho, somehow acting to disengage RNApII from the template and/or transcript using its ATP-dependent helicase activity.

Interestingly, Nrd1 complex is also associated with the exosome and the exosome-activating TRAMP complex containing noncanonical poly(A) polymerases Trf4 and Trf5 [43]. Since the exosome trims the 3′ end of snoRNAs for maturation [36, 37], this association may couple transcription termination and 3′ end processing at snoRNA genes. Recently, whole genome-scale microarray studies revealed a large number of cryptic unstable transcripts (CUTs) from cryptic promoters throughout the yeast genome [45–48]. Both Nrd1 and Sen1 are involved in the termination of these cryptic transcriptions [49–51]. However, unlike snoRNAs, the CUTs are completely degraded by nuclear exosome [47, 52, 53]. It is currently not known what regulates the extent of exosome degradation (3′ trimming vs. full degradation), but the Nrd1 complex and exosome provide a connection between termination and RNA surveillance at CUTs [54].

In addition to the Nrd1–Sen1 complex, subunits of the APT complex, which associate with the mRNA cleavage/polyadenylation factor Pta1, are also implicated in snoRNA termination [55–58]. A role for the Paf1 complex in snoRNA 3′ end formation was reported as well [59]. However, all these snoRNA termination factors, except for Pcf11 and Paf1C, do not normally function at mRNA genes, suggesting that mechanisms of termination at mRNA and snoRNA genes should be quite distinct.

The existence of proteins homologous to yeast Nrd1 and Sen1 in the mammalian genome suggests that higher eukaryotes may also have an alternative termination mechanism equivalent to the yeast snoRNA pathway. In mammals, however, snoRNAs are mostly processed from mRNA introns, indicating that mammalian snoRNA genes are not the target of this termination complex. Instead, considering widespread noncoding RNA transcription throughout the mammalian genome [60], this complex may primarily terminate CUTs and other noncoding RNAs in mammals. In higher eukaryotes, transcripts for snRNAs and replication-dependent histone mRNAs are also produced by RNApII but non-polyadenylated. Recently, CTD Ser7P and Thr4P were found to promote 3′ end processing of snRNAs and histone mRNAs, respectively [6, 61]. Since these CTD phosphorylations are not limited to these specific gene classes, protein factor(s) recognizing these marks must be recruited or functional only in the proper gene context. As in the case of the yeast snoRNA pathway, specific RNA sequences may contribute an additional layer of specificity (see Section 2.3).

2.3
Choice between the Two Termination Pathways

When termination fails at the 3′ end of snoRNA genes (e.g., snR13 and snR33), RNApII keeps transcribing until it reaches the poly(A) site of the downstream mRNA gene. Conversely, when the mRNA "torpedo" pathway fails, RNApII may utilize the alternative snoRNA termination pathway because RNApII eventually terminates even when polyadenylation factors or Rat1 is inactivated [34]. The non-polyadenylated mRNAs that are made under these conditions are retained at the site of transcription and degraded by the exosome [62]. These observations suggest that the two termination pathways function independently under different parameters, yet serve as "backup" systems to each other. Indeed, high-resolution studies of RNApII distribution across the yeast genome revealed that a Sen1 mutation (*sen1 E1597K*) exhibits defective termination at the majority of snoRNA genes but at less than 1% of mRNA genes [49]. Intriguingly, the affected mRNA genes are generally short (less than 600 bp), as are most snoRNA genes. Similarly, very few mRNA transcripts were affected in *nrd1-5* mutant [47].

As assayed by chromatin immunoprecipitation experiments in yeast, both mRNA and snoRNA termination/3′ processing factors are recruited to RNApII-transcribed genes, regardless of whether they encode mRNA or snoRNA [34]. Interestingly, the Nrd1–Nab3 complex shows strongest cross-linking to the 5′ ends of genes and polyadenylation factors to the 3′ ends [34]. Therefore, the choice of a termination pathway is not simply dependent on the recruitment of specific termination machinery to RNApII. These cross-linking experiments cannot distinguish whether the two termination complexes are simultaneously bound to a single RNApII or whether they are dynamically exchanging, but in either case, the elongating yeast RNApII apparently keeps both termination options available in preparation for the appropriate cues to emerge.

Given this overlap, how does RNApII choose which of the two termination pathways to use? Several things are likely to affect the choice. First, the identity of a gene as mRNA or snoRNA could be defined by the promoter. In mammalian cells, snRNA promoters are recognized by a unique TATA-binding protein

(TBP)-containing complex known as snRNA-activating protein complex (SNAPc) [63–65]. This complex may help to recruit Integrator, the complex that carries out snRNA 3′ processing [66]. *Saccharomyces cerevisiae* does not have either SNAPc or Integrator homologs; rather, sn/snoRNA promoters are thought to use the same transcription factor IID (TFIID) complex as mRNA promoters.

A second possibility is that sequences within the RNA could bind to proteins that confer identity on the transcript. Assembly of mRNPs by the hnRNP-like proteins Hrp1 and Npl3 may specify mRNA, while cotranscriptional assembly of snRNPs could lead RNApII to choose the snoRNA termination pathway. Supporting this idea, Hrp1 and Npl3 have been shown to affect mRNA 3′ end formation and/or termination [17, 67, 68]. Similarly, mutations of snoRNP components were shown to affect 3′ end formation and termination of snoRNA genes [69–71]. These results suggest that cotranscriptionally assembled RNA/protein complexes might interact with the downstream 3′ end processing machineries and guide RNApII to a specific termination pathway.

Another important feature that must affect the choice is specific RNA sequences at the 3′ end of genes. In fact, each termination pathway has sequence-specific RNA-binding proteins (Figure 2.3). For mRNA cleavage and polyadenylation, consensus sequences (e.g., AAUAAA and G/U-rich) have been identified and specific RRM (RNA recognition motif)-containing cleavage factors (e.g., Rna15) are known to bind them [72, 73]. Similarly, for snoRNA 3′ end maturation and termination,

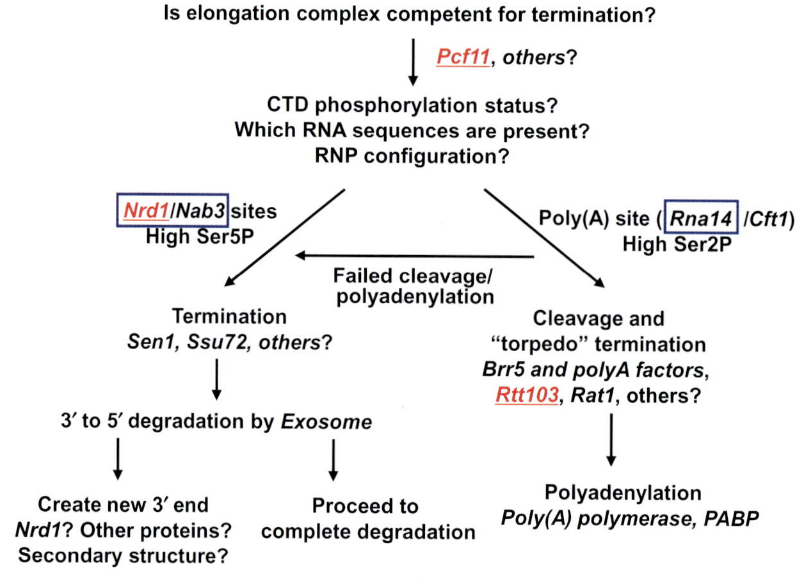

Figure 2.3 A choice of termination pathway by RNApII. When RNApII reaches the 3′ end, a "decision" may be made, depending on the CTD phosphorylation status and RNA sequences. If high CTD Ser5P level and/or multiple Nrd1–Nab3 binding sites are present, termination would occur by the Nrd1–Sen1 pathway. But if CTD Ser2P and/or cleavage/poly(A) signals are dominant, cleavage/poly(A) factors would lead the polymerase to terminate transcription by the Rat1–torpedo pathway. Note that each pathway has two distinct CID proteins (in red) and RNA-binding protein(s) (boxed in blue).

sequence-specific RRM-containing proteins Nrd1 and Nab3 are involved (e.g., Nrd1 interacting with the sequence GUAA) [38]. Therefore, the frequency and affinity of specific RNA sequences could determine the termination mechanism. Intriguingly, the dominant RNA sequences appearing at the 3′ end of mRNA genes in *Chlamydomonas reinhardtii* are UGUAA [74], instead of AAUAAA, raising the possibility that mRNA gene termination in this algae could involve an Nrd1-dependent mechanism or that the mRNA cleavage/polyadenylation machinery in this organism has a different binding specificity.

Finally, it is important to note that each termination pathway involves two CID proteins: Pcf11 and Rtt103 in the mRNA pathway, and Pcf11 and Nrd1 in the snoRNA pathway (Figure 2.3). Interestingly, these CID proteins differ in the specificity of CTD binding. The Rtt103 CID has a strong preference for Ser2P of the CTD [23], while Pcf11 has a preference for Ser2P CTD but also appreciable affinity for the nonphosphorylated CTD [75]. In contrast, the Nrd1 CID binds specifically to Ser5P of the CTD [44]. These striking differences in CTD binding suggest that one contributor to the choice between the mRNA and snoRNA pathways would be the CTD phosphorylation state of the polymerase, which is in turn related to how far it has transcribed from the promoter. Indeed, the Ser5P-specific binding of Nrd1 helps to explain why Nrd1–Sen1-dependent termination is preferentially used at short genes (less than 600 nt), where Ser5P predominates over Ser2P [44, 49, 76]. However, a CID deletion mutant of Nrd1 is viable and does not cause termination defects at many short and long genes tested [44], indicating that the binding of Nrd1 to Ser5P CTD is not absolutely required for the snoRNA termination pathway. On the other hand, the CID of Pcf11 is essential for viability and specifically affects only snoRNA termination [34], suggesting that Pcf11 CID might play a more crucial role in the choice. Since the level of Ser2P CTD is very low at snoRNA genes ([9–11] M. Kim, unpublished data), the ability of Pcf11 to bind nonphosphorylated or low level of Ser2P CTD might be particularly important to promote snoRNA termination in yeast.

2.4 Regulation of Transcription by Termination

2.4.1 Transcription Attenuation, Promoter Upstream/Associated Transcription, and Pausing of RNApII

RNA sequences for Nrd1–Sen1-dependent termination are mostly found at the 3′ end of snoRNA genes and small number of short mRNA genes. But in some yeast genes, an Nrd1–Sen1-dependent terminator exists at their 5′ UTR and/or early coding regions and regulates transcription by premature termination (i.e., attenuation). The *NRD1* gene has multiple Nrd1–Sen1 terminators at the 5′ end, so that its mRNA level is autoregulated by a negative feedback of premature termination [77]. The *IMD2* gene has two transcription start sites (TSS) separated by an Nrd1–Sen1 terminator within the same promoter. In high guanosine-5′-triphosphate

(GTP), transcription initiates at the upstream TSS with guanosine (G) but terminates soon due to adjacent Nrd1–Sen1 terminator, producing a CUT that is degraded by the exosome. However, when the GTP level is low, transcription initiates at the downstream TSS with adenosine (A), so that RNApII avoids attenuation and generates functional *IMD2* mRNAs [49, 78, 79]. This explains why the expression of *IMD2* mRNA is inversely correlated with cellular GTP concentration. Similar CUT-mediated premature termination was also observed at *URA2* and other nucleotide biosynthesis genes [80], implying that transcription attenuation is commonly utilized at many eukaryotic genes, as often seen in many bacterial operons [81].

Recent studies using ChIP-chip assays (chromatin immunoprecipitation followed by DNA microarray) or ChIP-seq (ChIP coupled to high-throughput sequencing) in the human and fly genomes revealed that RNApII is highly enriched at the 5′ promoter-proximal region of many genes relative to downstream regions [82–85], indicating that transition into productive elongation is another important regulatory step in transcription. Similarly, a nonactivated form of RNApII at the 5′ end of genes, distinct from activated RNApII reaching the 3′ end, has been reported in yeast [86]. Two protein complexes have been extensively studied to explain pausing of RNApII: DSIF (DRB sensitivity-inducing factor, Spt4/Spt5) [87] and NELF (negative elongation factor, NELF-A, B, C/D, E) [88]. However, depletion of NELF or DSIF does not lead to complete loss of paused RNApII or high constitutive expression [84, 89], suggesting that factors other than these two may also involve keeping RNApII close to the promoter.

Several termination factors are present at the 5′ promoter and early coding regions as well as the 3′ end of yeast genes [15, 23, 90], suggesting that these termination factors could play multiple roles during transcription. One possible function at the 5′ ends of genes might be to prematurely terminate transcription as observed in several yeast biosynthetic genes (see above). Along the same line, another termination factor, Pcf11 represses transcription of the HIV provirus by causing premature termination [91]. These findings raise the intriguing hypothesis that higher RNApII signals seen at the promoters of many genes might, at least in part, result from premature termination or cessation of elongation by termination machineries.

Recently, large-scale parallel sequencing of RNA transcripts associated with the transcriptionally engaged RNApII revealed unique peaks near the transcription start site (TSS) in both sense (+50 nt) and antisense (−250 nt) directions at ~30% of human genes [92]. This indicates that transcription initiates divergently from two separate preinitiation complexes at many promoters [93], but does not extend beyond the promoters, which is likely to be correlated to higher RNApII occupancy near promoters shown previously [82–85]. By analyzing short RNA cDNA libraries from mouse cell lines, Seila *et al.* also observed identical RNA peaks in both directions (+50 and −250 nt) [94]. Interestingly, generation of these TSS-associated RNAs (TSSa-RNAs) is not affected by lack of Dicer (an RNase III necessary for microRNA processing), and its 5′ end has a quite random nucleotide distribution instead of the preferred G residue seen at transcription initiation sites [94]. Therefore, these transcripts are likely decapped or 5′ processed by other ribonuclease(s). If RNAs are cleaved soon after initiation by an endonuclease, transcription could stop near the promoter by Rat1/Xrn2-dependent termination. Alternatively, given that 5′ capping protects RNAs from degradation by Rat1 [95], Rat1 might compete with capping enzymes, and begin to degrade the nascent RNAs even before the 5′ cap is formed. In either case, TSSa-RNAs (average 20 nt) could be protection products by RNApII from Rat1/Xrn2 degradation. Supporting a competition between Rat1/Xrn2 and capping enzymes, Rat1-mediated RNApII termination occurs just downstream of the promoter in capping-defective *rpb1-N488D* or *ceg1-63* mutants [96]. Furthermore, Rai1 (Rat1-interacting cofactor) appears to be a decapping endonuclease, removing the 5′ cap from RNA transcripts that is not methylated [97]. Interestingly, the cap methyltransferase (Abd1 in yeast) cross-links throughout the gene body in ChIP assays [3]. These mechanisms may comprise a quality control step to insure the 5′ integrity of nascent transcripts and to prevent premature termination by Rat1.

In addition to regulatory transcripts initiating from the same promoter, transcription from an adjacent promoter can also regulate downstream gene expression. The yeast *SER3* gene encodes an enzyme involved in serine biosynthesis, and its transcription is tightly repressed in the presence of serine by transcriptional interference initiated from the upstream noncoding *SRG1* promoter [98]. *SRG1* transcription reads through the *SER3* promoter, disrupting the formation of initiation complexes at the *SER3* promoter. Similarly, during zinc deficiency, an upstream intergenic transcript *ZRR1* interferes with the binding of activator to the *ADH1* and *ADH3* promoters, resulting in transcriptional repression of these two major zinc-requiring enzymes [99].

Using oligo-d(T)-priming and subsequent tiling microarray analysis of human transcripts, polyadenylated transcripts were found enriched upstream of promoters (−1 kb region) upon depletion of an exosome component [100]. Similarly, inactivation of exosome subunits led to accumulation of promoter-associated noncoding transcripts in *Arabidopsis* [101]. These exosome substrates in higher eukaryotes are likely similar to CUTs in yeast [45, 47, 52], suggesting that their termination might be mediated by yeast Nrd1 and Sen1 homologs in these organisms. It is not clear whether these transcripts are just noise or functional, but given that they are intimately linked to downstream gene transcription, defective termination of these upstream transcripts could severely interfere with downstream transcription initiated at the promoter.

2.4.2
Alternative Polyadenylation and Termination

In higher eukaryotes, proteins of different activities are often derived from a single gene by alternative polyadenylation and termination. A classic example is the IgM gene of B lymphocyte having two poly(A) sites: use of one poly(A) signal during development produces membrane-bound form of IgM, while use of the other weaker poly(A) signal generates secreted form [102]. Indeed, large-scale analyses revealed that more than half of mammalian genes have alternative poly(A) sites, and that their utilization patterns seem to be evolutionarily conserved [103–106]. In addition, recent discoveries of alternative polyadenylation in *Arabidopsis*, rice, and *Chlamydomonas* suggest that it may be a quite common mechanism for transcript diversity and gene regulation among eukaryotes [74, 107, 108].

The mechanisms for alternative polyadenylation and termination are not fully understood, but studies on several genes suggest multiple possible mechanisms. First, it could be mediated by a different promoter choice. In the brain of rats and mice, alternatively polyadenylated β-adductin mRNAs are driven by a brain-specific promoter [109], and alternative termination of the Opitz syndrome gene *MID1* in different tissues seems to be predetermined by the choice of alternative 5′ UTRs and promoters [110]. Second, structures of poly(A) sites may dictate their utilization. Retroviruses (including HIV-1) and retrotransposons often have two poly(A) sites located at their 5′ and 3′ LTRs (long terminal repeats), but only the 3′ poly(A) site normally directs cleavage and polyadenylation. Intriguingly, computational folding predictions of the poly(A) sites from multiple HIV-1 strains revealed that 5′ poly(A) sites are mostly in a hairpin loop while the 3′ poly(A) sites have a propensity for a linear structure [111]. The predicted hairpin structure at the 5′ poly(A) site may inhibit premature polyadenylation and termination, guiding RNApII to further elongate and use the 3′ poly(A) site for generating full-length viral transcripts. Alternatively, the strength of poly(A) signals would be important. In mouse primary B lymphocytes, the cellular level of an RNA cleavage factor CstF-64 is limiting, so that a poly(A) site with higher affinity for CstF-64 is preferentially used during the transcription of the IgM gene, generating the membrane-bound form. At the final stage of B-lymphocyte differentiation (plasma cells), however, the expression of CstF-64 is increased, and the weaker poly(A) site becomes recognized and utilized, resulting in the secreted form of IgM [112]. Additionally, direct competition or interference of the CstF-64 binding to RNA by other RNA-binding proteins recognizing similar RNA sequences (U-rich) was also shown to affect the choice of poly(A) sites [113, 114].

Interestingly, many tissue-specific alternative poly(A) sites have suboptimal AGUAAA polyadenylation sequences [109, 115], and a UCUU element was often found to associate with brain-specific poly(A) sites [116]. Coincidentally, these two sequence elements overlap the consensus binding sites of two yeast snoRNA termination factors Nrd1 and Nab3 (GUAA/G and UCUU, respectively) [38]. One might speculate that alternative polyadenylation choices might also be derived from alternative choice of termination pathways, although functional homologs of Nrd1 and Nab3 in mammals are not yet fully characterized.

2.5 Mechanisms of Termination by Other RNA Polymerases

RNA polymerases I and III (RNApI and RNApIII) share five subunits (Rpb5, Rpb6, Rpb8, Rpb10, and Rpb12) with RNApII, and recent structural studies elucidate that all three RNA polymerases have similar overall shapes (reviewed in [117]). However, detailed structural differences among them are likely to reflect differences in target gene recognition and transcription mechanisms.

Termination by RNApI, which transcribes rRNA genes, is known to be affected by a sequence-specific DNA-binding protein (Reb1 in yeast and transcription termination factor I (TTF-I) in mouse) [118–120]. Reb1 binds to DNA downstream of a T-rich sequence at the 3′ end of the 25S rRNA sequence. RNApI elongation is blocked by the factor, promoting termination at the upstream T-rich element [121, 122]. In yeast, the RNAse III endonuclease (Rnt1) cleaves a stem-loop structure at the 3′ end of the rRNA gene upstream of the Reb1-dependent termination site [123]. Deletion of Rnt1 leads to an RNApI termination defect [124, 125], suggesting that yeast RNApI termination might involve cotranscriptional RNA cleavage at the 3′ end, similar to poly(A) cleavage at the 3′ end of mRNA genes. Several groups tested the hypothesis that rRNA might terminate by a Rat1-dependent "torpedo" mechanism. Indeed, these studies confirmed that Rat1 and its cofactor Rai1 are required for efficient RNApI termination near the Reb1-binding site [126, 127]. Considering the close proximity (44 nt) between Rnt1 cleavage and the Reb1-dependent termination sites [123], it is likely that the role of Reb1 is to pause or slow down RNApI, allowing more time for Rat1 to catch up RNApI. No RNAse III site has been identified at the 3′ end of the mammalian 28S rRNA gene. Instead, in mice, an additional factor (polymerase I and transcript release factor (PTRF)) has been found to facilitate rRNA transcript release and termination along with TTF-I [128]. Considering that PTRF is not homologous to Rat1 and is not a helicase, the mechanism of RNApI termination in mammals may be different from that in yeast.

RNApIII transcribes tRNAs, 5S rRNA, and U6 snRNA as well as other noncoding RNAs such as 7SL RNA, 7SK RNA, some snoRNAs, at least one microRNA, and SINEs (short interspersed DNA elements)-encoded RNAs (reviewed in [129]). RNApIII is quite unique among other eukaryotic RNA polymerases in that it terminates without the need of any accessory factors upon recognizing a stretch of thymidine residues (4–6, varied among different organisms) on the nontranscribed strand [130–132]. Unlike the intrinsic terminators of bacterial RNA polymerase, RNA hairpin sequences are not seen upstream of these T clusters. Instead, flanking sequences beyond the T clusters affect the efficiency of termination [133]. Considering that T clusters are generally thought to pause RNA polymerases and lead to a weak RNA–DNA hybrid [134], RNApIII termination may involve extensive pausing and subsequent transcript release [135]. Recent add-back experiments of C53 and C37 subunits to an incomplete form of RNApIII revealed a key role of the C53/C37 subcomplex in the recognition of termination signals and slowing down the elongation rate of RNApIII [136]. These two subunits, which may be analogous to the RNApII initiation factor TFIIF, are positioned at the front end of the DNA-binding cleft, where they might be able to read the incoming termination signals and induce a conformational change of RNApIII into a termination-prone, slow elongation state [137].

2.6 Future Perspectives

The identification of the role of Rat1/Xrn2 5′–3′ exonuclease as a termination factor significantly improved our understanding of termination by eukaryotic RNApII (and RNApI as well). The existence of a distinct snoRNA termination pathway revealed that there can be multiple mechanisms for RNApII termination. Despite these advances, the molecular mechanisms that directly dissociate RNApII from the DNA template remain elusive. It may require additional, as yet unidentified, factors that interact with RNApII and/or key termination factors. Detailed *in vitro* termination assays will probably be necessary to elucidate this final step of termination.

Another area requiring more study is the mechanisms by which termination is regulated. There are several examples where termination represses transcription or produces alternative gene products in response to cellular signals and needs. Like transcriptional activators/repressors modulating the initiation process, there might be factors or signaling pathways activating or repressing termination at a particular site. Interestingly, sumoylation and phosphorylation of 3′ processing/termination factors have been detected [39, 138–140], suggesting that posttranslational modifications could regulate the activities of termination factors. Along this line, a genetic interaction between Nrd1–Nab3 and Ras/PKA pathways was recently reported, and dephosphorylation of Nrd1 and formation of nuclear speckles containing Nrd1–Nab3 was found upon glucose depletion [141]. Misregulation of termination would be detrimental. In fact, severe neurological disorders are correlated with mutations in Senataxin (SETX), the presumed human ortholog of yeast termination factor Sen1 [142, 143]. Finally, it has been evident that transcription is profoundly affected and regulated by chromatin. Various chromatin remodelers and histone modifications regulate the initiation and elongation stages of transcription. Either directly or indirectly, termination could also be affected by chromatin configuration. Tantalizingly, a few studies have suggested an effect of chromatin on termination [144, 145]. Future investigations may reveal more extensive connections between transcription termination and chromatin configurations.

Acknowledgments

Work in the authors' laboratories is funded by the WCU program (338-20110002) and the Basic Science Research program (0409-20100081) through the National Research Foundation of Korea to M.K., and by the US National Institutes of Health grants GM46498 and GM56663 to S.B.

References

1 Logan, J., Falck-Pedersen, E., Darnell, J.E., Jr., and Shenk, T. (1987) A poly(A) addition site and a downstream termination region are required for efficient cessation of transcription by RNA polymerase II in the mouse beta maj-globin gene. *Proc. Natl. Acad. Sci. U.S.A.*, **84**, 8306–8310.

2 Connelly, S. and Manley, J.L. (1988) A functional mRNA polyadenylation signal is required for transcription termination by RNA polymerase II. *Genes Dev.*, **2**, 440–452.

3 Komarnitsky, P., Cho, E.J., and Buratowski, S. (2000) Different phosphorylated forms of RNA

polymerase II and associated mRNA processing factors during transcription. *Genes Dev.*, **14**, 2452–2460.

4 Cho, E.J., Kobor, M.S., Kim, M., Greenblatt, J., and Buratowski, S. (2001) Opposing effects of Ctk1 kinase and Fcp1 phosphatase at Ser 2 of the RNA polymerase II C-terminal domain. *Genes Dev.*, **15**, 3319–3329.

5 Chapman, R.D., Heidemann, M., Albert, T.K., Maihammer, R., Flatly, A., Meisterernst, M., and Eick, D. (2007) Transcribing RNA polymerase II is phosphorylated at CTD residue serine-7. *Science*, **318**, 1780–1782.

6 Hsin, J.P., Sheth, A., and Manley, J.L. (2011) RNAP II CTD phosphorylated on threonine-4 is required for histone mRNA 3′ end processing. *Science*, **334**, 683–686.

7 Buratowski, S. (2003) The CTD code. *Nat. Struct. Biol.*, **10**, 679–680.

8 Buratowski, S. (2009) Progression through the RNA polymerase II CTD cycle. *Mol. Cell*, **36**, 541–546.

9 Kim, H., Erickson, B., Luo, W., Seward, D., Graber, J.H., Pollock, D.D., Megee, P.C., and Bentley, D.L. (2010) Gene-specific RNA polymerase II phosphorylation and the CTD code. *Nat. Struct. Mol. Biol.*, **17**, 1279–1286.

10 Mayer, A., Lidschreiber, M., Siebert, M., Leike, K., Soding, J., and Cramer, P. (2010) Uniform transitions of the general RNA polymerase II transcription complex. *Nat. Struct. Mol. Biol.*, **17**, 1272–1278.

11 Bataille, A.R., Jeronimo, C., Jacques, P.E., Laramee, L., Fortin, M.E., Forest, A., Bergeron, M., Hanes, S.D., and Robert, F. (2012) A universal RNA polymerase II CTD cycle is orchestrated by complex interplays between kinase, phosphatase, and isomerase enzymes along genes. *Mol. Cell*, **45**, 158–170.

12 Cho, E.J., Takagi, T., Moore, C.R., and Buratowski, S. (1997) mRNA capping enzyme is recruited to the transcription complex by phosphorylation of the RNA polymerase II carboxy-terminal domain. *Genes Dev.*, **11**, 3319–3326.

13 Qiu, H., Hu, C., and Hinnebusch, A.G. (2009) Phosphorylation of the Pol II CTD by KIN28 enhances BUR1/BUR2 recruitment and Ser2 CTD phosphorylation near promoters. *Mol. Cell*, **33**, 752–762.

14 Bartkowiak, B., Liu, P., Phatnani, H.P., Fuda, N.J., Cooper, J.J., Price, D.H., Adelman, K., Lis, J.T., and Greenleaf, A.L. (2010) CDK12 is a transcription elongation-associated CTD kinase, the metazoan ortholog of yeast Ctk1. *Genes Dev.*, **24**, 2303–2316.

15 Kim, M., Ahn, S.H., Krogan, N.J., Greenblatt, J.F., and Buratowski, S. (2004) Transitions in RNA polymerase II elongation complexes at the 3′ ends of genes. *EMBO J.*, **23**, 354–364.

16 Ahn, S.H., Kim, M., and Buratowski, S. (2004) Phosphorylation of serine 2 within the RNA polymerase II C-terminal domain couples transcription and 3′ end processing. *Mol. Cell*, **13**, 67–76.

17 Dermody, J.L., Dreyfuss, J.M., Villen, J., Ogundipe, B., Gygi, S.P., Park, P.J., Ponticelli, A.S., Moore, C.L., Buratowski, S., and Bucheli, M.E. (2008) Unphosphorylated SR-like protein Npl3 stimulates RNA polymerase II elongation. *PLoS One*, **3**, e3273.

18 Calvo, O. and Manley, J.L. (2001) Evolutionarily conserved interaction between CstF-64 and PC4 links transcription, polyadenylation, and termination. *Mol. Cell*, **7**, 1013–1023.

19 Osheim, Y.N., Sikes, M.L., and Beyer, A.L. (2002) EM visualization of Pol II genes in *Drosophila*: most genes terminate without prior 3′ end cleavage of nascent transcripts. *Chromosoma*, **111**, 1–12.

20 Sadowski, M., Dichtl, B., Hubner, W., and Keller, W. (2003) Independent functions of yeast Pcf11p in pre-mRNA 3′ end processing and in transcription termination. *EMBO J.*, **22**, 2167–2177.

21 Zhang, Z., Fu, J., and Gilmour, D.S. (2005) CTD-dependent dismantling of the RNA polymerase II elongation complex by the pre-mRNA 3′-end processing factor, Pcf11. *Genes Dev.*, **19**, 1572–1580.

22 Birse, C.E., Minvielle-Sebastia, L., Lee, B.A., Keller, W., and Proudfoot, N.J.

(1998) Coupling termination of transcription to messenger RNA maturation in yeast. *Science*, **280**, 298–301.

23 Kim, M., Krogan, N.J., Vasiljeva, L., Rando, O.J., Nedea, E., Greenblatt, J.F., and Buratowski, S. (2004) The yeast Rat1 exonuclease promotes transcription termination by RNA polymerase II. *Nature*, **432**, 517–522.

24 West, S., Gromak, N., and Proudfoot, N.J. (2004) Human 5′ --> 3′ exonuclease Xrn2 promotes transcription termination at co-transcriptional cleavage sites. *Nature*, **432**, 522–525.

25 Birse, C.E., Lee, B.A., Hansen, K., and Proudfoot, N.J. (1997) Transcriptional termination signals for RNA polymerase II in fission yeast. *EMBO J.*, **16**, 3633–3643.

26 Orozco, I.J., Kim, S.J., and Martinson, H.G. (2002) The poly(A) signal, without the assistance of any downstream element, directs RNA polymerase II to pause *in vivo* and then to release stochastically from the template. *J. Biol. Chem.*, **277**, 42899–42911.

27 Nag, A., Narsinh, K., and Martinson, H.G. (2007) The poly(A)-dependent transcriptional pause is mediated by CPSF acting on the body of the polymerase. *Nat. Struct. Mol. Biol.*, **14**, 662–669.

28 Luo, W., Johnson, A.W., and Bentley, D.L. (2006) The role of Rat1 in coupling mRNA 3′-end processing to transcription termination: implications for a unified allosteric-torpedo model. *Genes Dev.*, **20**, 954–965.

29 Dengl, S. and Cramer, P. (2009) Torpedo nuclease Rat1 is insufficient to terminate RNA polymerase II *in vitro*. *J. Biol. Chem.*, **284**, 21270–21279.

30 Ujvari, A., Pal, M., and Luse, D.S. (2002) RNA polymerase II transcription complexes may become arrested if the nascent RNA is shortened to less than 50 nucleotides. *J. Biol. Chem.*, **277**, 32527–32537.

31 Banerjee, S., Chalissery, J., Bandey, I., and Sen, R. (2006) Rho-dependent transcription termination: more questions than answers. *J. Microbiol.*, **44**, 11–22.

32 Deaconescu, A.M., Savery, N., and Darst, S.A. (2007) The bacterial transcription repair coupling factor. *Curr. Opin. Struct. Biol.*, **17**, 96–102.

33 Kiss, T. (2002) Small nucleolar RNAs: an abundant group of noncoding RNAs with diverse cellular functions. *Cell*, **109**, 145–148.

34 Kim, M., Vasiljeva, L., Rando, O.J., Zhelkovsky, A., Moore, C., and Buratowski, S. (2006) Distinct pathways for snoRNA and mRNA termination. *Mol. Cell*, **24**, 723–734.

35 Chanfreau, G., Rotondo, G., Legrain, P., and Jacquier, A. (1998) Processing of a dicistronic small nucleolar RNA precursor by the RNA endonuclease Rnt1. *EMBO J.*, **17**, 3726–3737.

36 Mitchell, P. and Tollervey, D. (2000) Musing on the structural organization of the exosome complex. *Nat. Struct. Biol.*, **7**, 843–846.

37 Houseley, J., LaCava, J., and Tollervey, D. (2006) RNA-quality control by the exosome. *Nat. Rev. Mol. Cell Biol.*, **7**, 529–539.

38 Carroll, K.L., Pradhan, D.A., Granek, J.A., Clarke, N.D., and Corden, J.L. (2004) Identification of cis elements directing termination of yeast nonpolyadenylated snoRNA transcripts. *Mol. Cell. Biol.*, **24**, 6241–6252.

39 Conrad, N.K., Wilson, S.M., Steinmetz, E.J., Patturajan, M., Brow, D.A., Swanson, M.S., and Corden, J.L. (2000) A yeast heterogeneous nuclear ribonucleoprotein complex associated with RNA polymerase II. *Genetics*, **154**, 557–571.

40 Steinmetz, E.J., Conrad, N.K., Brow, D.A., and Corden, J.L. (2001) RNA-binding protein Nrd1 directs poly(A)-independent 3′-end formation of RNA polymerase II transcripts. *Nature*, **413**, 327–331.

41 Steinmetz, E.J. and Brow, D.A. (1996) Repression of gene expression by an exogenous sequence element acting in concert with a heterogeneous nuclear ribonucleoprotein-like protein, Nrd1, and the putative helicase Sen1. *Mol. Cell. Biol.*, **16**, 6993–7003.

42 Steinmetz, E.J. and Brow, D.A. (1998) Control of pre-mRNA accumulation by

the essential yeast protein Nrd1 requires high-affinity transcript binding and a domain implicated in RNA polymerase II association. *Proc. Natl. Acad. Sci. U.S.A.*, **95**, 6699–6704.

43 Vasiljeva, L. and Buratowski, S. (2006) Nrd1 interacts with the nuclear exosome for 3′ processing of RNA polymerase II transcripts. *Mol. Cell*, **21**, 239–248.

44 Vasiljeva, L., Kim, M., Mutschler, H., Buratowski, S., and Meinhart, A. (2008) The Nrd1-Nab3-Sen1 termination complex interacts with the Ser5-phosphorylated RNA polymerase II C-terminal domain. *Nat. Struct. Mol. Biol.*, **15**, 795–804.

45 Davis, C.A. and Ares, M., Jr. (2006) Accumulation of unstable promoter-associated transcripts upon loss of the nuclear exosome subunit Rrp6p in *Saccharomyces cerevisiae*. *Proc. Natl. Acad. Sci. U.S.A.*, **103**, 3262–3267.

46 David, L., Huber, W., Granovskaia, M., Toedling, J., Palm, C.J., Bofkin, L., Jones, T., Davis, R.W., and Steinmetz, L.M. (2006) A high-resolution map of transcription in the yeast genome. *Proc. Natl. Acad. Sci. U.S.A.*, **103**, 5320–5325.

47 Houalla, R., Devaux, F., Fatica, A., Kufel, J., Barrass, D., Torchet, C., and Tollervey, D. (2006) Microarray detection of novel nuclear RNA substrates for the exosome. *Yeast*, **23**, 439–454.

48 Samanta, M.P., Tongprasit, W., Sethi, H., Chin, C.S., and Stolc, V. (2006) Global identification of noncoding RNAs in *Saccharomyces cerevisiae* by modulating an essential RNA processing pathway. *Proc. Natl. Acad. Sci. U.S.A.*, **103**, 4192–4197.

49 Steinmetz, E.J., Warren, C.L., Kuehner, J.N., Panbehi, B., Ansari, A.Z., and Brow, D.A. (2006) Genome-wide distribution of yeast RNA polymerase II and its control by Sen1 helicase. *Mol. Cell*, **24**, 735.

50 Arigo, J.T., Eyler, D.E., Carroll, K.L., and Corden, J.L. (2006) Termination of cryptic unstable transcripts is directed by yeast RNA-binding proteins Nrd1 and Nab3. *Mol. Cell*, **23**, 841.

51 Thiebaut, M., Kisseleva-Romanova, E., Rougemaille, M., Boulay, J., and Libri, D. (2006) Transcription termination and nuclear degradation of cryptic unstable transcripts: a role for the nrd1-nab3 pathway in genome surveillance. *Mol. Cell*, **23**, 853–864.

52 Wyers, F., Rougemaille, M., Badis, G., Rousselle, J.C., Dufour, M.E., Boulay, J., Regnault, B., Devaux, F., Namane, A., Seraphin, B., Libri, D., and Jacquier, A. (2005) Cryptic pol II transcripts are degraded by a nuclear quality control pathway involving a new poly(A) polymerase. *Cell*, **121**, 725–737.

53 Vanacova, S., Wolf, J., Martin, G., Blank, D., Dettwiler, S., Friedlein, A., Langen, H., Keith, G., and Keller, W. (2005) A new yeast poly(A) polymerase complex involved in RNA quality control. *PLoS Biol.*, **3**, e189.

54 Lykke-Andersen, S. and Jensen, T.H. (2006) CUT it out: silencing of noise in the transcriptome. *Nat. Struct. Mol. Biol.*, **13**, 860–861.

55 Dheur, S., Vo le, T.A., Voisinet-Hakil, F., Minet, M., Schmitter, J.M., Lacroute, F., Wyers, F., and Minvielle-Sebastia, L. (2003) Pti1p and Ref2p found in association with the mRNA 3′ end formation complex direct snoRNA maturation. *EMBO J.*, **22**, 2831–2840.

56 Ganem, C., Devaux, F., Torchet, C., Jacq, C., Quevillon-Cheruel, S., Labesse, G., Facca, C., and Faye, G. (2003) Ssu72 is a phosphatase essential for transcription termination of snoRNAs and specific mRNAs in yeast. *EMBO J.*, **22**, 1588–1598.

57 Steinmetz, E.J. and Brow, D.A. (2003) Ssu72 protein mediates both poly(A)-coupled and poly(A)-independent termination of RNA polymerase II transcription. *Mol. Cell. Biol.*, **23**, 6339–6349.

58 Dichtl, B., Aasland, R., and Keller, W. (2004) Functions for *S. cerevisiae* Swd2p in 3′ end formation of specific mRNAs and snoRNAs and global histone 3 lysine 4 methylation. *RNA*, **10**, 965–977.

59 Sheldon, K.E., Mauger, D.M., and Arndt, K.M. (2005) A requirement for the *Saccharomyces cerevisiae* Paf1 complex in snoRNA 3′ end formation. *Mol. Cell*, **20**, 225–236.

60 Kapranov, P., Cheng, J., Dike, S., Nix, D.A., Duttagupta, R., Willingham, A.T.,

Stadler, P.F., Hertel, J., Hackermuller, J., Hofacker, I.L., Bell, I., Cheung, E., Drenkow, J., Dumais, E., Patel, S., Helt, G., Ganesh, M., Ghosh, S., Piccolboni, A., Sementchenko, V., Tammana, H., and Gingeras, T.R. (2007) RNA maps reveal new RNA classes and a possible function for pervasive transcription. *Science*, **316**, 1484–1488.

61 Egloff, S., O'Reilly, D., Chapman, R.D., Taylor, A., Tanzhaus, K., Pitts, L., Eick, D., and Murphy, S. (2007) Serine-7 of the RNA polymerase II CTD is specifically required for snRNA gene expression. *Science*, **318**, 1777–1779.

62 Hilleren, P., McCarthy, T., Rosbash, M., Parker, R., and Jensen, T.H. (2001) Quality control of mRNA 3′-end processing is linked to the nuclear exosome. *Nature*, **413**, 538–542.

63 de Vegvar, H.E., Lund, E., and Dahlberg, J.E. (1986) 3′ end formation of U1 snRNA precursors is coupled to transcription from snRNA promoters. *Cell*, **47**, 259–266.

64 Hernandez, N. and Weiner, A.M. (1986) Formation of the 3′ end of U1 snRNA requires compatible snRNA promoter elements. *Cell*, **47**, 249–258.

65 Henry, R.W., Sadowski, C.L., Kobayashi, R., and Hernandez, N. (1995) A TBP-TAF complex required for transcription of human snRNA genes by RNA polymerase II and III. *Nature*, **374**, 653–656.

66 Baillat, D., Hakimi, M.A., Naar, A.M., Shilatifard, A., Cooch, N., and Shiekhattar, R. (2005) Integrator, a multiprotein mediator of small nuclear RNA processing, associates with the C-terminal repeat of RNA polymerase II. *Cell*, **123**, 265–276.

67 Kessler, M.M., Henry, M.F., Shen, E., Zhao, J., Gross, S., Silver, P.A., and Moore, C.L. (1997) Hrp1, a sequence-specific RNA-binding protein that shuttles between the nucleus and the cytoplasm, is required for mRNA 3′-end formation in yeast. *Genes Dev.*, **11**, 2545–2556.

68 Bucheli, M.E. and Buratowski, S. (2005) Npl3 is an antagonist of mRNA 3′ end formation by RNA polymerase II. *EMBO J.*, **24**, 2150–2160.

69 Morlando, M., Ballarino, M., Greco, P., Caffarelli, E., Dichtl, B., and Bozzoni, I. (2004) Coupling between snoRNP assembly and 3′ processing controls box C/D snoRNA biosynthesis in yeast. *EMBO J.*, **23**, 2392–2401.

70 Ballarino, M., Morlando, M., Pagano, F., Fatica, A., and Bozzoni, I. (2005) The cotranscriptional assembly of snoRNPs controls the biosynthesis of H/ACA snoRNAs in *Saccharomyces cerevisiae*. *Mol. Cell. Biol.*, **25**, 5396–5403.

71 Yang, P.K., Hoareau, C., Froment, C., Monsarrat, B., Henry, Y., and Chanfreau, G. (2005) Cotranscriptional recruitment of the pseudouridylsynthetase Cbf5p and of the RNA binding protein Naf1p during H/ACA snoRNP assembly. *Mol. Cell. Biol.*, **25**, 3295–3304.

72 Dichtl, B. and Keller, W. (2001) Recognition of polyadenylation sites in yeast pre-mRNAs by cleavage and polyadenylation factor. *EMBO J.*, **20**, 3197–3209.

73 Gross, S. and Moore, C.L. (2001) Rna15 interaction with the A-rich yeast polyadenylation signal is an essential step in mRNA 3′-end formation. *Mol. Cell. Biol.*, **21**, 8045–8055.

74 Shen, Y., Liu, Y., Liu, L., Liang, C., and Li, Q.Q. (2008) Unique features of nuclear mRNA poly(A) signals and alternative polyadenylation in *Chlamydomonas reinhardtii*. *Genetics*, **179**, 167–176.

75 Licatalosi, D.D., Geiger, G., Minet, M., Schroeder, S., Cilli, K., McNeil, J.B., and Bentley, D.L. (2002) Functional interaction of yeast Pre-mRNA 3′ end processing factors with RNA polymerase II. *Mol. Cell*, **9**, 1101–1111.

76 Gudipati, R.K., Villa, T., Boulay, J., and Libri, D. (2008) Phosphorylation of the RNA polymerase II C-terminal domain dictates transcription termination choice. *Nat. Struct. Mol. Biol.*, **15**, 786–794.

77 Arigo, J.T., Carroll, K.L., Ames, J.M., and Corden, J.L. (2006) Regulation of yeast NRD1 expression by premature transcription termination. *Mol. Cell*, **21**, 641–651.

78 Kuehner, J.N. and Brow, D.A. (2008) Regulation of a eukaryotic gene by

GTP-dependent start site selection and transcription attenuation. *Mol. Cell*, **31**, 201–211.

79 Jenks, M.H., O'Rourke, T.W., and Reines, D. (2008) Properties of an intergenic terminator and start site switch that regulate IMD2 transcription in yeast. *Mol. Cell. Biol.*, **28**, 3883–3893.

80 Thiebaut, M., Colin, J., Neil, H., Jacquier, A., Seraphin, B., Lacroute, F., and Libri, D. (2008) Futile cycle of transcription initiation and termination modulates the response to nucleotide shortage in *S. cerevisiae*. *Mol. Cell*, **31**, 671–682.

81 Merino, E. and Yanofsky, C. (2005) Transcription attenuation: a highly conserved regulatory strategy used by bacteria. *Trends Genet.*, **21**, 260–264.

82 Kim, T.H., Barrera, L.O., Zheng, M., Qu, C., Singer, M.A., Richmond, T.A., Wu, Y., Green, R.D., and Ren, B. (2005) A high-resolution map of active promoters in the human genome. *Nature*, **436**, 876–880.

83 Guenther, M.G., Levine, S.S., Boyer, L.A., Jaenisch, R., and Young, R.A. (2007) A chromatin landmark and transcription initiation at most promoters in human cells. *Cell*, **130**, 77–88.

84 Muse, G.W., Gilchrist, D.A., Nechaev, S., Shah, R., Parker, J.S., Grissom, S.F., Zeitlinger, J., and Adelman, K. (2007) RNA polymerase is poised for activation across the genome. *Nat. Genet.*, **39**, 1507–1511.

85 Zeitlinger, J., Stark, A., Kellis, M., Hong, J.W., Nechaev, S., Adelman, K., Levine, M., and Young, R.A. (2007) RNA polymerase stalling at developmental control genes in the *Drosophila melanogaster* embryo. *Nat. Genet.*, **39**, 1512–1516.

86 Akhtar, A., Faye, G., and Bentley, D.L. (1996) Distinct activated and non-activated RNA polymerase II complexes in yeast. *EMBO J.*, **15**, 4654–4664.

87 Wada, T., Takagi, T., Yamaguchi, Y., Ferdous, A., Imai, T., Hirose, S., Sugimoto, S., Yano, K., Hartzog, G.A., Winston, F., Buratowski, S., and Handa, H. (1998) DSIF, a novel transcription elongation factor that regulates RNA polymerase II processivity, is composed of human Spt4 and Spt5 homologs. *Genes Dev.*, **12**, 343–356.

88 Yamaguchi, Y., Takagi, T., Wada, T., Yano, K., Furuya, A., Sugimoto, S., Hasegawa, J., and Handa, H. (1999) NELF, a multisubunit complex containing RD, cooperates with DSIF to repress RNA polymerase II elongation. *Cell*, **97**, 41–51.

89 Lis, J.T. (2007) Imaging *Drosophila* gene activation and polymerase pausing *in vivo*. *Nature*, **450**, 198–202.

90 Nedea, E., He, X., Kim, M., Pootoolal, J., Zhong, G., Canadien, V., Hughes, T., Buratowski, S., Moore, C.L., and Greenblatt, J. (2003) Organization and function of APT, a subcomplex of the yeast cleavage and polyadenylation factor involved in the formation of mRNA and small nucleolar RNA 3′-ends. *J. Biol. Chem.*, **278**, 33000–33010.

91 Zhang, Z., Klatt, A., Henderson, A.J., and Gilmour, D.S. (2007) Transcription termination factor Pcf11 limits the processivity of Pol II on an HIV provirus to repress gene expression. *Genes Dev.*, **21**, 1609–1614.

92 Core, L.J., Waterfall, J.J., and Lis, J.T. (2008) Nascent RNA sequencing reveals widespread pausing and divergent initiation at human promoters. *Science*, **322**, 1845–1848.

93 Rhee, H.S. and Pugh, B.F. (2012) Genome-wide structure and organization of eukaryotic pre-initiation complexes. *Nature*, **483**, 295–301.

94 Seila, A.C., Calabrese, J.M., Levine, S.S., Yeo, G.W., Rahl, P.B., Flynn, R.A., Young, R.A., and Sharp, P.A. (2008) Divergent transcription from active promoters. *Science*, **322**, 1849–1851.

95 Stevens, A. and Poole, T.L. (1995) 5′-Exonuclease-2 of *Saccharomyces cerevisiae*. Purification and features of ribonuclease activity with comparison to 5′-exonuclease-1. *J. Biol. Chem.*, **270**, 16063–16069.

96 Jimeno-Gonzalez, S., Haaning, L.L., Malagon, F., and Jensen, T.H. (2010) The yeast 5′-3′ exonuclease Rat1p functions during transcription elongation by RNA polymerase II. *Mol. Cell*, **37**, 580–587.

97 Jiao, X., Xiang, S., Oh, C., Martin, C.E., Tong, L., and Kiledjian, M. (2010) Identification of a quality-control mechanism for mRNA 5′-end capping. *Nature*, **467**, 608–611.

98 Martens, J.A., Laprade, L., and Winston, F. (2004) Intergenic transcription is required to repress the *Saccharomyces cerevisiae* SER3 gene. *Nature*, **429**, 571–574.

99 Bird, A.J., Gordon, M., Eide, D.J., and Winge, D.R. (2006) Repression of ADH1 and ADH3 during zinc deficiency by Zap1-induced intergenic RNA transcripts. *EMBO J.*, **25**, 5726–5734.

100 Preker, P., Nielsen, J., Kammler, S., Lykke-Andersen, S., Christensen, M.S., Mapendano, C.K., Schierup, M.H., and Jensen, T.H. (2008) RNA exosome depletion reveals transcription upstream of active human promoters. *Science*, **322**, 1851–1854.

101 Chekanova, J.A., Gregory, B.D., Reverdatto, S.V., Chen, H., Kumar, R., Hooker, T., Yazaki, J., Li, P., Skiba, N., Peng, Q., Alonso, J., Brukhin, V., Grossniklaus, U., Ecker, J.R., and Belostotsky, D.A. (2007) Genome-wide high-resolution mapping of exosome substrates reveals hidden features in the *Arabidopsis* transcriptome. *Cell*, **131**, 1340–1353.

102 Peterson, M.L. and Perry, R.P. (1989) The regulated production of mu m and mu s mRNA is dependent on the relative efficiencies of mu s poly(A) site usage and the c mu 4-to-M1 splice. *Mol. Cell. Biol.*, **9**, 726–738.

103 Tian, B., Hu, J., Zhang, H., and Lutz, C.S. (2005) A large-scale analysis of mRNA polyadenylation of human and mouse genes. *Nucleic Acids Res.*, **33**, 201–212.

104 Ara, T., Lopez, F., Ritchie, W., Benech, P., and Gautheret, D. (2006) Conservation of alternative polyadenylation patterns in mammalian genes. *BMC Genomics*, **7**, 189.

105 Lee, J.Y., Yeh, I., Park, J.Y., and Tian, B. (2007) PolyA_DB 2: mRNA polyadenylation sites in vertebrate genes. *Nucleic Acids Res*, **35**, D165–D168.

106 Lee, J.Y., Ji, Z., and Tian, B. (2008) Phylogenetic analysis of mRNA polyadenylation sites reveals a role of transposable elements in evolution of the 3′-end of genes. *Nucleic Acids Res.*, **36**, 5581–5590.

107 Loke, J.C., Stahlberg, E.A., Strenski, D.G., Haas, B.J., Wood, P.C., and Li, Q.Q. (2005) Compilation of mRNA polyadenylation signals in *Arabidopsis* revealed a new signal element and potential secondary structures. *Plant Physiol.*, **138**, 1457–1468.

108 Shen, Y., Ji, G., Haas, B.J., Wu, X., Zheng, J., Reese, G.J., and Li, Q.Q. (2008) Genome level analysis of rice mRNA 3′-end processing signals and alternative polyadenylation. *Nucleic Acids Res.*, **36**, 3150–3161.

109 Costessi, L., Devescovi, G., Baralle, F.E., and Muro, A.F. (2006) Brain-specific promoter and polyadenylation sites of the beta-adducin pre-mRNA generate an unusually long 3′-UTR. *Nucleic Acids Res.*, **34**, 243–253.

110 Winter, J., Kunath, M., Roepcke, S., Krause, S., Schneider, R., and Schweiger, S. (2007) Alternative polyadenylation signals and promoters act in concert to control tissue-specific expression of the Opitz syndrome gene MID1. *BMC Mol. Biol.*, **8**, 105.

111 Gee, A.H., Kasprzak, W., and Shapiro, B.A. (2006) Structural differentiation of the HIV-1 polyA signals. *J. Biomol. Struct. Dyn.*, **23**, 417–428.

112 Takagaki, Y., Seipelt, R.L., Peterson, M.L., and Manley, J.L. (1996) The polyadenylation factor CstF-64 regulates alternative processing of IgM heavy chain pre-mRNA during B cell differentiation. *Cell*, **87**, 941–952.

113 Castelo-Branco, P., Furger, A., Wollerton, M., Smith, C., Moreira, A., and Proudfoot, N. (2004) Polypyrimidine tract binding protein modulates efficiency of polyadenylation. *Mol. Cell. Biol.*, **24**, 4174–4183.

114 Zhu, H., Zhou, H.L., Hasman, R.A., and Lou, H. (2007) Hu proteins regulate polyadenylation by blocking sites containing U-rich sequences. *J. Biol. Chem.*, **282**, 2203–2210.

115 Yu, M., Sha, H., Gao, Y., Zeng, H., Zhu, M., and Gao, X. (2006) Alternative 3′ UTR polyadenylation of Bzw1 transcripts display differential translation efficiency and tissue-specific expression. *Biochem. Biophys. Res. Commun.*, **345**, 479–485.

116 Zhang, H., Lee, J.Y., and Tian, B. (2005) Biased alternative polyadenylation in human tissues. *Genome Biol.*, **6**, R100.

117 Cramer, P., Armache, K.J., Baumli, S., Benkert, S., Brueckner, F., Buchen, C., Damsma, G.E., Dengl, S., Geiger, S.R., Jasiak, A.J., Jawhari, A., Jennebach, S., Kamenski, T., Kettenberger, H., Kuhn, C.D., Lehmann, E., Leike, K., Sydow, J.F., and Vannini, A. (2008) Structure of eukaryotic RNA polymerases. *Annu. Rev. Biophys.*, **37**, 337–352.

118 Lang, W.H. and Reeder, R.H. (1993) The REB1 site is an essential component of a terminator for RNA polymerase I in *Saccharomyces cerevisiae*. *Mol. Cell. Biol.*, **13**, 649–658.

119 Grummt, I., Rosenbauer, H., Niedermeyer, I., Maier, U., and Ohrlein, A. (1986) A repeated 18 bp sequence motif in the mouse rDNA spacer mediates binding of a nuclear factor and transcription termination. *Cell*, **45**, 837–846.

120 Kuhn, A., Bartsch, I., and Grummt, I. (1990) Specific interaction of the murine transcription termination factor TTF I with class-I RNA polymerases. *Nature*, **344**, 559–562.

121 Jeong, S.W., Lang, W.H., and Reeder, R.H. (1995) The release element of the yeast polymerase I transcription terminator can function independently of Reb1p. *Mol. Cell. Biol.*, **15**, 5929–5936.

122 Lang, W.H. and Reeder, R.H. (1995) Transcription termination of RNA polymerase I due to a T-rich element interacting with Reb1p. *Proc. Natl. Acad. Sci. U.S.A.*, **92**, 9781–9785.

123 Kufel, J., Dichtl, B., and Tollervey, D. (1999) Yeast Rnt1p is required for cleavage of the pre-ribosomal RNA in the 3′ ETS but not the 5′ ETS. *RNA*, **5**, 909–917.

124 Prescott, E.M., Osheim, Y.N., Jones, H.S., Alen, C.M., Roan, J.G., Reeder, R.H., Beyer, A.L., and Proudfoot, N.J. (2004) Transcriptional termination by RNA polymerase I requires the small subunit Rpa12p. *Proc. Natl. Acad. Sci. U.S.A.*, **101**, 6068–6073.

125 Catala, M., Tremblay, M., Samson, E., Conconi, A., and Abou Elela, S. (2008) Deletion of Rnt1p alters the proportion of open versus closed rRNA gene repeats in yeast. *Mol. Cell. Biol.*, **28**, 619–629.

126 El Hage, A., Koper, M., Kufel, J., and Tollervey, D. (2008) Efficient termination of transcription by RNA polymerase I requires the 5′ exonuclease Rat1 in yeast. *Genes Dev.*, **22**, 1069–1081.

127 Kawauchi, J., Mischo, H., Braglia, P., Rondon, A., and Proudfoot, N.J. (2008) Budding yeast RNA polymerases I and II employ parallel mechanisms of transcriptional termination. *Genes Dev.*, **22**, 1082–1092.

128 Jansa, P. and Grummt, I. (1999) Mechanism of transcription termination: PTRF interacts with the largest subunit of RNA polymerase I and dissociates paused transcription complexes from yeast and mouse. *Mol. Gen. Genet.*, **262**, 508–514.

129 Dieci, G., Fiorino, G., Castelnuovo, M., Teichmann, M., and Pagano, A. (2007) The expanding RNA polymerase III transcriptome. *Trends Genet.*, **23**, 614–622.

130 Cozzarelli, N.R., Gerrard, S.P., Schlissel, M., Brown, D.D., Bogenhagen, D.F., and Purified, R.N.A. (1983) Polymerase III accurately and efficiently terminates transcription of 5S RNA genes. *Cell*, **34**, 829–835.

131 Allison, D.S. and Hall, B.D. (1985) Effects of alterations in the 3′ flanking sequence on *in vivo* and *in vitro* expression of the yeast SUP4-o tRNATyr gene. *EMBO J.*, **4**, 2657–2664.

132 Hamada, M., Sakulich, A.L., Koduru, S.B., and Maraia, R.J. (2000) Transcription termination by RNA polymerase III in fission yeast. A genetic and biochemically tractable model system. *J. Biol. Chem.*, **275**, 29076–29081.

133 Braglia, P., Percudani, R., and Dieci, G. (2005) Sequence context effects on oligo(dT) termination signal recognition by *Saccharomyces cerevisiae* RNA polymerase III. *J. Biol. Chem.*, **280**, 19551–19562.

134 Martin, F.H. and Tinoco, I., Jr. (1980) DNA-RNA hybrid duplexes containing oligo(dA:rU) sequences are exceptionally unstable and may facilitate termination of transcription. *Nucleic Acids Res.*, **8**, 2295–2299.

135 Bogenhagen, D.F. and Brown, D.D. (1981) Nucleotide sequences in *Xenopus* 5S DNA required for transcription termination. *Cell*, **24**, 261–270.

136 Landrieux, E., Alic, N., Ducrot, C., Acker, J., Riva, M., and Carles, C. (2006) A subcomplex of RNA polymerase III subunits involved in transcription termination and reinitiation. *EMBO J.*, **25**, 118–128.

137 Fernandez-Tornero, C., Bottcher, B., Riva, M., Carles, C., Steuerwald, U., Ruigrok, R.W., Sentenac, A., Muller, C.W., and Schoehn, G. (2007) Insights into transcription initiation and termination from the electron microscopy structure of yeast RNA polymerase III. *Mol. Cell*, **25**, 813–823.

138 Panse, V.G., Hardeland, U., Werner, T., Kuster, B., and Hurt, E. (2004) A proteome-wide approach identifies sumoylated substrate proteins in yeast. *J. Biol. Chem.*, **279**, 41346–41351.

139 Vethantham, V., Rao, N., and Manley, J.L. (2007) Sumoylation modulates the assembly and activity of the pre-mRNA 3′ processing complex. *Mol. Cell. Biol.*, **27**, 8848–8858.

140 Vethantham, V., Rao, N., and Manley, J.L. (2008) Sumoylation regulates multiple aspects of mammalian poly(A) polymerase function. *Genes Dev.*, **22**, 499–511.

141 Darby, M.M., Serebreni, L., Pan, X., Boeke, J.D., and Corden, J.L. (2012) The *S. cerevisiae* Nrd1-Nab3 transcription termination pathway acts in opposition to ras signaling and mediates response to nutrient depletion. *Mol. Cell. Biol.*, **32**, 1762–1775.

142 Moreira, M.C., Klur, S., Watanabe, M., Nemeth, A.H., Le Ber, I., Moniz, J.C., Tranchant, C., Aubourg, P., Tazir, M., Schols, L., Pandolfo, M., Schulz, J.B., Pouget, J., Calvas, P., Shizuka-Ikeda, M., Shoji, M., Tanaka, M., Izatt, L., Shaw, C.E., M'Zahem, A., Dunne, E., Bomont, P., Benhassine, T., Bouslam, N., Stevanin, G., Brice, A., Guimaraes, J., Mendonca, P., Barbot, C., Coutinho, P., Sequeiros, J., Durr, A., Warter, J.M., and Koenig, M. (2004) Senataxin, the ortholog of a yeast RNA helicase, is mutant in ataxia-ocular apraxia 2. *Nat. Genet.*, **36**, 225–227.

143 Chen, Y.Z., Hashemi, S.H., Anderson, S.K., Huang, Y., Moreira, M.C., Lynch, D.R., Glass, I.A., Chance, P.F., and Bennett, C.L. (2006) Senataxin, the yeast Sen1p orthologue: characterization of a unique protein in which recessive mutations cause ataxia and dominant mutations cause motor neuron disease. *Neurobiol. Dis.*, **23**, 97–108.

144 Alen, C., Kent, N.A., Jones, H.S., O'Sullivan, J., Aranda, A., and Proudfoot, N.J. (2002) A role for chromatin remodeling in transcriptional termination by RNA polymerase II. *Mol. Cell*, **10**, 1441–1452.

145 Jones, H.S., Kawauchi, J., Braglia, P., Alen, C.M., Kent, N.A., and Proudfoot, N.J. (2007) RNA polymerase I in yeast transcribes dynamic nucleosomal rDNA. *Nat. Struct. Mol. Biol.*, **14**, 123–130.

3
Posttranscriptional Gene Regulation by an Editor: ADAR and its Role in RNA Editing

Louis Valente, Yukio Kawahara, Boris Zinshteyn, Hisashi Iizasa, and Kazuko Nishikura

3.1
Introduction

The process of RNA editing is a means by which greater diversity can be achieved from the DNA-encoded genome, generating a greater set of RNA transcripts to produce a multitude of proteins. With the advent of the genome sequencing era, it is now clear that there are no more than 25 000 human genes and that the complexity of higher eukaryotes must be derived from other sources such as noncoding RNA and RNA modifications [1]. RNA editing differs from other posttranscriptional processes in that it is a site-specific alteration in order to fine-tune gene products [2]. This is unlike splicing and polyadenylation, which are mechanisms that affect large stretches of sequence. Specific changes in the coding region by editing of mRNA can lead to functional alterations of the protein product, while modification of the noncoding regions may globally affect such processes as splicing, RNA stability, translational efficiency, and RNA-based gene silencing mechanisms [3–5]. The phenomenon of RNA editing was first revealed 20 years ago in the kinetoplastid protozoa where uridine nucleotides of its mitochondrial mRNA were inserted or deleted, and this editing is necessary to generate functional proteins for this trypanosome [6]. Shortly afterward, RNA editing was detected in a nuclear-encoded mammalian apolipoprotein mRNA that is edited by APOBEC1, a cytidine deaminase family member that modifies cytidine to uridine (C-to-U editing) [7–10]. Of the various classes of nuclear-encoded RNA editing, the most widespread type is the modification of adenosine to inosine (A-to-I) in higher eukaryotes [5, 11–13].

During RNA editing, a base alteration of a ribonucleotide takes place. A-to-I editing just as C-to-U editing involves a base deamination reaction of RNA (Figure 3.1a) [11]. The deamination reactions of C-to-U or A-to-I are the best characterized examples of base conversion and are the major RNA editing events in metazoans [14]. Editing by A-to-I base modification occurs by a hydrolytic deamination reaction (Figure 3.1a) [15, 16]. This hydrolytic attack takes place on carbon 6 of the adenine base by removal of the exocyclic amine with oxygen serving as the nucleophile and is proposed to continue through a tetravalent intermediate, which then releases ammonia.

Posttranscriptional Gene Regulation: RNA Processing in Eukaryotes, First Edition. Edited by Jane Wu.

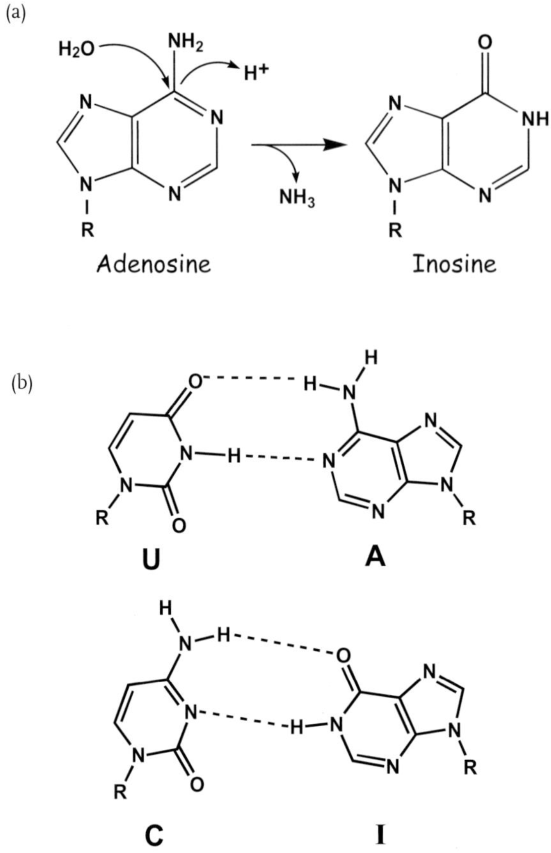

(a)

(b)

Figure 3.1 ADAR edits adenosine residues in dsRNA to inosines that can consequently function as guanosines. (a) The hydrolytic deamination reaction at position C_6 of the adenine ring converts adenosine to inosine. (b) Inosine converted from adenosine preferentially base pairs with cytosine in a Watson–Crick hydrogen-bonding configuration, as if it were guanosine. Inosine is recognized as a guanosine by the splicing machinery, translational apparatus, and also by reverse transcriptase.

The adenosine deaminases acting on RNA (ADARs), which provide for the A-to-I editing activity, are believed to have evolved from the cytidine deaminases [11, 17–19]. ADARs were first identified as a cellular helicase or RNA unwinding activity resulting in the destabilization of double-stranded (ds) RNA molecules by the introduction of I:U mismatches [15, 20]. Upon editing of an adenosine in an A:U base pair, the newly created structure makes an I:U wobble base pair. An inosine preferentially base pairs with cytidine (Figure 3.1b). The ADAR proteins can alter the structure of its target RNAs by melting the double strandedness of RNA duplexes (Figure 3.2). The homologs of ADAR have been cloned and characterized from many organism since their discovery in *Xenopus laevis* such as the

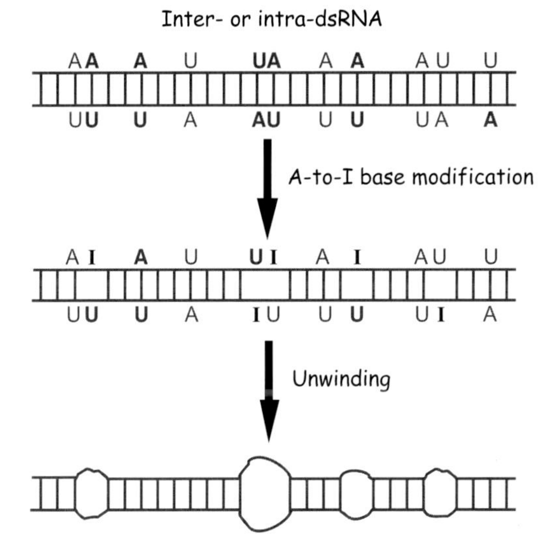

Figure 3.2 Reduced double strandedness by modification of multiple adenosines in a long dsRNA. A-to-I modification can occur in inter- (as shown) or intra-dsRNA. The resultant dsRNA containing multiple I:U mismatched base pairs can become destabilized and subsequently unwound. This relatively random attack of many adenosines in long completely base paired dsRNA substrates by ADARs contrast to their precise activity in site-selective editing of natural target RNAs.

mammalian family members ADAR1 [17, 21], ADAR2 [22–24], and ADAR3 [25, 26] (Figure 3.3).

The biological significance and molecular mechanism of A-to-I RNA editing has been intensely studied with many examples of A-to-I editing altering the decoding of mRNA transcripts [27]. These changes in the RNA allow the inosine to be read as a guanosine during splicing and protein synthesis, as well as by the polymerases such as reverse transcriptase. Thus, the recoding of the mRNA can have functional consequences in the cell by translating these changes into a diverse set of proteins with altered functions [28, 29]. Mostly all the initial examples of protein recoding were discovered serendipitously when a discrepancy between a genomic sequence and its cDNA was noticed. It has been widely viewed that there should be many more examples of protein recoding by A-to-I RNA editing that remain to be indentified [30].

Recent bioinformatic studies have revealed that the most common targets of A-to-I RNA editing is found within the noncoding regions of RNA that can form inverted repeats [29, 31–37]. Although the biological significance of noncoding RNA editing remains to be established, it may regulate many different processes that also utilize dsRNA. In addition, A-to-I RNA editing appears to intersect with other pathways that exploit dsRNA to act as a host antiviral defense mechanism [38, 39] and modulate dsRNA of the RNA interference (RNAi)-mediated gene silencing pathway [5, 40–46]. Furthermore, dsRNA formed between sense and

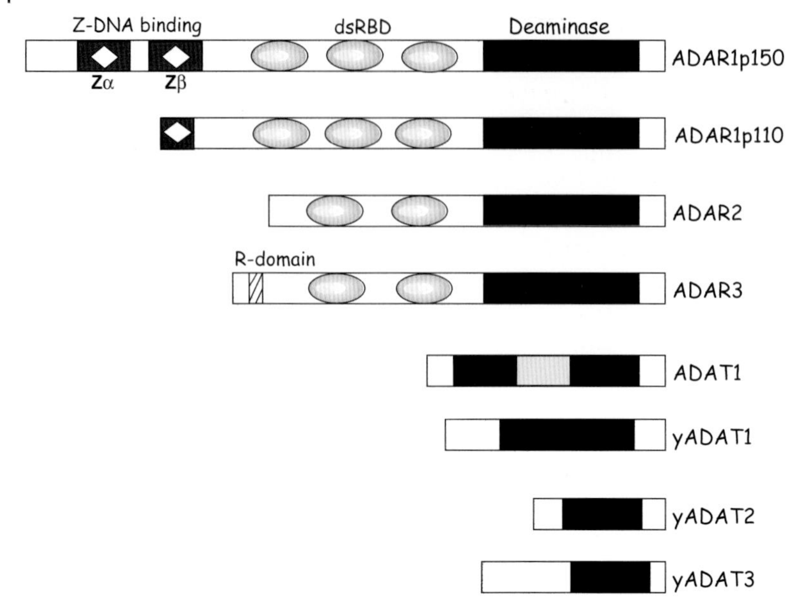

Figure 3.3 A-to-I RNA editing is performed by the ADAR gene family and ADAT subfamily members. Three ADARs and single ADAT1 are known to exist in mammals. The yeast homologs of the ADAT family are also represented. Shown are the Z-DNA-binding subdomains (diamonds), double-stranded RNA-binding domains (dsRBD, circles), deaminase domain (black box), an arginine/lysine-rich domain (R-domain) of ADAR3 that binds ssRNA (striped box), and a unique sequence to mammalian ADAT1 that is located within the deaminase domain (gray box). ADARs target dsRNA, and tRNA is targeted by ADATs despite the fact that they lack any known RNA-binding motifs. yADAT2 and yADAT3 form an active heterodimers, while ADAR1 and ADAR2 form active homodimers.

antisense transcripts might undergo A-to-I editing [47–50]. The global processes and pathways that ADAR gene family members are interconnected with indicate the importance of A-to-I RNA editing. The purpose of this chapter is to highlight recent studies on A-to-I RNA editing that have revealed new roles for ADAR activity and does not include a comprehensive citation list of all the references and recent reviews on this subject by others.

3.2 The RNA Editing Kinship

From bacteria to man, RNA editing has been reported in mRNAs, tRNAs, and ribosomal RNA. Since the discovery of RNA editing in trypanosomes, where uridine nucleotides of its mitochondrial mRNA are inserted or deleted, for the generation of functional proteins [6], many examples have emerged. The next form of RNA editing to be found was the mammalian nuclear-encoded apolipoprotein mRNA, which is edited by APOBEC1, a family member of the cytidine deaminases acting on RNA (CDAR) involved in C-to-U editing [7–10]. It is held that the predecessor to the adenosine deaminases acting on tRNA (ADATs) and consequently ADARs (see Section 3.3) are the cytidine deaminases acting on mononucleotides (CDAs) or on RNA (CDAR), but not the adenosine deaminases acting on mononucleotides (ADAs) [11, 17, 18, 51]. An *Escherichia coli* CDA forms a homodimer [52], as well as the CDAR APOBEC1 [53–55]. This may provide an interesting evolutionary link to the ADAT and ADAR families of proteins, which also require hetero- or homodimerization for activity [56, 57] (see Section 3.3).

The ADAT family members edit tRNA to change adenosine to inosine at specific locations and are the closest relatives to the ADAR proteins [18] (Figure 3.3). ADAT members are conserved in eukaryotes from yeast to man and were found based on sequence homology searches to ADAR [58–61]. Recently, a bacterial ortholog of the ADAT family was discovered, tRNA adenosine deaminase (TadA), indicating conservation of this deaminase function between prokaryotes and eukaryotes [62]. Recent, structural studies of prokaryotic TadA (an active homodimer) have shed light on the possible evolutionary origin and mechanism of the eukaryotic ADAT2/3 (an active heterodimer) [63, 64]. Interestingly, the trypanosome homologs of yeast ADAT, which can perform A-to-I editing on tRNA can also perform C-to-U editing on single-stranded (ss) DNA, thus providing strong evidence for the evolution of the editing deaminases [65].

ADATs resemble ADARs with a high degree of homology in their deaminase domains that are located in the C-terminal portion of these proteins (Figure 3.3). It is believed that the ADAR proteins evolved from the ADAT family after acquiring domains that provided for RNA binding, such as the dsRNA-binding domains (dsRBDs) [18]. ADAT sites of action on tRNA are on or near the anticodon position, which can further modulate codon recognition during the decoding of mRNA by the translation machinery [51, 58]. In yeast, the ADAT family consists of three proteins ADAT1, 2, and 3 with nucleotide-specific enzymatic activity. ADAT1 edits position A37 [58], while in yeast, ADAT3 forms a heterodimer with ADAT2 to produce an enzymatically active complex that edits the wobble position of an anticodon at nucleotide A34, indicating the importance of this substitution [51]. ADAT2 is the catalytic subunit of the heterodimer complex due to the lack of a conserved glutamate residue in the deaminase domain of ADAT3 [51] (Figure 3.3). Besides in yeast where ADAR is not present, ADAT is also found in metazoans, thus overlapping in expression with the ADAR proteins for A-to-I function in higher eukaryotes [18]. ADAR and ADAT families do not overlap in substrate targets and are highly specific for either dsRNA or tRNA, respectively [18, 51, 58, 61, 66].

3.3
The ADAR Gene Family

In mammals, three ADAR gene family members have been identified, and they are conserved in their dsRBDs as well as in their C-terminal deaminase domain

(Figure 3.3) [17, 21–26, 67]. The ADAR1, 2, and 3 proteins are highly conserved from fish to humans, according to sequence homology alignments [68, 69]. There is also a well-conserved invertebrate member for the single *Drosophila* (dADAR) that is similar to mammalian ADAR2 [70]. *Caenorhabditis elegans* contains two members (c.e.ADAR1 and c.e.ADAR2) that are conserved to a lesser extent [17, 71]. The expression pattern of mammalian ADAR1 and ADAR2 is ubiquitous in many tissues with a high level of expression in the brain [17, 21–24]. This contrast to the inactive mammalian ADAR3 whose expression is limited to the brain [25, 26], but similar to the confinement of dADAR and c.e.ADAR1, which are also restricted to the nervous system [70, 72]. This expression pattern correlates to the targets of recoding by ADAR, which are mainly ion channels and receptors of neurotransmission. The critical role of these ADAR genes is illustrated by knock-out mutations and their essential phenotypes displayed, indicating a vital function of A-to-I editing in posttranscriptional gene regulation.

The similar structural features shared by mammalian ADARs include dsRBDs repeated two or three times that are located in the N-terminus and a C-terminal deaminase domain that supplies the enzymatic activity [17, 21]. Nonetheless, there are also unique characteristics that set these ADARs apart such as the Z-DNA-binding domain of ADAR1 [73] and the arginine/lysine-rich domain (R-domain) of ADAR3 at the N-terminus preceding the dsRBDs [25, 26] (Figure 3.3). The R-domain of ADAR3 has been shown to be important for ssRNA binding and is perhaps necessary for the association with a specific subset dsRNA substrates that have distinctive single-stranded character [26]. Although the function of the Z-domain in ADAR1 is not understood yet, it appears to bind the left-handed form of negatively supercoiled helical DNA [74]. Sites actively undergoing transcription generate Z-DNA by action of the polymerase, and this may perhaps localize ADAR1 to the newly transcribed RNA before splicing [73]. Indeed, there is newer experimental evidence to suggest that RNA Pol II helps to ensure that editing precedes splicing [75]. Interestingly, recent structural studies of the Z_α and Z_β subdomains of ADAR1 indicate that the two similar domains evolved into distinctive functions, even on the same polypeptide [76, 77]. With a 50 base paired (bp) substrate of Z-form dsRNA, editing efficiency is enhanced and a modulation in editing pattern is observed compared with the same context RNA without Z-formation [78]. The Z-DNA domain of ADAR1 is important for the editing of small 15 bp dsRNA segments; however, editing efficiency increases with longer substrate dsRNA, indicating the significant function of the dsRBDs [79].

ADARs' dsRBDs resemble those of the RNase III ribonuclease complexes of Drosha and Dicer that are part of the RNAi pathway, as well as the dsRNA-activated protein kinase (PKR) that is involved in antiviral mechanisms, placing ADAR in this dsRNA-binding protein superfamily [80, 81]. Evidence suggests that dsRBDs provide general binding with little selectivity, although the number and distance between ADAR dsRBDs may provide some specificity for its substrates [23] (Figure 3.3). Two separate structures of a protein's dsRBD interacting with dsRNA indicate that dsRBDs bind in a similar manner [82, 83]. The dsRBD of *Xenopus laevis* RNA-binding protein A (Xlrbpa) show that binding is presumably nonsequence specific, interacting with two successive minor grooves with an intervening major groove on one face of the dsRNA helix [82]. Similarly, the structure of the *Drosophila* Staufen protein's third dsRBD displayed binding over a minor and major groove like Xlrbpa [83]. These dsRBDs consisting of approximately 70 amino acids (aa) make contact with dsRNA spanning 16 bp in a sequence and nucleotide-independent fashion while interacting with the phosphate oxygen [82]. A common theme of these domains appear to be a KKxxK motif in the dsRBD that is essential for this association, since mutagenesis of this motif reduces binding [82, 83]. Likewise, ADARs contain this KKxxK motif in their dsRBDs (Figures 3.4 and 3.5), and deletions or site-directed mutagenesis also indicates that they are important for function [57, 84–88].

Studies on serial deletion constructs of a natural ADAR2 substrate indicate that the dsRBDs of ADAR2 bind in a distinct manner from the dsRBDs of PKR. Examination of PKR has revealed that dsRBDs can select specific sites on dsRNA, and the site selectivity differs among the dsRBD superfamily members [80]. An

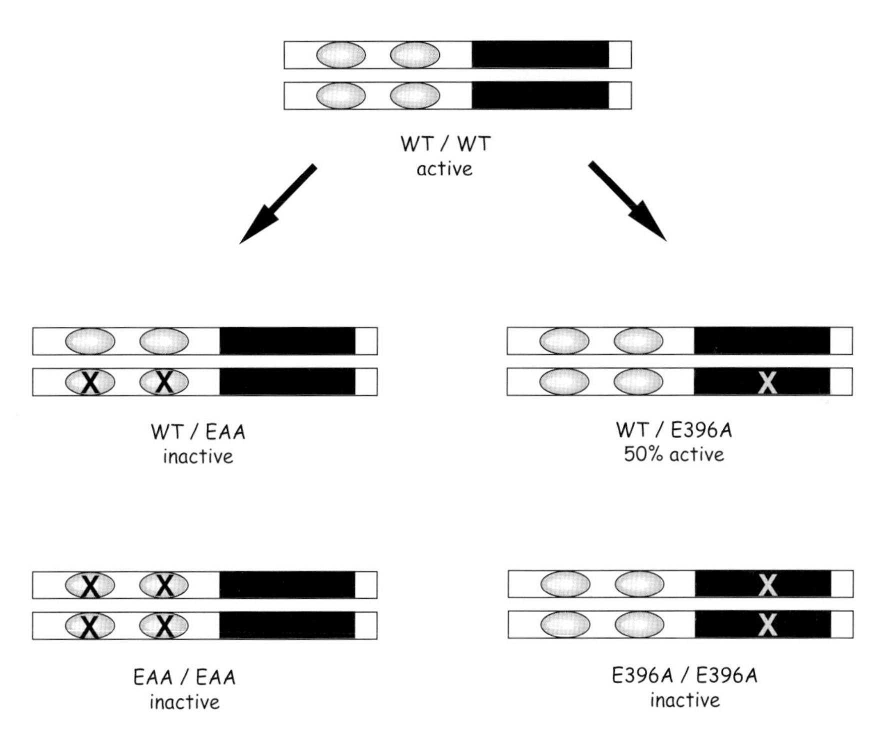

Figure 3.4 Model for ADAR dimerization and activity. Mutation of the dsRBDs in the KKxxK motif to EAxxA (denoted as EAA, black X) eliminates RNA binding. A dominant negative effect occurs due to one monomer having the dsRBD mutations and is inactive. The mutation in the deaminase domain E396A (white X) of ADAR2 allows for 50% activity when one monomer is mutated. Catalytic function is totally destroyed when both monomers contain the mutation in the deaminase domain. ADAR2 is represented here, but the same occurs for ADAR1.

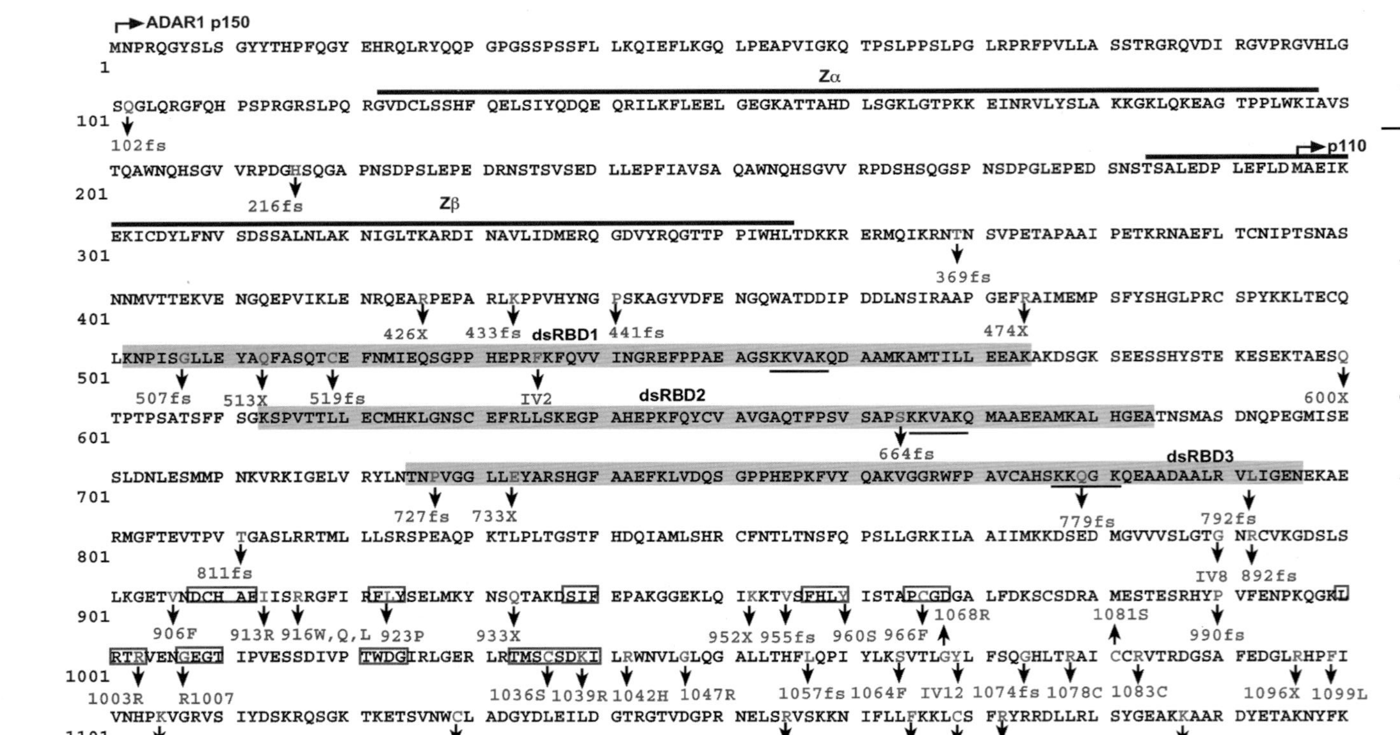

Figure 3.5 Mutations of the ADAR1 gene identified in DSH patients. A total of 41 mutations (arrow) associated with DSH have been identified in Japanese and Chinese populations. Indicated are Zα and Zβ domains (overline), three dsRNA-binding domains dsRBD1–3 (gray-filled box), conserved KKxxK motifs (underline), and nine stretches of core deaminase subdomains, highly conserved in all three mammalian ADAR gene family members ADAR1–3 (box). Two methionines (M1 and M296) used for alternative translation initiation of ADAR1p150 and ADAR1p110, respectively, are also shown with arrows. X; nonsense, fs; frameshift, IV; splice site mutations.

analysis utilizing footprinting of a full-length ADAR2 and also of its dsRBDs lacking the catalytic domain indicates that the dsRBDs may be the discriminating factor for selective recognition [89, 90]. Newer studies have indicated that there may be differences even among the dsRBDs on the same protein where individual dsRBDs do not have sequence specificity but the combinations of these dsRBDs may then impart the substrate specificity [91]. On the other hand, there is also evidence that dsRBD1 in the N-terminus of ADAR2 can inhibit its own activity on short 15 bp substrates, indicating the unique function of a specific dsRBD [92]. These differences of the dsRBDs on the same ADAR protein may convey the specificity of dsRNA binding and activity, as suggested by studies on ADAR2, which indicate a more important role of dsRBD1 for general binding, whereas dsRBD2 is crucial for editing activity [93, 94]. Structural studies of the dsRBDs of ADAR2 indicate that its two dsRBDs have different structure-based recognition for the dsRNA substrates [87]. Interestingly, ADAR3 binds dsRNA with high affinity even though it is not active on known substrates of the other two ADARs or on long dsRNA substrates *in vitro* and may need other factors in the brain for catalytic activity [25, 26]. The extent of binding cooperativity among ADARs' dsRBDs may provide for binding selectivity and subsequent activity, while evidence suggests that binding and catalysis are separate events [95].

The catalytic site of action for the cytidine deaminases is a conserved HAE tripeptide amino acid sequence, which is also contained in the ADAR deaminase domain. Because cytidine deaminases coordinate zinc, it was assumed that ADAR having these homologous residues performs a similar catalytic activity by utilizing zinc for the hydrolytic deamination reaction [12, 17, 84]. The HAE tripeptide amino acid residues have been proposed to coordinate zinc to activate a water molecule for nucleophilic attack (Figure 3.1a). Site-directed mutagenesis of some of these conserved residues in the ADAR deaminase domain abolished catalytic activity while still retaining its dsRNA-binding function [84]. It is viewed that ADAR is a metalloenzyme that uses a base-flipping mechanism to pull out the targeted adenosine from the helix into the enzyme active site [96]. The ADAR2 dsRBDs change the conformational flexibility of nucleotides (nt) surrounding the targeted adenosine to provide a lower activation energy for base flipping to occur [90, 97]. The edited adenosine to be flipped into the ADAR active site can also be affected by its surrounding nucleotide neighbors and can affect the efficiency of A-to-I conversion [95]. Recently, the structure of the ADAR2 deaminase domain has been solved and provides new models to how the catalytic core functions. Surprisingly, the ADAR active site binds a molecule of inositol hexakisphosphate (IP_6), which is pretty well buried in this catalytic site of action [98]. Many lysines in the catalytic pocket form a three-dimensional structure that is stabilized by IP_6, allowing the core amino acids to coordinate with a zinc ion and water molecule that is necessary for enzymatic activity. As predicted, the HAE triplet is making important contacts to the zinc and the nucleophilic water molecule that is proposed for proton shuttling. Interestingly, this IP_6 molecule is also vital for the functional activity of ADAT1, indicating its evolutionary significance for A-to-I editing [98].

More recently, the requirement of ADAR dimerization on A-to-I editing activity has become known. ADAR homologs act as a dimer complex in flies and mammals, showing the importance of this oligomerization for enzymatic activity [56, 57]. By contrast, ADAR3 does not dimerize [56], which is indicative of its lack of activity on various substrates even though it binds dsRNA [25, 26]. The cellular dimerization of the mammalian ADARs has been verified *in vivo* through studies utilizing resonance energy transfer methods [94, 99]. Although initial *in vitro* experiments on ADAR1 and ADAR2 indicate that they form only homodimers, it has now been determined that they can as well form heterodimers *in vivo* [94, 99, 100]. It appears that the dimerization ability may be due to the N-terminal dsRBD region and not the C-terminal deaminase domain as experimentally shown using truncations of ADAR [94, 98] (Figure 3.4). It is proposed that this dimer interaction allows for the proper formation of active site alignment to deaminate the adenosine moiety [56, 57]. It was initially determined that mammalian ADAR dimerization was not mediated via RNA by the inclusion of RNases into the assays, while fly dADAR indicated that this association was RNA dependent [56, 57, 99]. Since RNAs could have been protected during the RNase treatment and provide for the bridging of the two monomer subunits, several mutations in the critical lysines (KKxxK to EAxxA) of each dsRBD were analyzed. This mutagenesis study has revealed that ADAR dimerization is independent of RNA [88]. Most interesting is that dominant negative effects are seen on dimer functions if one subunit of an ADAR complex is mutated in the dsRBDs, then its wild-type partner is unable to bind the dsRNA and therefore incapable to catalyze the editing reaction [88] (Figure 3.4). This clearly indicates the cooperative nature with respect to their dsRBDs for the interplay of the monomers, thus defining their functional significance in activity.

The ADAR1, 2, and 3 family members are primarily nuclear localized with the exception of the longer p150 form of ADAR1, which contains an N-terminal extension (Figures 3.3 and 3.5). The nuclear localized shorter ADAR1p110 form is expressed under constitutive promoters, while the mainly cytoplasmic p150 form is under the control of an interferon (IFN) inducible promoter [67, 101–103]. ADAR1p110 initiates at a methionine at position 296 due to alternative exons at the 5′ end of the mRNA and lacks most of the N-terminal Z-DNA-binding domain [67, 101]. Recent experimental evidence suggests that ADAR1-specific regulation of alternative splicing is controlled in a tissue-specific manner [104]. ADAR1p110 is primary nuclear due to the lack of the nuclear export signal (NES) at the N-terminus in the Z_α subdomain, which is present within the p150 isoform [105]. This cytoplasmic export of ADAR1p150 is mediated by the CRM1 export factor and Ran-GTP [105]. The third dsRBD of ADAR1 was shown to contain a nuclear localization signal (NLS), indicating that a dsRBD can have dual roles [106]. Furthermore, it was found that ADAR1 localization is determined by a dynamic balance of the NES in the amino-terminus, the NLS containing a nucleolar localization signal at dsRBD III, a regulatory element in the deaminase domain, and a predominant NLS at the carboxy-terminus [107]. ADAR is found to associate with transcriptionally active chromosomes within the nucleus, although this is attributed to the dsRBDs and not mediated by the Z-DNA-binding domain [108, 109]. Closer inspection of the nucleus revealed that ADAR1 and ADAR2 are localized to the nucleolus and mutagenesis of the dsRBDs resulted in translocation to the nucleoplasm [86, 110]. The dynamic relationship of ADAR1 and ADAR2 to the nucleolar compartment is reversed upon active editing substrates becoming present in the nucleoplasm. Transient sequestration may keep enzymatic activity away from potential RNA targets [86, 110], thus mediating ADAR localization.

3.4 The Role of RNA in the A-to-I Editing Mechanism

RNA can form duplex structures such as hairpins interrupted by bulges and loops that the ADAR proteins recognize for the base modification reaction to occur. The A-to-I editing specificity for the ADAR enzymes is influenced by the RNA topography (see Figure 3.7). RNA secondary structural elements consisting of hairpins containing mismatches, bulges, and loops are edited more selectively than completely base paired duplexed RNA. These loops divide the adjacent helices into separate units for ADAR to recognize, dividing long substrates into a series of smaller components for more selective editing of its targets [111, 112]. Completely long complementary dsRNA will be edited in a nonspecific manner. ADAR1 and ADAR2 will edit about 50% of the adenosines present in a nonselective manner for a perfect RNA duplex of greater than ~50 bp [96, 113] (Figure 3.2).

The structural aspects of RNA are not the only determinants for editing efficiency. ADARs must first identify its dsRNA substrate and then differentiate which adenosine to deaminate among the other adenosines present in an RNA molecule. A sequence bias surrounding an edited site has shown that ADAR does indeed have nucleotide preference for neighboring residues. The sequences encompassing an edited site provide proper context for the deamination of selected adenosine residues that will be modified to inosine. *In vitro* studies have shown that ADAR1 has a 5′ nearest-neighbor sequence preference ($U = A > G > C$) for the targeted adenosine [96]. ADAR2 is very similar to ADAR1 in its 5′ nearest-nucleotide preference ($U = A > C = G$), but dissimilar in that it has a suggested 3′ nearest-neighbor preference ($U = G > C = A$) as opposed to ADAR1 [114]. Other analysis indicates that additional nucleotides surrounding the targeted adenosine besides its nearest sequence neighbors may be just as important for the deamination reaction. In one study, at least for the case of ADAR2, it was suggested that the sequence context of several nucleotides surrounding the adenosine moiety to be edited defines a consensus sequence for preference and activity [112]. Site-specific editing of dsRNA can be altered by mutagenesis surrounding the targeted adenosine, while not affecting the binding of ADAR [90, 95]. Substrate recognition and/or catalysis by ADAR could involve the nucleotide that is on the opposite strand of RNA since the base that pairs with the adenosine to be edited is in close proximity to the ADAR active site. It was found that substrates having an A:C mismatch at the catalytic site of deamination had a greater tendency for editing (will generate an I:C matched base pair) (Figure 3.1b) as compared with when A:A or A:G mismatches or even A:U base pairs resided at the same position [115].

As editing commences, it decreases the double strandedness of the RNA molecule, thus imparting selectivity that is determined as the reaction ceases. As further adenosines become deaminated, the RNA structure becomes less duplexed due to I:U mismatches and more single stranded (Figure 3.2), as studies have shown that I:U mismatches decrease the stability of dsRNA [116]. Since active ADARs bind dsRNA and not ssRNA [15, 117], their substrate pool becomes diminished upon ADAR active deamination. ADAR catalysis terminates when ≥50% of the adenosines are modified in a long dsRNA [96, 113]. The reduction in double strandedness of an RNA molecule can be correlated with the thermodynamic stability of a dsRNA molecule that decreases with every inosine conversion, creating greater amounts of I:U mismatches and less A:U base pairs [116]. This may explain why ADAR selectively edits RNA helical structures separated by loops that divide adjacent helices [111, 112]. For these structures, after a deamination event, the stability is reduced more rapidly than in long perfectly base paired dsRNA providing for less recognition by ADAR, granting it more specificity on this kind of substrate. This implies that an RNA duplex undergoing a subsequent deamination reaction may significantly change in its stability, thus altering editing specificity of this target.

The differences in site-selective editing between ADAR1 and ADAR2 are best illustrated in several experiments on the RNA of the glutamate receptor B-subunit (GluR-B) and the serotonin receptor subtype 2C (5-HT_{2C}R). *In vitro,* recombinant ADAR1 and ADAR2 proteins display distinctive editing patterns for site selectivity on these known substrates [22–24, 118, 119]. For example, ADAR1 selectively edits the A and B sites of 5-HT_{2C}R and the intronic hotspot +60 site of the B-subunit GluR RNAs, while ADAR2 does not significantly target these sites. However, ADAR2 displays site selectivity for the D site of 5-HT_{2C}R and the Q/R site of GluR-B RNAs, whereas ADAR1 barely edits these sites. This clearly indicates that ADAR1 and ADAR2, although they are family homologs with similar A-to-I activity, have distinct site selectivity that is conveyed from the differences in RNA substrates.

3.5 Splice Site Alterations

The editing of exons often utilizes intronic sequences of complementarity for which to base pair with. An inosine is interpreted by the splicing machinery as a guanosine (Figure 3.1b); therefore, alternative splicing can be another effect of A-to-I RNA editing, which can vary a protein to generate more isoforms with distinctive behavior. Editing precedes splicing, and A-to-I changes can potentially affect the introns and processing of pre-mRNA. A highly conserved canonical 3′ splice site dinucleotide recognition sequence (AG) can be edited as to remove this identifying sequence ($\underline{A}G \rightarrow \underline{I}G \rightarrow \underline{G}G$), or a nonsplice site can be altered to create a new acceptor site for splicing to occur (AA → AI → AG) (Figure 3.6). It is also possible to generate a 5′ splice donor site (AU → IU → GU) by ADAR action. It appears that editing and splicing are coordinated events for the efficient processing of pre-mRNA in that editing often overlaps with these intronic sequences that are important for splicing [120, 121]. The effect of extensive alternative splicing in gene regulation, potentially modulated by ADARs to some extent, provides a mechanism to expand the subset of gene-encoded proteins.

ADAR1 and ADAR2 are found to be complexed with large nuclear ribonucleoprotein (lnRNP) particles that contain spliceosomal components, and these associated ADARs are enzymatically active [120]. To help ensure that editing precedes splicing, a recent study found that there is coordination between the RNA polymerase II and site-selective RNA editing by ADAR2 on GluR-B pre-mRNAs for regulated splicing [75]. Interestingly, ADAR1 and hUPF1, the mRNA surveillance protein for decay, are found associated with nuclear RNA splicing complexes and suggest a regulatory pathway by A-to-I editing and RNA degradation [122]. Furthermore, it is believed that ADARs may mark RNAs in the nucleus with inosine, which are then retained in the nucleus, thereby preventing these flagged RNAs from being prematurely exported to the cytoplasm [123].

It is conceivable that multiple editing events encompassing a duplexed region containing a splice signal can become more accessible as this region is destabilized (Figure 3.2), allowing for splicing to occur. The kinetics of splicing have been shown to change by a 10-fold reduction in *ADAR2*$^{-/-}$ mice brains for the GluR-B transcript, due to the almost complete lack of editing at the Q/R site of this RNA, which is a target for the ADAR2 editing enzyme [124]. In support of this, preferential splicing is observed as a consequence of Q/R site editing of GluR-B transcripts, as compared with 10% in the intron-containing pre-mRNAs to that of 40% in processed mRNAs [124]. For *Drosophila*, mutation in a specific ATP-dependent dsRNA helicase A confers a temperature-sensitive paralytic phenotype similar to *para* Na^+ channel mutants in this organism. The mutated helicase is incapable of resolving the dsRNA structure of this channel's mRNA and consequently results in exon-skipping events by aberrant splicing and frequent editing in this area of RNA [125]. The model of editing involvement is that this region of dsRNA contains a 5′ splice site that is occluded and is unwound by the helicase for splicing to occur, which is also impacted by editing efficiency.

A classic example of splice site alteration is of the mammalian ADAR2 pre-mRNA transcript that is edited at the −1 position by its own protein to produce an alternative 3′ splice site acceptor (AA → AG) in a gene-encoded intron [4] (Figure 3.6). This new splice site forms a frameshift producing a nonfunctional truncated protein lacking the dsRBD as well as the catalytic domain. It is also observed that preferential splicing occurs when ADAR2 transcripts are self-edited at the −1 position for this acceptor site, which correlates well with ADAR2 editing activity [126]. This suggests that auto-editing may provide a source of negative feedback regulation, presumably to restrict active levels of ADAR2 editing action from indiscriminate targeting of wrong RNAs due to a high abundance of ADAR2 protein. Interestingly, it was revealed that ADAR1 may modulate ADAR2

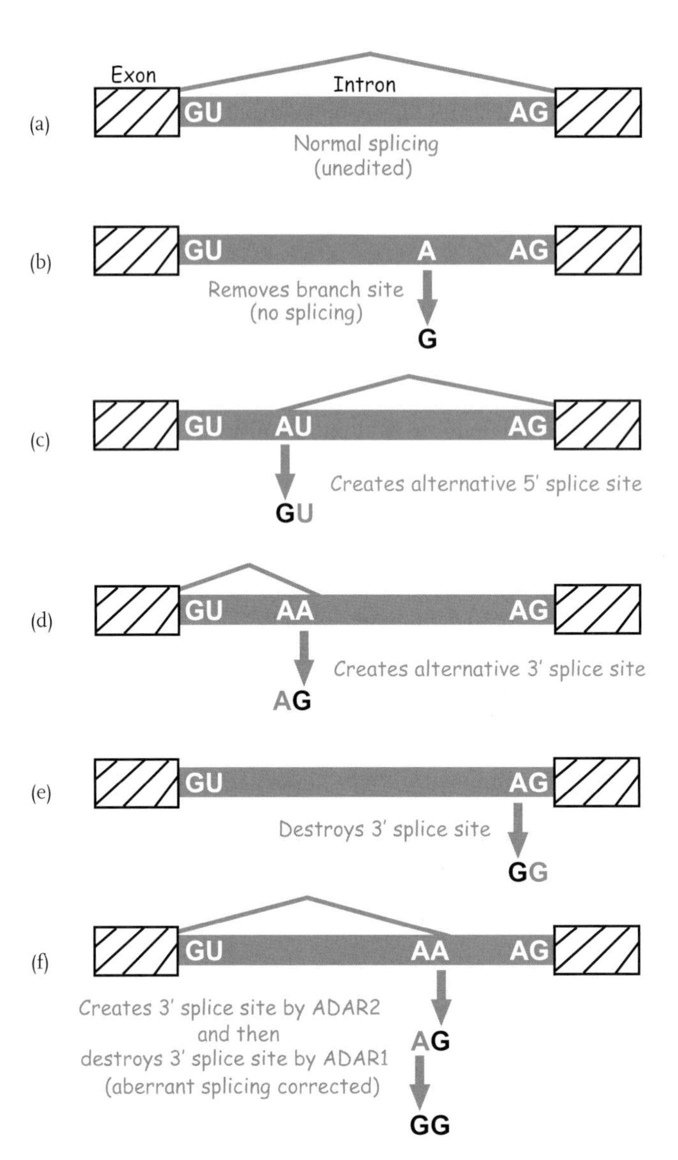

Figure 3.6 Splicing is altered by A-to-I modification and changes the coding regions in mRNA. In the normal case, (a) the splicing machinery utilizes a 5′ donor GU and a 3′ acceptor AG, with a branch site adenosine that is needed between the two sites. Obligatory adenosines can be converted into inosine (guanosine) and are no longer recognized by the splicing machinery, while on the other hand, modified adenosines can be read as the requisite guanosines that are needed for splicing to occur. RNA editing by ADARs can alter the branch site adenosine not allowing for splicing to proceed (b), can create alternative 5′ (c) or 3′ (d) splice sites, or can abolish a 3′ acceptor site (e). An interesting case of splice site-selective editing is shown (f) where ADAR2 edits its own pre-mRNA to create an atypical 3′ acceptor site in a probable feedback loop, but the action of ADAR1 can reverse this negative constraint on ADAR2 transcripts by eliminating this noncanonical splice site. The bent line above the introns indicates the section that would be removed due to splicing. The A-to-I changes are shown as a black "G" for simplicity.

pre-mRNA splicing patterns by editing the −2 position to destroy this potential splice signal created by ADAR2 self-editing (AA → AG → GG) [112] (Figure 3.6). Therefore, after the ADAR2 protein has modified the −1 site for splicing to occur, ADAR1 may subsequently override the ADAR2 negative feedback loop by not allowing this splicing to occur. In a similar scenario, it seems that *Drosophila* dADAR, which is more like the ADAR2 protein, is also self-edited within a highly conserved residue in the catalytic domain proposed to downregulate activity [127].

There are a few other recent examples of editing affecting splice site recognition in non-ADAR transcripts. An intronic branched site adenosine that is a typical distance from the 3′ acceptor splice site has been proposed to be edited by ADAR in the tyrosine phosphatase PTPN6 pre-mRNA [128] (Figure 3.6). This editing removes the branched site adenosine, thus retaining the intron that should be excised in this splice variant, which has been implicated in leukemogenesis [128]. In another case, for the human serotonin receptor 5-HT_{2C}R, splicing has been shown to be directly modulated by ADAR function [129]. Editing of 5-HT_{2C}R occurs in five exonic positions (termed A–E sites) in a stem loop that is formed with an intron, which contains an additional editing site F [129]. Depending on the different editing combinations of these six sites, it was shown that they can have profound effects on alternative splice site selection within this 5′ donor site at the intron/exon boundary. In another analysis, a 5′ splice site signal was reported to be generated by producing the consensus donor sequence in the pre-mRNA of the G-protein coupled receptor GPR81, creating an alternative splice site that is mediated by Alu inverted repeats in the noncoding regions [36] (Figure 3.6). Splice site editing is more difficult to detect than in the coding regions, and we anticipate that many more cases will arise in these noncoding sites that are significant to exonic sequence formation, providing for a posttranscriptional regulatory effect.

3.6
A-to-I RNA Recoding Modifies Proteins Such As Neurotransmitters

For editing of protein coding sequences, a fairly duplexed pre-mRNA secondary structure formed between an exon sequence harboring the editing sites and complementary intronic sequence is necessary [130–132]. This more or less duplexed structure is difficult to detect because intron/exon pairings can have over 1000 nucleotides separating them before the relatively small base paired RNA is formed [4, 130–133]. When the modified adenosine is located in the coding region, a change arises within an mRNA codon as to read A-to-I as a guanosine change by the translational machinery (Figure 3.1b), thus altering the protein primary sequence and/or structure [134]. This altered codon meaning allows for more than one protein isoform to be produced from a single gene. In this sense, the proteins encoded by the genome itself can expand the repertoire of genes expressed that diversifies the proteome [2, 12, 18, 135].

Besides coding region alterations, it is possible to have adenosine modifications remove or add start and stop codons within the messenger RNA. Analysis of

C-to-U RNA editing sites in higher plant chloroplasts indicates that it is possible to introduce modifications that will create a start codon [136], and such discoveries are likely to be found for A-to-I editing. Thus far, no examples have been identified for A-to-I editing recoding a nontranslational start site as to generate a new translation initiation codon (AU<u>A</u> → AU<u>G</u>). At the other end of an mRNA, a stop codon change has been identified for a *Drosophila* G-protein coupled receptor in the rhodopsin family similar to the β_2-adrenergic receptors found in mammals [137], but the physiological significance of this editing site generating a tryptophan (U<u>A</u>G → U<u>G</u>G) is perplexing as it only produces two extra amino acids that immediately hit another stop codon [137]. In a more crucial case of stop codon modification, the antigenome RNA of the hepatitis delta virus also replaces a translational stop signal with a tryptophan codon; this is an essential step in this viral life cycle [138, 139]. Additionally, the polyadenylation signal (AAUAAA) important for the proper processing of an mRNA can be a potential target, but no examples have been identified of A-to-I editing removing this consensus signal.

Polypeptides expressed in the central nervous system such as neurotransmitter receptors and ion channels are often targets of A-to-I editing at the RNA level [134], and a few examples are described in Sections 3.6.1 and 3.6.2. It is presumed that proteins as exemplified by neurotransmitters acquired an advantage in function by an A-to-I editing and consequent recoding event that subsequently underwent positive selection and has resulted in increased editing at this site [12, 140]. Furthermore, a general theme is emerging that few mRNAs are left to be identified that are marked for editing within the coding regions, which will ultimately lead to protein sequence variation. It could possibly be that coding region editing events have a higher prevalence than what has been identified, but it occurs at a low frequency for a specific site to almost near background levels, making it difficult to detect. This kind of editing can provide more diversity of protein isoforms to fine-tune neurotransmitter function, especially in mammalian channels and receptors [13, 134, 141].

To identify novel proteins that contained modifications due to A-to-I RNA editing, a different approach from previous screens was taken. These investigations utilized a comparative genomic methodology to systematically detect editing within conserved portions of known neurotransmitter proteins that are themselves edited by ADAR [29]. This coding region screen for A-to-I editing yielded many ADAR-modified proteins. Most significantly, these altered targets were all involved in neurotransmission, and many of the edited locations are in conserved amino acid clusters demonstrating the importance of phylogenetic conservation of editing among different species [29]. Notably, a new human target protein has emerged from this analysis alluding to the editing of the K^+ channel gene KCNA1, an ortholog to *Drosophila* shaker gene Kv1.1 and is also found in rodents [29]. These as well as the squid Kv1.1 are edited throughout evolution indicating the functional importance of editing these ancient potassium channels [29, 142, 143]. The voltage-gated K^+ channels of squid has alterations in channel closure rates and its tetramerization upon being edited [142, 144]. The levels of human Kv1.1 (hKv1.1)-specific editing in the nervous system is prominent displaying spatial regulation within the different regions compared [29]. hKv1.1 is edited in a particularly interesting location creating an amino acid change at the conserved isoleucine 400 to a conserved edited valine (I/V) [29]. Ile400 is positioned in the pore of this K^+ channel, and the edited form generating Val400 has selective functional consequences on the process of fast inactivation [143]. Residues in close proximity to hKv1.1 Ile400 have been implicated in the autosomal dominant neurological disorder of episodic ataxia type-1 (EA-1) that results from missense mutations at the DNA level [145]. It is intriguing to speculate that RNA editing near the mutated locations may perhaps become influenced by the affected dsRNA structure surrounding the Ile400 codon, or the RNA editing event itself may be involved in this disease state, directly or indirectly.

Some other newly indentified channels and receptors that contain editing sites in the protein-coding regions are FLNA, BLCAP, CYFIP2, and IGFBP7 [146], but the physiological significance of these editing events remains to be determined. Recent analysis of these proteins indicate that BLCAP and IGFBP7 are edited by only ADAR1, while CYFIP2 and FLNA are more strongly edited by ADAR2 as compared with ADAR1 [147]. In another similar recent study, it was confirmed that editing of BLCAP and FLNA can be done by ADAR1, and determined that ADAR2 edits CYFIP2 and FLNA as well [148]. These new target differences between ADAR proteins may further establish the distinctions for each protein in RNA editing as well as explain the phenotypic dissimilarity of ADAR1 and ADAR2 knockout mice models. Additionally, editing was found in Gabra-3, which codes for the alpha3 subunit of the GABA(A) receptor [149]. An amino acid change of isoleucine to methionine (I/M site) is catalyzed by both ADAR1 and ADAR2. The editing frequency is low at birth, but increases during development, reaching close to 100% in the adult mouse brain [149]. The physiological significance remains to be established but might have importance for gating and inactivation of the channel. The best studied mammalian neurotransmitters examples are the GluR-B and the serotonin 5-HT_{2C}R receptors, which contain editing events that significantly alter membrane protein functions.

3.6.1 Glutamate Receptor Editing–GluR-B

The first example found in a coding region of mRNA was for GluR-B, a subtype of the AMPA (α-amino-3-hydroxy-5-methyl-4-isoxazolepropionic acid) receptors [27]. ADAR2 has been implicated in this major mRNA target *in vivo* for the editing of the GluR-B Q/R site, where it is edited to ~100% [124, 150]. The postnatal lethality of an *ADAR2*$^{-/-}$ null mouse is rescued if the edited adenosine at the Q/R site of GluR-B is homozygously substituted by a guanosine at the genomic level, thus translating the amino acid at this site as an obligatory arginine [124]. The results indicate that Q/R site editing of GluR-B by ADAR2 is essential for viability [124]. It is interesting that nature selected for this edited site to such a high degree and did not just simply encode for it at the genomic level. Perhaps, this GluR-B Q/R site editing provided a critical advantage that this change became a prerequisite for proper channel function, which mediates fast excitatory neurotransmission in the brain.

The glutamate receptors consist of four subunits to form an inner channel lining defining a critical pore loop structure where the Q/R site is located [150, 151]. This site is important in determining the ion flow of the channel, and the functional consequence of Q/R editing is voltage-independent gating with decreased calcium (Ca^{2+}) permeability [27, 150, 152]. A role for GluR-B Q/R site subunit editing in cellular trafficking and assembly of its heteromeric channel receptors has been reported [153, 154]. The edited Q/R site provides for retention within the endoplasmic reticulum (ER) and determines the proper assembly of the AMPA receptor at the step of tetramerization [153, 154]. It appears that the Q/R edited site is critical for different aspects of GluR maturation.

3.6.2
Serotonin Receptor Editing – 5-$HT_{2C}R$

Of the at least 14 distinct members of the serotonin (5-hydroxytryptamine or 5-HT) receptors expressed within the central nervous system, the 5-$HT_{2C}R$ subtype is the only one known that undergoes A-to-I RNA editing. The G-protein coupling functions of the receptor changes upon editing of the A–E sites that are located at the intracellular loop II domain [130, 155–158]. The different amino acid combinations produced as a result of editing at these five sites can modify up to three residues I156, N158, and I160 in the unedited receptor (denoted "INI") and can result in the expression of 24 different edited isoforms within regions of the brain. Pharmacological studies of receptor isoforms revealed a substantial decrease in the basal G-protein coupling activity, agonist affinity, and 5-HT potency due to RNA editing at these sites [130, 155–158]. Greater editing for sites A–E of 5-$HT_{2C}R$ results in a lower response to serotonin, which in turn, decreases phospholipase C activation and a cascade of downstream pathways. An affect was also seen for the coupling of different G-protein α-subunits for the edited versions [159]. Significant conformational changes of the intracellular loop II domain observed in the fully edited VGV isoform as opposed to the unedited INI amino acid residues have been proposed as a cause for the alterations in G-protein coupling activity for 5-$HT_{2C}R$ [160].

It has been revealed that the pattern of 5-$HT_{2C}R$ editing is significantly different among inbred strains of mice [161]. In C57BL/6 and 129SV, more than 80% of forebrain neocortical 5-$HT_{2C}R$ mRNAs are edited at A, B, and D sites or A, B, C, and D sites. C57BL/6 mice show no alteration of the editing pattern against acute stress or chronic treatment of fluoxetine, an antidepressant. In contrast, in the BALB/c strain, about 80% of 5-$HT_{2C}R$ mRNAs are not edited at all and encode receptors with the highest constitutive activity and the highest agonist affinity and potency. However, exposure of BALB/c mice to acute stress or chronic treatment with fluoxetine elicits significant increases of RNA editing at sites A–D [161]. Interestingly, BALB/c mice have a deficiency in the 5-HT biosynthesis pathway and thus may need to maintain a high level of unedited 5-$HT_{2C}R$ mRNAs to compensate for their naturally low concentration of 5-HT. These observations imply that editing responsiveness to stress and medication is modulated by genetic background, as well as behavioral state. Human psychiatry studies utilizing post-mortem brains have limitations and development of animal models, such as 5-HT$_{2C}$R editing-deficient mice, will be needed to gain substantial insights into the relationships between RNA editing of 5-HTRs and mental disorders.

3.7
Cellular Effects and *in Vivo* Phenotypes of ADAR Gene Inactivation

The physiological consequences of A-to-I editing mediated by a specific ADAR have been validated in various species. A *C. elegans* strain containing double homozygous deletions for both *c.e.ADAR1* and *c.e.ADAR2* genes is viable, but it displays defects in chemotaxis and has abnormal development of the vulva in a subset of worms lacking only *c.e.ADAR1* [72]. *Drosophila* engineered with a homozygous deletion in the lone *dADAR* gene are also viable but exhibit defective locomotion and behavior connected to a variety of anatomical and neurological alterations in the brain [70]. The range of neurological defects include obsessive grooming, abnormal mating behavior, tremors, sluggish recovery from hypoxia, and age-dependent neurodegeneration [70, 162, 163]. These fly abnormalities are presumed to be caused by the lack of editing for mRNAs linked to several known targets such as the *cac* Ca^{2+} channel and *para* Na^{+} channel [70, 162, 163]. ADAR knockout phenotypes of these lower organisms indicate that A-to-I editing is non-essential for life but hint at clues toward neurological disorders.

In mammals, ADAR genes are required for life in contrast to their fly and worm counterparts. Initially, *ADAR2*$^{-/-}$ mice are viable but die shortly after birth, post-natally by day 20 of repeated episodes of epileptic seizures [124]. This phenotype is completely reversed upon mutagenic targeting of a known ADAR2 substrate that generates the purely edited form of the protein from the genome, mimicking as if it were edited [124]. The expression of ADAR2 and the editing frequency of ADAR2 targets were increased, but no obvious phenotype was detected in mutant mice generated in which the ability of ADAR2 to edit its own pre-mRNA has been selectively ablated [164]. Although the transgenic mice expressing ADAR2b with an hGH (human growth hormone) 3′ untranslated region (UTR) display a phenotype of hyperphagia and obesity [165], the mechanism underlying hyperphagia remains to be known. However, it is independent of the editing activity, because expression of deaminase-deficient ADAR2b resulted in the same phenotype [165].

ADAR1 function appears to be more important during embryonic development. Heterozygous *ADAR1*$^{+/-}$ mouse chimeras die at embryonic day 14.5 at the midgestation stage with erythropoietic dysfunction [166]. The embryonic lethality observed in this ADAR1 heterozygote might have been due to antisense affects generated by transcripts derived from the targeted allele and has not been ruled out [166]. The same group engineered a new *ADAR1*$^{-/-}$ null mutation in mice, and analysis revealed lethality for the embryos at days 11.0–12.5 [167]. Widespread cellular death of apoptotic cells was detected in many tissues of the *ADAR1*$^{-/-}$ embryos collected at days 10.5–11.5 [167]. Similarly, another group reported congruent findings in *ADAR1*$^{-/-}$ mice that died at embryonic days 11.5–12.5 and observed severe defects in liver structure along with a hematopoietic deficiency [168].

Consistent with animal models is the correlation of ADAR A-to-I editing involvement in human diseases and disorders. In its most benign case, ADAR1 gene locus mutations in humans has been directly associated to cause dyschromatosis symmetrica hereditaria (DSH), a hereditary pigmentation disorder [169, 170]. Intriguingly, many of these point mutations in human ADAR1 are located in the C-terminal portion of the protein [170] and most likely results in full-length proteins but with altered function (Figure 3.5). Interestingly, it was hypothesized that missense mutations are rarely found in the dsRBDs or the critical KKxxK motif contained within, because these alterations would have a more dominant effect when paired with a wild-type partner, thus drastically reducing ADAR function [88] (Figures 3.4 and 3.5). Since the initial reports [169, 170], 12 missense, 18 frameshift, 8 nonsense, and 3 splice site mutations have been identified for ADAR1 in DSH patients [141]. Among these mutations identified in total, the two mutations (Q102fs and H216fs) are specifically interesting, since these frameshift mutations are located upstream of the methionine codon 296, which is used as the initiation codon for translation of the p110 form of ADAR1 (Figures 3.3 and 3.5). Thus, no p150 form of ADAR1 is synthesized from the mutated allele, but synthesis of p110 proteins still occurs in patients carrying this particular mutation. This mutation is clearly linked to the DSH phenotype, indicating that the dosage of functional p150 proteins is a central denominator of DSH [141]. In a previous study, no pigmentation abnormality was detected on the limbs of *ADAR1*$^{+/-}$ mice [167]. It will be interesting to see if an ADAR1-specific RNA editing target will emerge in the future that can explain the observed DSH phenotype and disease mechanism.

Evidence in mice of epileptic seizures are a major consequence of the GluR-B Q/R site underediting, with increased AMPA receptor conductance at the macroscopic level [124, 171, 172]. Also, mice that have been eliminated for Q/R site editing in the GluR-6 kainate receptor subunit display a greater susceptibility to kainite-induced seizures [173]. Lack of normal editing at the GluR-B Q/R site may in part be responsible for the occurrence of epileptic seizures in patients with malignant gliomas [126]. It is not known whether reduced RNA editing rates cause the disease state or are a consequence of the tumor itself. The critical role of ADAR2 is underscored by its A-to-I editing activity at the Q/R site of GluR-B.

The editing of the serotonin receptor 5-$HT_{2C}R$ RNA raises the possibility that this may be to some degree a relevant source of neuropsychiatric dysfunction [134]. This idea has become more relevant in that editing of 5-$HT_{2C}R$ is considerably altered in the prefrontal cortex of suicide victims [174, 175] as well as in schizophrenic individuals [176]. The use of interferons (IFNs) for clinical treatment of chronic hepatitis virus and other ailments have had positive outcomes on the immune system, but these cytokines have adverse effects including depression

[177–179]. The antiserotonergic effects in IFN-induced depression have been implicated as a causative mechanism of this illness [177–179]. The possibility that 5-HT_{2C}R mRNA editing can be the underlying basis for this disorder is intriguing. As mentioned earlier ADAR1 contains an IFN inducible promoter as well as constitutive promoters [67, 101–103]. Utilizing glioblastoma cell lines, it has been demonstrated that the expression of ADAR1 and pattern of 5-HT_{2C}R mRNA editing is rapidly changed in response to IFN-α treatment [180]. These results support the hypothesis that induced suicidal depression from cytokines may have associated affects from the editing of 5-HT_{2C}R mRNA by ADAR1.

Different cases of A-to-I editing dysregulation are beginning to surface and implicate ADAR function in human disease [141]. Just like in other organisms studied, ADAR's activity in protein recoding appears to be directed toward neurotransmitter receptors of the central nervous system. Neurological disorders in which ADAR2 has been implicated are epilepsy, Alzheimer's disease, Huntington's disease, schizophrenia, and amyotrophic lateral sclerosis (ALS), which may be a result of glutamate receptor channel editing in human brain tissue [141, 181–187]. Furthermore, IFN-induced editing by ADAR1 may have causative relevance in the pathophysiology of depression, schizophrenia, and the propensity to become suicidal that is linked to the serotonin receptor [141, 174–176, 180]. With new classes of A-to-I RNA editing targets recently discovered, it is anticipated that more potential diseases may be attributed to ADAR function.

3.8 Noncoding RNA and Repetitive Sequences

Only a few A-to-I changes have been identified in target genes from discrepancies between the mRNA (cDNA) and genomic sequences, which have been found totally by chance. These serendipitous occurrences provided optimism that many more coding region editing events will be uncovered. The level of inosine present in poly$(A)^+$ RNA from various mammalian tissues, especially in the brain, was reported to be estimated at one inosine for every 17 000 ribonucleotides [30]. This substantial amount of editing particularly in the central nervous system correlated with the few known ADAR targets that are involved in neurotransmission and suggested that many more genes may undergo A-to-I editing. This created much interest in developing methods to thoroughly identify novel targets of A-to-I editing. In recent years, there has been much in the development of new approaches to uncover targets of ADAR.

An initial study created a system in *C. elegans* for the recovery and cloning of inosine-containing RNAs that offered a broad search tool to identify new ADAR substrates [31]. This first round of analysis in search of recoding events yielded five A-to-I targets in worms, but it was not the case as these events where found in purely noncoding regions of mRNA and also in a noncoding RNA that was hyperedited [31]. A subsequent analysis utilizing an identical method confirmed other noncoding A-to-I activities, not only in *C. elegans* but also in human brain tissue [3]. This new tool identified 10 novel target RNAs in worm and 19 from human, which suggested a general theme throughout evolution that A-to-I editing is perhaps mainly limited to the UTRs and introns of mRNA as well as noncoding RNA. More importantly, these initial studies opened the door to show that editing intersected with repetitive elements that are contained within the RNA sequences that originate from the chromosomal DNA. These repetitive elements, such as the Alu and LINE1 sequences of the human genome are dispersed in the noncoding portions of DNA, some as inverted repeats and many are capable of folding back to generate hairpin structures at the RNA level [188] (Figure 3.7). These initial studies provided the first mechanism to get a better understanding of the targets of ADAR A-to-I editing.

The method developed based on the sequence and dsRNA structure required for editing of glutamate and serotonin receptor mRNAs was utilized to detect a first round of new A-to-I targets based on unbiased screening [3, 31]. This technique could readily detect inosine incorporation in the coding region of these

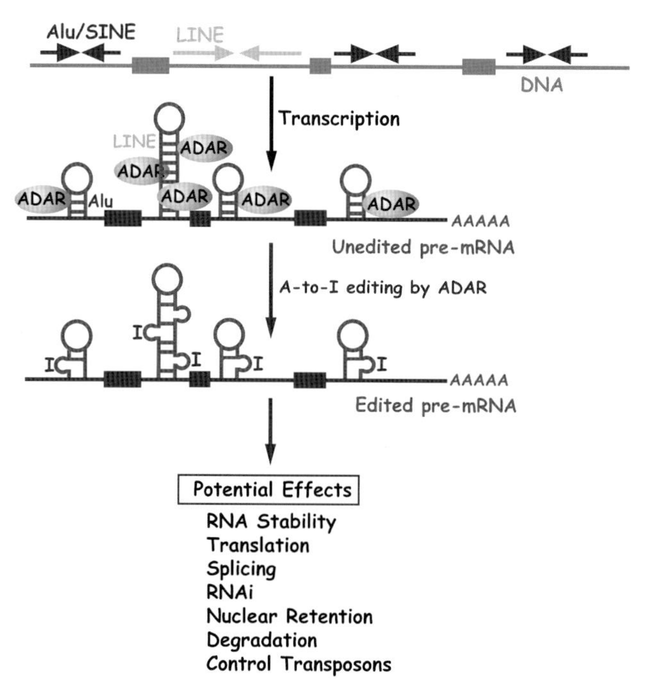

Figure 3.7 Editing of noncoding and repetitive RNAs and its potential significance. Long duplexed RNA structures formed by inverted repeats of transcribed retrotransposable Alu and Line1 (L1) sequences pose as targets for ADAR A-to-I RNA editing activity. The consequent destabilization of this dsRNA may have effects on a variety of cellular functions, in turn affecting gene regulation at the posttranscriptional level. Horizontal arrows represent repetitive sequences in the genome, boxes are exons, lines are introns and UTRs, and "I" (inosine) indicates sites of ADAR activity.

transcripts under defined experiments, but it did not yield any new coding region targets from this screen. In the posthuman genome sequencing era, it is now possible to get a more realistic amount to the degree of A-to-I targets in the human transcriptome. Stringent analysis employing bioinformatic tools along with experimental verification of putative editing sites has increased the known number of editing sites by several orders of magnitude [32, 33, 35–37]. The newly identified sites for editing have amplified our small amount of ADAR substrates to now include greater than 2500 target RNAs, especially within *Homo sapiens*. Finally, it seems that it is well established that the majority of all editing sites are contained within the noncoding regions composed of UTRs, introns, or noncoding RNAs. Primarily, repetitive elements such as the abundant Alu and LINE1 transposons of primates are in effect the major targets of ADAR A-to-I editing [32, 33, 35–37] (Figure 3.7). In parallel experiments of the mouse transcriptome, which does not contain these abundant Alus, only 91 transcripts were found to be edited [35]. However, this still dramatically increases the number of editing sites for rodents. These analyses shift the focus from editing of the coding regions to editing of the noncoding repetitive elements that form inverted repeats as to create long hairpin structures [189]. Editing frequency in human is much higher than that in other organisms such as mouse, rat, chicken, and fly [190]. This difference can be explained by the specific properties of different repeat families such as abundance, length, and divergence.

By utilizing cDNA sequencing data, the first bioinformatic methodologies generated only small amounts of ADAR A-to-I activity that was mainly positioned within the repeat sequences but provided initial clues as to where this abundant editing might occur [32]. However, a genomic wide comprehensive analysis of abundant A-to-I editing sites utilizing millions of expressed sequence tags (ESTs) from the human transcriptome revealed that Alu sequences are indeed the major target of this modification [33]. This computational analysis mapped 12 723 A-to-I events in 1637 different genes, thus amplifying greatly the known targets to be edited. Nearly all reported sites consisted of inversely oriented repetitive sequences such as the Alu (92%) and LINE1 (1%) elements and were located in the 5′ UTR (12%), 3′ UTR (54%), and in the introns (33%) [33].

Thereafter, other groups confirmed that the Alu repeats in humans are indeed edited, and this expanded our knowledge of A-to-I targets by utilization of new bioinformatic searches [35–37]. These computational data compiled together indicate that editing is much more widespread than previously thought with thousands of transcripts being affected and may occur at the greater frequency of one edit per 2000 ribonucleotides [37]. It was determined that over 2600 human mRNAs (~2%) are subjected to RNA editing events [35]. Furthermore, 88% of the A-to-I editing events were found to be located in the Alu sequences even though they only comprise 20% of the total length of transcripts [35]. Interestingly, A-to-I editing was found to be most prevalent in the brain as opposed to other human tissues [33, 35–37], correlating with ADAR expression. These newer studies indicate that the majority of these editing events occurred in the introns of pre-mRNA [36, 37]. However, A-to-I RNA editing is not detected in the regions outside of repetitive sequences [191]. Both sense and antisense RNA containing repetitive sequences are edited. A-to-I changes can be observed only in regions encompassing repetitive sequences, indicating that editing occurs independently in each strand and less likely of formation of an intermolecular duplex between sense and antisense RNA containing an inverted repeat of Alus [192]. These editing events can affect splice site selection via modification [36]. Also, these posttranscriptional modifications have the potential to affect alternatively spliced Alu-containing exons by changes in these splice signals [36] (Figure 3.7). Alu-derived exons are possibly selected against due to the harmful effects that they may produce, as previous evidence suggests that alternatively spliced exons are not found as constitutive exons nor in the coding regions of functionally mature mRNAs [193].

It is of great interest to understand the functional consequences of Alu editing by ADAR, which may represent a probable housekeeping gene. Or it is also possible that A-to-I modification of Alu sequences may not serve any function and signify a system that just coincidently uses these highly prevalent dsRNA structures. On the contrary, these primate Alu elements may mark the RNA to affect processing, stability, transport, localization, or translation of the message that can conceivably be modulated by ADAR function (Figure 3.7). The presence of an intronic dsRNA and its stability have been shown to influence the kinetics of splicing or even result in alternative splicing [125, 193]. Human nuclear prelamin A recognition factor contains a primate-specific Alu exon that exclusively depends on RNA editing for its exonization [194]. Or perhaps, the editing of Alu elements act as a cellular marker of unspliced mRNA, as evidence suggests that hyperedited RNAs are retained in the nucleus, although mainly an antiviral mechanism [195]. Furthermore, editing may affect the transposon activities of Alu and LINE1 elements [196, 197], and it may interfere with suppressive effects of transposon transcriptional elongation [198, 199]. Editing frequency of Alu repetitive elements as well as the expression level of ADARs is reduced in human brain cancers [200]. The extent of the reduction is correlated with the grade of malignancy, suggesting that reduced editing may be involved in the pathogenesis of cancer. The wide variety of potential functions performed by noncoding RNA suggests that ADARs may play regulatory roles in many processes.

3.9 Effects on the RNA Interference Silencing Pathway

RNAi is the process by which ~22 bp dsRNAs called small interfering RNAs (siRNAs) direct the degradation of homologous mRNA transcripts [201, 202]. In addition to being an important viral defense and gene regulatory mechanism in the cell, RNAi has become an immensely powerful and widely used experimental tool. RNAi can be induced by a long dsRNA, which is cleaved by the ribonuclease Dicer into siRNAs (Figure 3.8a) that direct the destruction of the original transcript and any other identical sequences [203]. Long dsRNA is also a preferred substrate for ADARs, which can nonspecifically edit more than 50% of adenosines in such

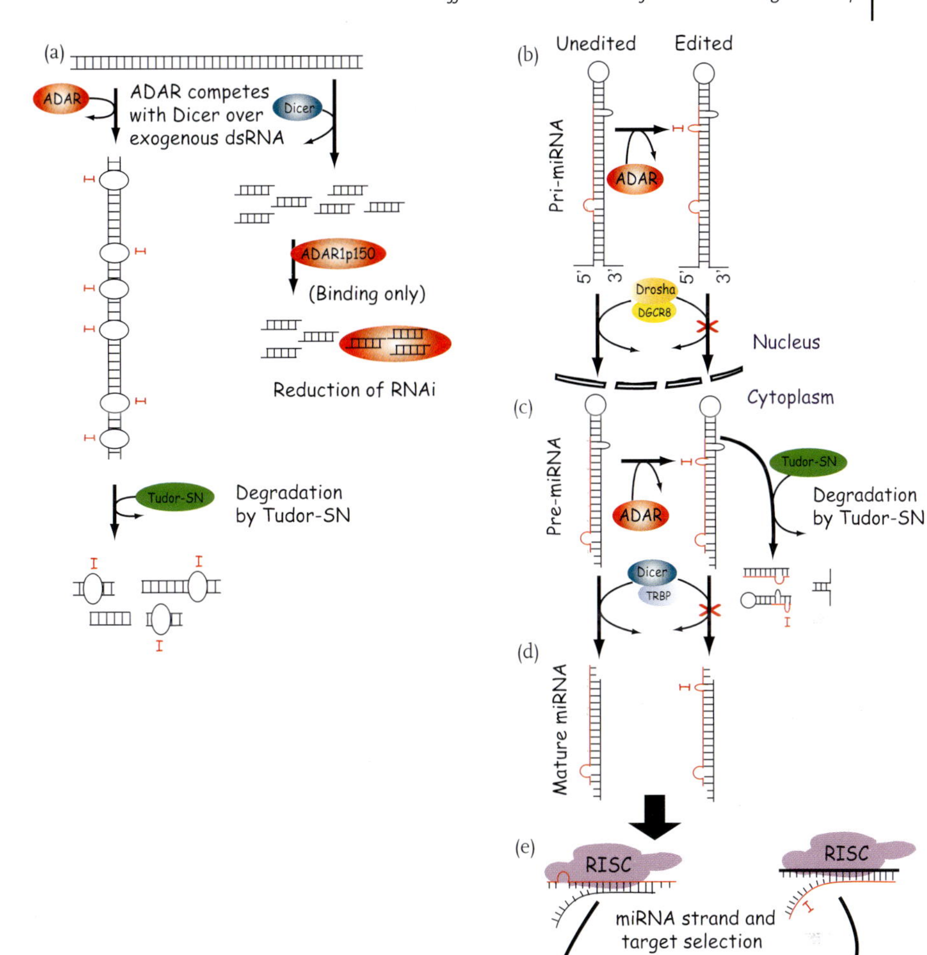

Figure 3.8 Involvement of ADAR in RNA-mediated gene regulation pathways. (a) ADAR and Dicer compete for long dsRNA substrates of viral or synthetic origin. If ADAR succeeds in hyperediting the RNA, Dicer activity, and thus RNAi, is suppressed and the resulting RNA may be degraded by Tudor-SN. If allowed, Dicer will process the long dsRNA into ~22 bp siRNAs to begin the RNAi process. ADAR1p150 can sequester some of these siRNAs, reducing RNAi efficiency. (b) Pri-miRNAs are subject to editing by ADARs, which may block cleavage by Drosha. (c) Pre-miRNAs may also be subject to ADAR activity, inhibiting cleavage by Dicer. Edited pri- and pre-miRNAs can thus build up and may be degraded by Tudor-SN. If processing is unaffected by editing, edited mature miRNAs will be expressed (d), which can silence a different set of targets from their unedited counterparts (e). Editing may also alter miRNA stability, potentially affecting active strand selection by RISC.

molecules [12, 113]. Such extensive editing has been shown *in vitro* to suppress dicing of these RNA targets, and thus to suppress RNAi induced by these RNAs [204] (Figure 3.8a). In addition to competing with Dicer for long dsRNA substrates, ADAR1p150 has been shown to bind directly to siRNAs via its dsRBDs. This tight association decreases the effective concentration of siRNAs in the cytoplasm, thereby decreasing the efficiency of RNAi (Figure 3.8a) [42]. Strains of *C. elegans* in which the active ADAR genes have been deleted show defects in chemotaxis (the ability to seek out or avoid certain substances). These defects are eliminated in strains that have gene mutations affecting the RNAi machinery [205]. This suggests that the chemotaxis phenotype is a result of hyperactivity in an RNAi pathway, which is normally inhibited by A-to-I editing. This antagonistic relationship suggests that RNA editing may have evolved to protect organisms from nonspecific effects of RNAi.

3.10
Effects on MicroRNA Processing and Target Selection

MicroRNAs (miRNAs) are small dsRNAs, encoded by eukaryotic genomes, which regulate gene expression via an RNAi-like pathway. There are currently over 500 known miRNAs in humans [206]. They need not be perfectly complementary to their targets; therefore, a single miRNA can affect a large number of mRNAs, and a single mRNA may be targeted by many miRNAs. MicroRNAs are transcribed into long primary transcripts called primary miRNAs (pri-miRNAs), which form a stem-loop secondary structure. The pri-miRNAs are cleaved by the nuclear Drosha-DGCR8 complex to remove the flanking ssRNA sequence and yield ~60–70 nt miRNA precursors (pre-miRNAs) (Figure 3.8b,c). Pre-miRNAs are transported into the cytoplasm by Exportin 5-Ran-GTP and then cleaved by the cytoplasmic Dicer-TRBP complex into mature miRNAs: dsRNAs ~22 bp in length, with two-nucleotide 3′ overhangs (Figure 3.8c,d). These mature miRNAs (miRs) are subsequently loaded into the RNA-induced silencing complex (RISC), which selects the active miRNA strand and unwinds the miRNA (Figure 3.8e). The miRNA guides the RISC complex to its target site, usually in the 3′ UTR of an mRNA, resulting in translational repression or mRNA degradation [201, 207].

The dsRNA regions of pri- and pre-miRNAs allow ADARs to interact with intermediates in the miRNA biogenesis pathway. A-to-I editing has been detected in numerous endogenous pri-miRNAs [41, 44]. Editing can affect miRNA processing, by inhibiting the Drosha [43] or Dicer [45] cleavage steps (Figure 3.8b,c), thereby reducing levels of the mature miRNA. In the case of miR-142, edited pri-miRNAs are not processed and are subsequently degraded by the inosine-specific nuclease Tudor-SN [43]. Editing of miR-151 precursors blocks cleavage by Dicer, resulting in accumulation of edited pre-miRNAs [45]. Since the transcription of miRNAs is often controlled by the promoters of other genes [207, 208], such modulation of processing is thought to be the main method of regulating miRNA levels. If processing is not affected by these editing events, then mature miRNAs with A-to-I substitutions are expressed (Figure 3.8d) [46, 208]. These miRNAs can silence a set of gene targets different from that of their unedited counterpart (Figure 3.8e). In the case of miR-376a-5p, the edited form of the mature miRNA gains the ability to downregulate the mRNA of the PRPS1 gene [46]. PRPS1 is involved in purine metabolism, whose end product is uric acid. Its overexpression in humans is associated with the buildup of purines and uric acid, resulting in gout and, in some cases, neurodevelopmental impairment [209]. Mice lacking ADAR2, the enzyme responsible for editing of miR-376a-5p, cannot express the edited form of this mature miRNA. As a result, they have increased levels of the PRPS1 protein and uric acid in their brains compared with wild-type mice. The effect of uric acid on the brain is unclear, but some studies have shown reduced levels in patients with multiple sclerosis, suggesting that it may have a neuroprotective effect [210].

In most cases, only one strand of the miRNA is thought to be active, while the other is degraded. Selection of the active miRNA strand is based on the thermodynamic stability of the 5′ ends of the miRNA [207]. The structural changes caused by editing could change the effective strand [5], which is likely to drastically change the genes silenced by the miRNA (Figure 3.8e). The 3′ UTRs of mRNAs are known to be frequently edited [3, 33, 140], so it is possible that editing of miRNA target sites in these UTRs might enhance or suppress the silencing of specific mRNAs [211]. Similarly, editing may destabilize secondary structures in 3′ UTRs, allowing the RISC complex access to previously inaccessible target sites. Alternatively, such editing may prevent Drosha and Dicer from cleaving dsRNA regions of mRNAs. Through all of these mechanisms, A-to-I RNA editing has the potential to rapidly change gene expression levels in response to cellular stress or other stimuli.

Misregulation of miRNAs may play a role in a number of human ailments. Downregulation of miR-133 makes mice more susceptible to cardiac hypertrophy, while an increase in either miR-133 or miR-1 may reduce occurrence of the disease [212]. MicroRNAs are also likely to play a role in cancer. Numerous studies have found changes in miRNA expression in tumors when compared with normal tissue [213]. Specific miRNAs have been shown to have both oncogenic and tumor-suppressive effects [214]. In addition, pediatric astrocytomas show a reduced level of ADAR2 editing activity [100]. Increasing ADAR2 levels in these tumor cells has been shown to reduce their malignant behavior *in vitro* [100], but the editing targets responsible for this effect remain unknown. This suppressive effect may be the result of editing of an oncogenic miRNA, suggesting a link between the editing of noncoding RNAs and cancer.

3.11 RNA Editing Role as an Antiviral Mechanism

In some cases, the viral RNA genome can form dsRNA, which may be edited by ADARs [215]. The brain of patients infected latently with measles viruses, viral sense, and antisense RNA genome is hyperedited [216]. In the liver of woodchucks, hepatitis delta virus antisense genomic RNA is also edited by ADARs and this editing converts a stop codon (U<u>A</u>G) to a tryptophan codon (U<u>G</u>G) that inhibits viral replication [139, 215]. ADAR1 can also edit the hepatitis C virus (HCV) RNA genome and inhibit the replication of HCV [217]. It was reported that inosine-containing RNA-specific nuclease exists in the cell [38, 218]. These edited dsRNAs may stimulate their degradation by Tudor-SN, which is a micrococcal nuclease homolog and can bind to inosine-containing dsRNA [39] (Figure 3.8a). Moreover, the antiviral cytokine IFN-α/β induces expression of cytoplasmic ADAR1p150 [67, 219, 220]. A key regulator for expression of IFN-α/β is IKKε. It is reported that the RNA expression level of ADAR1 is dramatically reduced in IKKε-deficient mice with influenza virus infection [221]. ADAR1 may participate in the antiviral activity of IFN-α/β through RNA editing of the viral RNA genome. On the other hand, viral genomic DNA can code for pri-miRNA and is processed by the miRNA machinery [222, 223]. Human herpesvirus 8 has 13 pri-miRNA-like structures, and miR-K12 may be edited by ADAR1 [224]. miR-K12 is located in the protein coding region of the K12 transcript, and editing changes a serine to a glycine, which inhibits its transforming activity [224]. It is unknown whether the edited miR-K12 silences a unique set of target genes of its own. Interestingly, the cytidine deaminase APOBEC3 family has been shown to be able to reduce HIV-1 replication *in vivo* [225]. These examples suggest that RNA editing may inhibit viral replication and tumorgenesis as an innate immunity. However, it has also been reported that ADAR1p150 enhanced the vesicular stomatitis virus's replication through inhibition of PKR activation by an RNA editing-independent manner [226]. Moreover, ADAR1 inhibits the cytosolic DNA-mediated induction of IFN-β mRNA [227]. ADAR1 binds to cytosolic DNA via the Z-DNA-binding domain. These data also suggest that ADAR1 may stimulate the DNA virus's replication. Most of the antiviral effect of ADARs is largely unclear.

3.12 Conclusions

Almost two decades ago, ADARs were originally discovered as a mysterious dsRNA unwinding activity. Soon after that, they were identified as the enzymes responsible for A-to-I editing to recode crucial mammalian genes involved in neurotransmission. But the last five years of ADAR analysis by several research groups has indicated new and exciting targets of A-to-I RNA editing. It appears that the recoding of protein genes is a minor activity of ADAR action, although the targets are usually neurotransmitters that can have drastic affects on an organism's physiology and disease state. These novel targets are widespread and contained within the noncoding regions of RNA that are formed by repetitive elements, which can generate "fold-back" structures that form dsRNA. Furthermore, the intersection of ADAR A-to-I editing with the RNAi pathway provides another whole subset of dsRNA targets including miRNAs for the regulation of posttranscriptional gene silencing.

Lower organisms do not have ADAR genes, while for the more complex metazoans from flies to mammals and primates, it appears that editing has evolved with RNA substrates. It is possible that the ADAR proteins could have evolved synergistically with its substrates to become a critical mediator in the cell. This is indicated by the increased levels of editing targets in mammals; specifically, in humans, A-to-I editing activity is required for life, as demonstrated by embryonic lethal phenotypes of knockout mouse models of ADAR1. ADAR2 knockout mouse models also indicate postnatal lethality, but this has more to do with the synergistic evolution of a neurotransmitter, which is 100% edited at a critical residue. Interestingly, the specific cause of the lethal phenotype of the ADAR1 null mutation is yet to be determined. Perhaps this new era in the edited target landscape will provide better hints as to the importance of ADAR1 during development.

Another exciting direction for future studies is the potential of ADAR and A-to-I RNA editing to be involved in cancer development and progression. Recently, many more studies are implicating altered epigenetic control and noncoding RNA, such as repetitive elements and miRNAs, in the progression of cancer and other diseases. It is interesting to speculate that ADAR will be involved in these processes since it has the potential to be an upstream regulator of RNA sequence/function, expression, biogenesis, and activity. The concerted effort by ADAR action may allow for the posttranscriptional regulation of many processes to affect cellular activities on a global scale. Future studies directed toward this elucidation should allow for a greater understanding of posttranscriptional gene regulation within the cell.

Acknowledgments

Support for K.N. includes funds from the National Institutes of Health, the Juvenile Diabetes Research Foundation, and the Commonwealth Universal Research Enhancement Program, Pennsylvania Department of Health. Funds received to L.V. are the National Institutes of Health Postdoctoral Supplemental Award R01 HL070045 and the NCI National Institutes of Health Postdoctoral Training Grant T32 CA09171. Educational funding to B.Z. is provided by the Vagelos MLS Program – University of Pennsylvania.

References

1 International Human Genome Sequencing Consortium (2004) Finishing the euchromatic sequence of the human genome. *Nature*, **431**, 931–945.

2 Gott, J.M. and Emeson, R.B. (2000) Functions and mechanisms of RNA editing. *Annu. Rev. Genet.*, **34**, 499–531.

3 Morse, D.P., Aruscavage, P.J., and Bass, B.L. (2002) RNA hairpins in noncoding regions of human brain and *Caenorhabditis elegans* mRNA are edited by adenosine deaminases that act on RNA. *Proc. Natl. Acad. Sci. U.S.A.*, **99**, 7906–7911.

4 Rueter, S.M., Dawson, T.R., and Emeson, R.B. (1999) Regulation of alternative splicing by RNA editing. *Nature*, **399**, 75–80.
5 Nishikura, K. (2006) Editor meets silencer: crosstalk between RNA editing and RNA interference. *Nat. Rev. Mol. Cell Biol.*, **7**, 919–931.
6 Benne, R., Van den Burg, J., Brakenhoff, J.P., Sloof, P., Van Boom, J.H., and Tromp, M.C. (1986) Major transcript of the frameshifted coxII gene from trypanosome mitochondria contains four nucleotides that are not encoded in the DNA. *Cell*, **46**, 819–826.
7 Powell, L.M., Wallis, S.C., Pease, R.J., Edwards, Y.H., Knott, T.J., and Scott, J. (1987) A novel form of tissue-specific RNA processing produces apolipoprotein-B48 in intestine. *Cell*, **50**, 831–840.
8 Chen, S.H., Habib, G., Yang, C.Y., Gu, Z.W., Lee, B.R., Weng, S.A., Silberman, S.R., Cai, S.J., *et al.* (1987) Apolipoprotein B-48 is the product of a messenger RNA with an organ-specific in-frame stop codon. *Science*, **238**, 363–366.
9 Teng, B., Burant, C.F., and Davidson, N.O. (1993) Molecular cloning of an apolipoprotein B messenger RNA editing protein. *Science*, **260**, 1816–1819.
10 Navaratnam, N. and Sarwar, R. (2006) An overview of cytidine deaminases. *Int. J. Hematol.*, **83**, 195–200.
11 Gerber, A.P. and Keller, W. (2001) RNA editing by base deamination: more enzymes, more targets, new mysteries. *Trends Biochem. Sci.*, **26**, 376–384.
12 Bass, B.L. (2002) RNA editing by adenosine deaminases that act on RNA. *Annu. Rev. Biochem.*, **71**, 817–846.
13 Jepson, J.E. and Reenan, R.A. (2008) RNA editing in regulating gene expression in the brain. *Biochim. Biophys. Acta*, **1779**, 459–470.
14 Maas, S. and Rich, A. (2000) Changing genetic information through RNA editing. *Bioessays*, **22**, 790–802.
15 Wagner, R.W., Smith, J.E., Cooperman, B.S., and Nishikura, K. (1989) A double-stranded RNA unwinding activity introduces structural alterations by means of adenosine to inosine conversions in mammalian cells and *Xenopus* eggs. *Proc. Natl. Acad. Sci. U.S.A.*, **86**, 2647–2651.
16 Polson, A.G., Crain, P.F., Pomerantz, S.C., McCloskey, J.A., and Bass, B.L. (1991) The mechanism of adenosine to inosine conversion by the double-stranded RNA unwinding/modifying activity: a high-performance liquid chromatography-mass spectrometry analysis. *Biochemistry*, **30**, 11507–11514.
17 Kim, U., Wang, Y., Sanford, T., Zeng, Y., and Nishikura, K. (1994) Molecular cloning of cDNA for double-stranded RNA adenosine deaminase, a candidate enzyme for nuclear RNA editing. *Proc. Natl. Acad. Sci. U.S.A.*, **91**, 11457–11461.
18 Schaub, M. and Keller, W. (2002) RNA editing by adenosine deaminases generates RNA and protein diversity. *Biochimie*, **84**, 791–803.
19 Bass, B.L., Nishikura, K., Keller, W., Seeburg, P.H., Emeson, R.B., O'Connell, M.A., Samuel, C.E., and Herbert, A. (1997) A standardized nomenclature for adenosine deaminases that act on RNA. *RNA*, **3**, 947–949.
20 Bass, B.L. and Weintraub, H. (1988) An unwinding activity that covalently modifies its double-stranded RNA substrate. *Cell*, **55**, 1089–1098.
21 O'Connell, M.A., Krause, S., Higuchi, M., Hsuan, J.J., Totty, N.F., Jenny, A., and Keller, W. (1995) Cloning of cDNAs encoding mammalian double-stranded RNA-specific adenosine deaminase. *Mol. Cell. Biol.*, **15**, 1389–1397.
22 Melcher, T., Maas, S., Herb, A., Sprengel, R., Seeburg, P.H., and Higuchi, M. (1996) A mammalian RNA editing enzyme. *Nature*, **379**, 460–464.
23 Lai, F., Chen, C.X., Carter, K.C., and Nishikura, K. (1997) Editing of glutamate receptor B subunit ion channel RNAs by four alternatively spliced DRADA2 double-stranded RNA adenosine deaminases. *Mol. Cell. Biol.*, **17**, 2413–2424.
24 Gerber, A., O'Connell, M.A., and Keller, W. (1997) Two forms of human double-stranded RNA-specific editase 1 (hRED1) generated by the insertion of an Alu cassette. *RNA*, **3**, 453–463.

25 Melcher, T., Maas, S., Herb, A., Sprengel, R., Higuchi, M., and Seeburg, P.H. (1996) RED2, a brain-specific member of the RNA-specific adenosine deaminase family. *J. Biol. Chem.*, **271**, 31795–31798.

26 Chen, C.X., Cho, D.S., Wang, Q., Lai, F., Carter, K.C., and Nishikura, K. (2000) A third member of the RNA-specific adenosine deaminase gene family, ADAR3, contains both single- and double-stranded RNA binding domains. *RNA*, **6**, 755–767.

27 Sommer, B., Kohler, M., Sprengel, R., and Seeburg, P.H. (1991) RNA editing in brain controls a determinant of ion flow in glutamate-gated channels. *Cell*, **67**, 11–19.

28 Seeburg, P.H. (2002) A-to-I editing: new and old sites, functions and speculations. *Neuron*, **35**, 17–20.

29 Hoopengardner, B., Bhalla, T., Staber, C., and Reenan, R. (2003) Nervous system targets of RNA editing identified by comparative genomics. *Science*, **301**, 832–836.

30 Paul, M.S. and Bass, B.L. (1998) Inosine exists in mRNA at tissue-specific levels and is most abundant in brain mRNA. *EMBO J.*, **17**, 1120–1127.

31 Morse, D.P. and Bass, B.L. (1999) Long RNA hairpins that contain inosine are present in *Caenorhabditis elegans* poly(A)+ RNA. *Proc. Natl. Acad. Sci. U.S.A.*, **96**, 6048–6053.

32 Kikuno, R., Nagase, T., Waki, M., and Ohara, O. (2002) HUGE: a database for human large proteins identified in the Kazusa cDNA sequencing project. *Nucleic Acids Res.*, **30**, 166–168.

33 Levanon, E.Y., Eisenberg, E., Yelin, R., Nemzer, S., Hallegger, M., Shemesh, R., Fligelman, Z.Y., Shoshan, A., *et al.* (2004) Systematic identification of abundant A-to-I editing sites in the human transcriptome. *Nat. Biotechnol.*, **22**, 1001–1005.

34 Nishikura, K. (2004) Editing the message from A to I. *Nat. Biotechnol.*, **22**, 962–963.

35 Kim, D.D., Kim, T.T., Walsh, T., Kobayashi, Y., Matise, T.C., Buyske, S., and Gabriel, A. (2004) Widespread RNA editing of embedded Alu elements in the human transcriptome. *Genome Res.*, **14**, 1719–1725.

36 Athanasiadis, A., Rich, A., and Maas, S. (2004) Widespread A-to-I RNA editing of Alu-containing mRNAs in the human transcriptome. *PLoS Biol.*, **2**, e391.

37 Blow, M., Futreal, P.A., Wooster, R., and Stratton, M.R. (2004) A survey of RNA editing in human brain. *Genome Res.*, **14**, 2379–2387.

38 Scadden, A.D. and Smith, C.W. (2001) Specific cleavage of hyper-edited dsRNAs. *EMBO J.*, **20**, 4243–4252.

39 Scadden, A.D. (2005) The RISC subunit Tudor-SN binds to hyper-edited double-stranded RNA and promotes its cleavage. *Nat. Struct. Mol. Biol.*, **12**, 489–496.

40 Knight, S.W. and Bass, B.L. (2002) The role of RNA editing by ADARs in RNAi. *Mol. Cell*, **10**, 809–817.

41 Luciano, D.J., Mirsky, H., Vendetti, N.J., and Maas, S. (2004) RNA editing of a miRNA precursor. *RNA*, **10**, 1174–1177.

42 Yang, W., Wang, Q., Howell, K.L., Lee, J.T., Cho, D.S., Murray, J.M., and Nishikura, K. (2005) ADAR1 RNA deaminase limits short interfering RNA efficacy in mammalian cells. *J. Biol. Chem.*, **280**, 3946–3953.

43 Yang, W., Chendrimada, T.P., Wang, Q., Higuchi, M., Seeburg, P.H., Shiekhattar, R., and Nishikura, K. (2006) Modulation of microRNA processing and expression through RNA editing by ADAR deaminases. *Nat. Struct. Mol. Biol.*, **13**, 13–21.

44 Blow, M.J., Grocock, R.J., van Dongen, S., Enright, A.J., Dicks, E., Futreal, P.A., Wooster, R., and Stratton, M.R. (2006) RNA editing of human microRNAs. *Genome Biol.*, **7**, R27.

45 Kawahara, Y., Zinshteyn, B., Chendrimada, T.P., Shiekhattar, R., and Nishikura, K. (2007) RNA editing of the microRNA-151 precursor blocks cleavage by the Dicer-TRBP complex. *EMBO Rep.*, **8**, 763–769.

46 Kawahara, Y., Zinshteyn, B., Sethupathy, P., Iizasa, H., Hatzigeorgiou, A.G., and Nishikura, K. (2007) Redirection of silencing targets by adenosine-to-inosine editing of miRNAs. *Science*, **315**, 1137–1140.

47 Carmichael, G.G. (2003) Antisense starts making more sense. *Nat. Biotechnol.*, **21**, 371–372.

48 Yelin, R., Dahary, D., Sorek, R., Levanon, E.Y., Goldstein, O., Shoshan, A., Diber, A., Biton, S., *et al.* (2003) Widespread occurrence of antisense transcription in the human genome. *Nat. Biotechnol.*, **21**, 379–386.

49 Kampa, D., Cheng, J., Kapranov, P., Yamanaka, M., Brubaker, S., Cawley, S., Drenkow, J., Piccolboni, A., *et al.* (2004) Novel RNAs identified from an in-depth analysis of the transcriptome of human chromosomes 21 and 22. *Genome Res.*, **14**, 331–342.

50 Lavorgna, G., Dahary, D., Lehner, B., Sorek, R., Sanderson, C.M., and Casari, G. (2004) In search of antisense. *Trends Biochem. Sci.*, **29**, 88–94.

51 Gerber, A.P. and Keller, W. (1999) An adenosine deaminase that generates inosine at the wobble position of tRNAs. *Science*, **286**, 1146–1149.

52 Betts, L., Xiang, S., Short, S.A., Wolfenden, R., and Carter, C.W., Jr. (1994) Cytidine deaminase. The 2.3 A crystal structure of an enzyme: transition-state analog complex. *J. Mol. Biol.*, **235**, 635–656.

53 Lau, P.P., Zhu, H.J., Baldini, A., Charnsangavej, C., and Chan, L. (1994) Dimeric structure of a human apolipoprotein B mRNA editing protein and cloning and chromosomal localization of its gene. *Proc. Natl. Acad. Sci. U.S.A.*, **91**, 8522–8526.

54 MacGinnitie, A.J., Anant, S., and Davidson, N.O. (1995) Mutagenesis of apobec-1, the catalytic subunit of the mammalian apolipoprotein B mRNA editing enzyme, reveals distinct domains that mediate cytosine nucleoside deaminase, RNA binding, and RNA editing activity. *J. Biol. Chem.*, **270**, 14768–14775.

55 Navaratnam, N., Fujino, T., Bayliss, J., Jarmuz, A., How, A., Richardson, N., Somasekaram, A., Bhattacharya, S., *et al.* (1998) *Escherichia coli* cytidine deaminase provides a molecular model for ApoB RNA editing and a mechanism for RNA substrate recognition. *J. Mol. Biol.*, **275**, 695–714.

56 Cho, D.S., Yang, W., Lee, J.T., Shiekhattar, R., Murray, J.M., and Nishikura, K. (2003) Requirement of dimerization for RNA editing activity of adenosine deaminases acting on RNA. *J. Biol. Chem.*, **278**, 17093–17102.

57 Gallo, A., Keegan, L.P., Ring, G.M., and O'Connell, M.A. (2003) An ADAR that edits transcripts encoding ion channel subunits functions as a dimer. *EMBO J.*, **22**, 3421–3430.

58 Gerber, A., Grosjean, H., Melcher, T., and Keller, W. (1998) Tad1p, a yeast tRNA-specific adenosine deaminase, is related to the mammalian pre-mRNA editing enzymes ADAR1 and ADAR2. *EMBO J.*, **17**, 4780–4789.

59 Keegan, L.P., Gerber, A.P., Brindle, J., Leemans, R., Gallo, A., Keller, W., and O'Connell, M.A. (2000) The properties of a tRNA-specific adenosine deaminase from *Drosophila melanogaster* support an evolutionary link between pre-mRNA editing and tRNA modification. *Mol. Cell. Biol.*, **20**, 825–833.

60 Maas, S., Kim, Y.G., and Rich, A. (2000) Sequence, genomic organization and functional expression of the murine tRNA-specific adenosine deaminase ADAT1. *Gene*, **243**, 59–66.

61 Maas, S., Gerber, A.P., and Rich, A. (1999) Identification and characterization of a human tRNA-specific adenosine deaminase related to the ADAR family of pre-mRNA editing enzymes. *Proc. Natl. Acad. Sci. U.S.A.*, **96**, 8895–8900.

62 Wolf, J., Gerber, A.P., and Keller, W. (2002) tadA, an essential tRNA-specific adenosine deaminase from *Escherichia coli*. *EMBO J.*, **21**, 3841–3851.

63 Elias, Y. and Huang, R.H. (2005) Biochemical and structural studies of A-to-I editing by tRNA:A34 deaminases at the wobble position of transfer RNA. *Biochemistry*, **44**, 12057–12065.

64 Losey, H.C., Ruthenburg, A.J., and Verdine, G.L. (2006) Crystal structure of *Staphylococcus aureus* tRNA adenosine deaminase TadA in complex with RNA. *Nat. Struct. Mol. Biol.*, **13**, 153–159.

65 Rubio, M.A., Pastar, I., Gaston, K.W., Ragone, F.L., Janzen, C.J., Cross, G.A., Papavasiliou, F.N., and Alfonzo, J.D.

(2007) An adenosine-to-inosine tRNA-editing enzyme that can perform C-to-U deamination of DNA. *Proc. Natl. Acad. Sci. U.S.A.*, **104**, 7821–7826.

66 Keller, W., Wolf, J., and Gerber, A. (1999) Editing of messenger RNA precursors and of tRNAs by adenosine to inosine conversion. *FEBS Lett.*, **452**, 71–76.

67 Patterson, J.B. and Samuel, C.E. (1995) Expression and regulation by interferon of a double-stranded-RNA-specific adenosine deaminase from human cells: evidence for two forms of the deaminase. *Mol. Cell. Biol.*, **15**, 5376–5388.

68 Slavov, D., Clark, M., and Gardiner, K. (2000) Comparative analysis of the RED1 and RED2 A-to-I RNA editing genes from mammals, pufferfish and zebrafish. *Gene*, **250**, 41–51.

69 Slavov, D., Crnogorac-Jurcevic, T., Clark, M., and Gardiner, K. (2000) Comparative analysis of the DRADA A-to-I RNA editing gene from mammals, pufferfish and zebrafish. *Gene*, **250**, 53–60.

70 Palladino, M.J., Keegan, L.P., O'Connell, M.A., and Reenan, R.A. (2000) A-to-I pre-mRNA editing in *Drosophila* is primarily involved in adult nervous system function and integrity. *Cell*, **102**, 437–449.

71 Hough, R.F., Lingam, A.T., and Bass, B.L. (1999) *Caenorhabditis elegans* mRNAs that encode a protein similar to ADARs derive from an operon containing six genes. *Nucleic Acids Res.*, **27**, 3424–3432.

72 Tonkin, L.A., Saccomanno, L., Morse, D.P., Brodigan, T., Krause, M., and Bass, B.L. (2002) RNA editing by ADARs is important for normal behavior in *Caenorhabditis elegans*. *EMBO J.*, **21**, 6025–6035.

73 Herbert, A., Alfken, J., Kim, Y.G., Mian, I.S., Nishikura, K., and Rich, A. (1997) A Z-DNA binding domain present in the human editing enzyme, double-stranded RNA adenosine deaminase. *Proc. Natl. Acad. Sci. U.S.A.*, **94**, 8421–8426.

74 Rich, A. and Zhang, S. (2003) Timeline: Z-DNA: the long road to biological function. *Nat. Rev. Genet.*, **4**, 566–572.

75 Ryman, K., Fong, N., Bratt, E., Bentley, D.L., and Ohman, M. (2007) The C-terminal domain of RNA Pol II helps ensure that editing precedes splicing of the GluR-B transcript. *RNA*, **13**, 1071–1078.

76 Athanasiadis, A., Placido, D., Maas, S., Brown, B.A., 2nd, Lowenhaupt, K., and Rich, A. (2005) The crystal structure of the Zbeta domain of the RNA-editing enzyme ADAR1 reveals distinct conserved surfaces among Z-domains. *J. Mol. Biol.*, **351**, 496–507.

77 Placido, D., Brown, B.A., 2nd, Lowenhaupt, K., Rich, A., and Athanasiadis, A. (2007) A left-handed RNA double helix bound by the Z alpha domain of the RNA-editing enzyme ADAR1. *Structure*, **15**, 395–404.

78 Koeris, M., Funke, L., Shrestha, J., Rich, A., and Maas, S. (2005) Modulation of ADAR1 editing activity by Z-RNA *in vitro*. *Nucleic Acids Res.*, **33**, 5362–5370.

79 Herbert, A. and Rich, A. (2001) The role of binding domains for dsRNA and Z-DNA in the *in vivo* editing of minimal substrates by ADAR1. *Proc. Natl. Acad. Sci. U.S.A.*, **98**, 12132–12137.

80 Carlson, C.B., Stephens, O.M., and Beal, P.A. (2003) Recognition of double-stranded RNA by proteins and small molecules. *Biopolymers*, **70**, 86–102.

81 Saunders, L.R. and Barber, G.N. (2003) The dsRNA binding protein family: critical roles, diverse cellular functions. *FASEB J.*, **17**, 961–983.

82 Ryter, J.M. and Schultz, S.C. (1998) Molecular basis of double-stranded RNA-protein interactions: structure of a dsRNA-binding domain complexed with dsRNA. *EMBO J.*, **17**, 7505–7513.

83 Ramos, A., Grunert, S., Adams, J., Micklem, D.R., Proctor, M.R., Freund, S., Bycroft, M., St Johnston, D., *et al.* (2000) RNA recognition by a Staufen double-stranded RNA-binding domain. *EMBO J.*, **19**, 997–1009.

84 Lai, F., Drakas, R., and Nishikura, K. (1995) Mutagenic analysis of double-stranded RNA adenosine deaminase, a candidate enzyme for RNA editing of glutamate-gated ion channel transcripts. *J. Biol. Chem.*, **270**, 17098–17105.

85 Liu, Y. and Samuel, C.E. (1996) Mechanism of interferon action: functionally distinct RNA-binding and catalytic domains in the interferon-inducible, double-stranded RNA-specific adenosine deaminase. *J. Virol.*, **70**, 1961–1968.

86 Sansam, C.L., Wells, K.S., and Emeson, R.B. (2003) Modulation of RNA editing by functional nucleolar sequestration of ADAR2. *Proc. Natl. Acad. Sci. U.S.A.*, **100**, 14018–14023.

87 Stefl, R., Xu, M., Skrisovska, L., Emeson, R.B., and Allain, F.H. (2006) Structure and specific RNA binding of ADAR2 double-stranded RNA binding motifs. *Structure*, **14**, 345–355.

88 Valente, L. and Nishikura, K. (2007) RNA binding-independent dimerization of adenosine deaminases acting on RNA and dominant negative effects of nonfunctional subunits on dimer functions. *J. Biol. Chem.*, **282**, 16054–16061.

89 Ohman, M., Kallman, A.M., and Bass, B.L. (2000) *In vitro* analysis of the binding of ADAR2 to the pre-mRNA encoding the GluR-B R/G site. *RNA*, **6**, 687–697.

90 Yi-Brunozzi, H.Y., Stephens, O.M., and Beal, P.A. (2001) Conformational changes that occur during an RNA-editing adenosine deamination reaction. *J. Biol. Chem.*, **276**, 37827–37833.

91 Hallegger, M., Taschner, A., and Jantsch, M.F. (2006) RNA aptamers binding the double-stranded RNA-binding domain. *RNA*, **12**, 1993–2004.

92 Macbeth, M.R., Lingam, A.T., and Bass, B.L. (2004) Evidence for auto-inhibition by the N terminus of hADAR2 and activation by dsRNA binding. *RNA*, **10**, 1563–1571.

93 Xu, M., Wells, K.S., and Emeson, R.B. (2006) Substrate-dependent contribution of double-stranded RNA-binding motifs to ADAR2 function. *Mol. Biol. Cell*, **17**, 3211–3220.

94 Poulsen, H., Jorgensen, R., Heding, A., Nielsen, F.C., Bonven, B., and Egebjerg, J. (2006) Dimerization of ADAR2 is mediated by the double-stranded RNA binding domain. *RNA*, **12**, 1350–1360.

95 Kallman, A.M., Sahlin, M., and Ohman, M. (2003) ADAR2 A-->I editing: site selectivity and editing efficiency are separate events. *Nucleic Acids Res.*, **31**, 4874–4881.

96 Polson, A.G. and Bass, B.L. (1994) Preferential selection of adenosines for modification by double-stranded RNA adenosine deaminase. *EMBO J.*, **13**, 5701–5711.

97 Stephens, O.M., Yi-Brunozzi, H.Y., and Beal, P.A. (2000) Analysis of the RNA-editing reaction of ADAR2 with structural and fluorescent analogues of the GluR-B R/G editing site. *Biochemistry*, **39**, 12243–12251.

98 Macbeth, M.R., Schubert, H.L., Vandemark, A.P., Lingam, A.T., Hill, C.P., and Bass, B.L. (2005) Inositol hexakisphosphate is bound in the ADAR2 core and required for RNA editing. *Science*, **309**, 1534–1539.

99 Chilibeck, K.A., Wu, T., Liang, C., Schellenberg, M.J., Gesner, E.M., Lynch, J.M., and MacMillan, A.M. (2006) FRET analysis of *in vivo* dimerization by RNA-editing enzymes. *J. Biol. Chem.*, **281**, 16530–16535.

100 Cenci, C., Barzotti, R., Galeano, F., Corbelli, S., Rota, R., Massimi, L., Di Rocco, C., O'Connell, M.A., *et al.* (2008) Down-regulation of RNA editing in pediatric astrocytomas: ADAR2 editing activity inhibits cell migration and proliferation. *J. Biol. Chem.*, **283**, 7251–7260.

101 George, C.X. and Samuel, C.E. (1999) Human RNA-specific adenosine deaminase ADAR1 transcripts possess alternative exon 1 structures that initiate from different promoters, one constitutively active and the other interferon inducible. *Proc. Natl. Acad. Sci. U.S.A.*, **96**, 4621–4626.

102 George, C.X. and Samuel, C.E. (1999) Characterization of the 5'-flanking region of the human RNA-specific adenosine deaminase ADAR1 gene and identification of an interferon-inducible ADAR1 promoter. *Gene*, **229**, 203–213.

103 Kawakubo, K. and Samuel, C.E. (2000) Human RNA-specific adenosine deaminase (ADAR1) gene specifies transcripts that initiate from a

constitutively active alternative promoter. *Gene*, **258**, 165–172.

104 Lykke-Andersen, S., Pinol-Roma, S., and Kjems, J. (2007) Alternative splicing of the ADAR1 transcript in a region that functions either as a 5'-UTR or an ORF. *RNA*, **13**, 1732–1744.

105 Poulsen, H., Nilsson, J., Damgaard, C.K., Egebjerg, J., and Kjems, J. (2001) CRM1 mediates the export of ADAR1 through a nuclear export signal within the Z-DNA binding domain. *Mol. Cell. Biol.*, **21**, 7862–7871.

106 Eckmann, C.R., Neunteufl, A., Pfaffstetter, L., and Jantsch, M.F. (2001) The human but not the *Xenopus* RNA-editing enzyme ADAR1 has an atypical nuclear localization signal and displays the characteristics of a shuttling protein. *Mol. Biol. Cell.*, **12**, 1911–1924.

107 Nie, Y., Zhao, Q., Su, Y., and Yang, J.H. (2004) Subcellular distribution of ADAR1 isoforms is synergistically determined by three nuclear discrimination signals and a regulatory motif. *J. Biol. Chem.*, **279**, 13249–13255.

108 Eckmann, C.R. and Jantsch, M.F. (1999) The RNA-editing enzyme ADAR1 is localized to the nascent ribonucleoprotein matrix on *Xenopus* lampbrush chromosomes but specifically associates with an atypical loop. *J. Cell Biol.*, **144**, 603–615.

109 Doyle, M. and Jantsch, M.F. (2003) Distinct *in vivo* roles for double-stranded RNA-binding domains of the *Xenopus* RNA-editing enzyme ADAR1 in chromosomal targeting. *J. Cell Biol.*, **161**, 309–319.

110 Desterro, J.M., Keegan, L.P., Lafarga, M., Berciano, M.T., O'Connell, M., and Carmo-Fonseca, M. (2003) Dynamic association of RNA-editing enzymes with the nucleolus. *J. Cell Sci.*, **116**, 1805–1818.

111 Lehmann, K.A. and Bass, B.L. (1999) The importance of internal loops within RNA substrates of ADAR1. *J. Mol. Biol.*, **291**, 1–13.

112 Dawson, T.R., Sansam, C.L., and Emeson, R.B. (2004) Structure and sequence determinants required for the RNA editing of ADAR2 substrates. *J. Biol. Chem.*, **279**, 4941–4951.

113 Nishikura, K., Yoo, C., Kim, U., Murray, J.M., Estes, P.A., Cash, F.E., and Liebhaber, S.A. (1991) Substrate specificity of the dsRNA unwinding/modifying activity. *EMBO J.*, **10**, 3523–3532.

114 Lehmann, K.A. and Bass, B.L. (2000) Double-stranded RNA adenosine deaminases ADAR1 and ADAR2 have overlapping specificities. *Biochemistry*, **39**, 12875–12884.

115 Wong, S.K., Sato, S., and Lazinski, D.W. (2001) Substrate recognition by ADAR1 and ADAR2. *RNA*, **7**, 846–858.

116 Serra, M.J., Smolter, P.E., and Westhof, E. (2004) Pronounced instability of tandem IU base pairs in RNA. *Nucleic Acids Res.*, **32**, 1824–1828.

117 Bass, B.L. and Weintraub, H. (1987) A developmentally regulated activity that unwinds RNA duplexes. *Cell*, **48**, 607–613.

118 Dabiri, G.A., Lai, F., Drakas, R.A., and Nishikura, K. (1996) Editing of the GLuR-B ion channel RNA *in vitro* by recombinant double-stranded RNA adenosine deaminase. *EMBO J.*, **15**, 34–45.

119 Maas, S., Melcher, T., Herb, A., Seeburg, P.H., Keller, W., Krause, S., Higuchi, M., and O'Connell, M.A. (1996) Structural requirements for RNA editing in glutamate receptor pre-mRNAs by recombinant double-stranded RNA adenosine deaminase. *J. Biol. Chem.*, **271**, 12221–12226.

120 Raitskin, O., Cho, D.S., Sperling, J., Nishikura, K., and Sperling, R. (2001) RNA editing activity is associated with splicing factors in lnRNP particles: the nuclear pre-mRNA processing machinery. *Proc. Natl. Acad. Sci. U.S.A.*, **98**, 6571–6576.

121 Bratt, E. and Ohman, M. (2003) Coordination of editing and splicing of glutamate receptor pre-mRNA. *RNA*, **9**, 309–318.

122 Agranat, L., Raitskin, O., Sperling, J., and Sperling, R. (2008) The editing enzyme ADAR1 and the mRNA surveillance protein hUpf1 interact in the cell nucleus. *Proc. Natl. Acad. Sci. U.S.A.*, **105**, 5028–5033.

123 DeCerbo, J. and Carmichael, G.G. (2005) Retention and repression: fates of hyperedited RNAs in the nucleus. *Curr. Opin. Cell Biol.*, **17**, 302–308.

124 Higuchi, M., Maas, S., Single, F.N., Hartner, J., Rozov, A., Burnashev, N., Feldmeyer, D., Sprengel, R., *et al.* (2000) Point mutation in an AMPA receptor gene rescues lethality in mice deficient in the RNA-editing enzyme ADAR2. *Nature*, **406**, 78–81.

125 Reenan, R.A., Hanrahan, C.J., and Barry, G. (2000) The mle(napts) RNA helicase mutation in *Drosophila* results in a splicing catastrophe of the para Na+ channel transcript in a region of RNA editing. *Neuron*, **25**, 139–149.

126 Maas, S., Patt, S., Schrey, M., and Rich, A. (2001) Underediting of glutamate receptor GluR-B mRNA in malignant gliomas. *Proc. Natl. Acad. Sci. U.S.A.*, **98**, 14687–14692.

127 Palladino, M.J., Keegan, L.P., O'Connell, M.A., and Reenan, R.A. (2000) dADAR, a *Drosophila* double-stranded RNA-specific adenosine deaminase is highly developmentally regulated and is itself a target for RNA editing. *RNA*, **6**, 1004–1018.

128 Beghini, A., Ripamonti, C.B., Peterlongo, P., Roversi, G., Cairoli, R., Morra, E., and Larizza, L. (2000) RNA hyperediting and alternative splicing of hematopoietic cell phosphatase (PTPN6) gene in acute myeloid leukemia. *Hum. Mol. Genet.*, **9**, 2297–2304.

129 Flomen, R., Knight, J., Sham, P., Kerwin, R., and Makoff, A. (2004) Evidence that RNA editing modulates splice site selection in the 5-HT2C receptor gene. *Nucleic Acids Res.*, **32**, 2113–2122.

130 Burns, C.M., Chu, H., Rueter, S.M., Hutchinson, L.K., Canton, H., Sanders-Bush, E., and Emeson, R.B. (1997) Regulation of serotonin-2C receptor G-protein coupling by RNA editing. *Nature*, **387**, 303–308.

131 Higuchi, M., Single, F.N., Kohler, M., Sommer, B., Sprengel, R., and Seeburg, P.H. (1993) RNA editing of AMPA receptor subunit GluR-B: a base-paired intron-exon structure determines position and efficiency. *Cell*, **75**, 1361–1370.

132 Lomeli, H., Mosbacher, J., Melcher, T., Hoger, T., Geiger, J.R., Kuner, T., Monyer, H., Higuchi, M., *et al.* (1994) Control of kinetic properties of AMPA receptor channels by nuclear RNA editing. *Science*, **266**, 1709–1713.

133 Herb, A., Higuchi, M., Sprengel, R., and Seeburg, P.H. (1996) Q/R site editing in kainate receptor GluR5 and GluR6 pre-mRNAs requires distant intronic sequences. *Proc. Natl. Acad. Sci. U.S.A.*, **93**, 1875–1880.

134 Seeburg, P.H. and Hartner, J. (2003) Regulation of ion channel/neurotransmitter receptor function by RNA editing. *Curr. Opin. Neurobiol.*, **13**, 279–283.

135 Keegan, L.P., Gallo, A., and O'Connell, M.A. (2001) The many roles of an RNA editor. *Nat. Rev. Genet.*, **2**, 869–878.

136 Tsudzuki, T., Wakasugi, T., and Sugiura, M. (2001) Comparative analysis of RNA editing sites in higher plant chloroplasts. *J. Mol. Evol.*, **53**, 327–332.

137 Stapleton, M., Carlson, J., Brokstein, P., Yu, C., Champe, M., George, R., Guarin, H., Kronmiller, B., *et al.* (2002) A *Drosophila* full-length cDNA resource. *Genome Biol.*, **3**, RESEARCH0080.

138 Luo, G.X., Chao, M., Hsieh, S.Y., Sureau, C., Nishikura, K., and Taylor, J. (1990) A specific base transition occurs on replicating hepatitis delta virus RNA. *J. Virol.*, **64**, 1021–1027.

139 Polson, A.G., Bass, B.L., and Casey, J.L. (1996) RNA editing of hepatitis delta virus antigenome by dsRNA-adenosine deaminase. *Nature*, **380**, 454–456.

140 Maas, S., Rich, A., and Nishikura, K. (2003) A-to-I RNA editing: recent news and residual mysteries. *J. Biol. Chem.*, **278**, 1391–1394.

141 Maas, S., Kawahara, Y., Tamburro, K.M., and Nishikura, K. (2006) A-to-I RNA editing and human disease. *RNA Biol.*, **3**, 1–9.

142 Rosenthal, J.J. and Bezanilla, F. (2002) Extensive editing of mRNAs for the squid delayed rectifier K+ channel regulates subunit tetramerization. *Neuron*, **34**, 743–757.

143 Bhalla, T., Rosenthal, J.J., Holmgren, M., and Reenan, R. (2004) Control of human potassium channel inactivation by editing of a small mRNA hairpin. *Nat. Struct. Mol. Biol.*, **11**, 950–956.

144 Patton, D.E., Silva, T., and Bezanilla, F. (1997) RNA editing generates a diverse array of transcripts encoding squid Kv2 K+ channels with altered functional properties. *Neuron*, **19**, 711–722.

145 Maylie, B., Bissonnette, E., Virk, M., Adelman, J.P., and Maylie, J.G. (2002) Episodic ataxia type 1 mutations in the human Kv1.1 potassium channel alter hKvbeta 1-induced N-type inactivation. *J. Neurosci.*, **22**, 4786–4793.

146 Levanon, E.Y., Hallegger, M., Kinar, Y., Shemesh, R., Djinovic-Carugo, K., Rechavi, G., Jantsch, M.F., and Eisenberg, E. (2005) Evolutionarily conserved human targets of adenosine to inosine RNA editing. *Nucleic Acids Res.*, **33**, 1162–1168.

147 Nishimoto, Y., Yamashita, T., Hideyama, T., Tsuji, S., Suzuki, N., and Kwak, S. (2008) Determination of editors at the novel A-to-I editing positions. *Neurosci. Res*, **61**, 201–206.

148 Riedmann, E.M., Schopoff, S., Hartner, J.C., and Jantsch, M.F. (2008) Specificity of ADAR-mediated RNA editing in newly identified targets. *RNA*, **14**, 1110–1118.

149 Ohlson, J., Pedersen, J.S., Haussler, D., and Ohman, M. (2007) Editing modifies the GABA(A) receptor subunit alpha3. *RNA*, **13**, 698–703.

150 Seeburg, P.H., Single, F., Kuner, T., Higuchi, M., and Sprengel, R. (2001) Genetic manipulation of key determinants of ion flow in glutamate receptor channels in the mouse. *Brain Res.*, **907**, 233–243.

151 Kuner, T., Seeburg, P.H., and Guy, H.R. (2003) A common architecture for K+ channels and ionotropic glutamate receptors? *Trends Neurosci.*, **26**, 27–32.

152 Egebjerg, J. and Heinemann, S.F. (1993) Ca2+ permeability of unedited and edited versions of the kainate selective glutamate receptor GluR6. *Proc. Natl. Acad. Sci. U.S.A.*, **90**, 755–759.

153 Greger, I.H., Khatri, L., and Ziff, E.B. (2002) RNA editing at arg607 controls AMPA receptor exit from the endoplasmic reticulum. *Neuron*, **34**, 759–772.

154 Greger, I.H., Khatri, L., Kong, X., and Ziff, E.B. (2003) AMPA receptor tetramerization is mediated by Q/R editing. *Neuron*, **40**, 763–774.

155 Fitzgerald, L.W., Iyer, G., Conklin, D.S., Krause, C.M., Marshall, A., Patterson, J.P., Tran, D.P., Jonak, G.J., *et al.* (1999) Messenger RNA editing of the human serotonin 5-HT2C receptor. *Neuropsychopharmacology*, **21**, 82S–90S.

156 Herrick-Davis, K., Grinde, E., and Niswender, C.M. (1999) Serotonin 5-HT2C receptor RNA editing alters receptor basal activity: implications for serotonergic signal transduction. *J. Neurochem.*, **73**, 1711–1717.

157 Niswender, C.M., Copeland, S.C., Herrick-Davis, K., Emeson, R.B., and Sanders-Bush, E. (1999) RNA editing of the human serotonin 5-hydroxytryptamine 2C receptor silences constitutive activity. *J. Biol. Chem.*, **274**, 9472–9478.

158 Wang, Q., O'Brien, P.J., Chen, C.X., Cho, D.S., Murray, J.M., and Nishikura, K. (2000) Altered G protein-coupling functions of RNA editing isoform and splicing variant serotonin2C receptors. *J. Neurochem.*, **74**, 1290–1300.

159 Price, R.D., Weiner, D.M., Chang, M.S., and Sanders-Bush, E. (2001) RNA editing of the human serotonin 5-HT2C receptor alters receptor-mediated activation of G13 protein. *J. Biol. Chem.*, **276**, 44663–44668.

160 Visiers, I., Hassan, S.A., and Weinstein, H. (2001) Differences in conformational properties of the second intracellular loop (IL2) in 5HT(2C) receptors modified by RNA editing can account for G protein coupling efficiency. *Prot. Eng.*, **14**, 409–414.

161 Englander, M.T., Dulawa, S.C., Bhansali, P., and Schmauss, C. (2005) How stress and fluoxetine modulate serotonin 2C receptor pre-mRNA editing. *J. Neurosci.*, **25**, 648–651.

162 O'Farrell, P.H. (2001) Conserved responses to oxygen deprivation. *J. Clin. Invest.*, **107**, 671–674.

163 Reenan, R.A. (2001) The RNA world meets behavior: a-->I pre-mRNA editing in animals. *Trends Genet.*, **17**, 53–56.

164 Feng, Y., Sansam, C.L., Singh, M., and Emeson, R.B. (2006) Altered RNA editing in mice lacking ADAR2 autoregulation. *Mol. Cell. Biol.*, **26**, 480–488.

165 Singh, M., Kesterson, R.A., Jacobs, M.M., Joers, J.M., Gore, J.C., and Emeson, R.B. (2007) Hyperphagia-mediated obesity in transgenic mice misexpressing the RNA-editing enzyme ADAR2. *J. Biol. Chem.*, **282**, 22448–22459.

166 Wang, Q., Khillan, J., Gadue, P., and Nishikura, K. (2000) Requirement of the RNA editing deaminase ADAR1 gene for embryonic erythropoiesis. *Science*, **290**, 1765–1768.

167 Wang, Q., Miyakoda, M., Yang, W., Khillan, J., Stachura, D.L., Weiss, M.J., and Nishikura, K. (2004) Stress-induced apoptosis associated with null mutation of ADAR1 RNA editing deaminase gene. *J. Biol. Chem.*, **279**, 4952–4961.

168 Hartner, J.C., Schmittwolf, C., Kispert, A., Muller, A.M., Higuchi, M., and Seeburg, P.H. (2004) Liver disintegration in the mouse embryo caused by deficiency in the RNA-editing enzyme ADAR1. *J. Biol. Chem.*, **279**, 4894–4902.

169 Miyamura, Y., Suzuki, T., Kono, M., Inagaki, K., Ito, S., Suzuki, N., and Tomita, Y. (2003) Mutations of the RNA-specific adenosine deaminase gene (DSRAD) are involved in dyschromatosis symmetrica hereditaria. *Am. J. Hum. Genet.*, **73**, 693–699.

170 Zhang, X.J., He, P.P., Li, M., He, C.D., Yan, K.L., Cui, Y., Yang, S., Zhang, K.Y., *et al.* (2004) Seven novel mutations of the ADAR gene in Chinese families and sporadic patients with dyschromatosis symmetrica hereditaria (DSH). *Hum. Mutat.*, **23**, 629–630.

171 Brusa, R., Zimmermann, F., Koh, D.S., Feldmeyer, D., Gass, P., Seeburg, P.H., and Sprengel, R. (1995) Early-onset epilepsy and postnatal lethality associated with an editing-deficient GluR-B allele in mice. *Science*, **270**, 1677–1680.

172 Feldmeyer, D., Kask, K., Brusa, R., Kornau, H.C., Kolhekar, R., Rozov, A., Burnashev, N., Jensen, V., *et al.* (1999) Neurological dysfunctions in mice expressing different levels of the Q/R site-unedited AMPAR subunit GluR-B. *Nat. Neurosci.*, **2**, 57–64.

173 Vissel, B., Royle, G.A., Christie, B.R., Schiffer, H.H., Ghetti, A., Tritto, T., Perez-Otano, I., Radcliffe, R.A., *et al.* (2001) The role of RNA editing of kainate receptors in synaptic plasticity and seizures. *Neuron*, **29**, 217–227.

174 Gurevich, I., Tamir, H., Arango, V., Dwork, A.J., Mann, J.J., and Schmauss, C. (2002) Altered editing of serotonin 2C receptor pre-mRNA in the prefrontal cortex of depressed suicide victims. *Neuron*, **34**, 349–356.

175 Niswender, C.M., Herrick-Davis, K., Dilley, G.E., Meltzer, H.Y., Overholser, J.C., Stockmeier, C.A., Emeson, R.B., and Sanders-Bush, E. (2001) RNA editing of the human serotonin 5-HT2C receptor. Alterations in suicide and implications for serotonergic pharmacotherapy. *Neuropsychopharmacology*, **24**, 478–491.

176 Sodhi, M.S., Burnet, P.W., Makoff, A.J., Kerwin, R.W., and Harrison, P.J. (2001) RNA editing of the 5-HT(2C) receptor is reduced in schizophrenia. *Mol. Psychiatry*, **6**, 373–379.

177 Malek-Ahmadi, P. (2001) Mood disorders associated with interferon treatment: theoretical and practical considerations. *Ann. Pharmacother.*, **35**, 489–495.

178 Menkes, D.B. and MacDonald, J.A. (2000) Interferons, serotonin and neurotoxicity. *Psychol. Med.*, **30**, 259–268.

179 Schaefer, M., Engelbrecht, M.A., Gut, O., Fiebich, B.L., Bauer, J., Schmidt, F., Grunze, H., and Lieb, K. (2002) Interferon alpha (IFNalpha) and psychiatric syndromes: a review. *Prog. Neuropsychopharmacol. Biol. Psychiatry*, **26**, 731–746.

180 Yang, W., Wang, Q., Kanes, S.J., Murray, J.M., and Nishikura, K. (2004) Altered RNA editing of serotonin 5-HT2C receptor induced by interferon: implications for depression associated

with cytokine therapy. *Brain Res. Mol. Brain Res.*, **124**, 70–78.

181 Paschen, W., Hedreen, J.C., and Ross, C.A. (1994) RNA editing of the glutamate receptor subunits GluR2 and GluR6 in human brain tissue. *J. Neurochem.*, **63**, 1596–1602.

182 Kawahara, Y., Ito, K., Sun, H., Kanazawa, I., and Kwak, S. (2003) Low editing efficiency of GluR2 mRNA is associated with a low relative abundance of ADAR2 mRNA in white matter of normal human brain. *Eur. J. Neurosci.*, **18**, 23–33.

183 Kawahara, Y., Ito, K., Sun, H., Ito, M., Kanazawa, I., and Kwak, S. (2004) Regulation of glutamate receptor RNA editing and ADAR mRNA expression in developing human normal and Down's syndrome brains. *Brain Res. Dev. Brain Res.*, **148**, 151–155.

184 Vollmar, W., Gloger, J., Berger, E., Kortenbruck, G., Kohling, R., Speckmann, E.J., and Musshoff, U. (2004) RNA editing (R/G site) and flip-flop splicing of the AMPA receptor subunit GluR2 in nervous tissue of epilepsy patients. *Neurobiol. Dis.*, **15**, 371–379.

185 Akbarian, S., Smith, M.A., and Jones, E.G. (1995) Editing for an AMPA receptor subunit RNA in prefrontal cortex and striatum in Alzheimer's disease, Huntington's disease and schizophrenia. *Brain Res.*, **699**, 297–304.

186 Kawahara, Y., Kwak, S., Sun, H., Ito, K., Hashida, H., Aizawa, H., Jeong, S.Y., and Kanazawa, I. (2003) Human spinal motoneurons express low relative abundance of GluR2 mRNA: an implication for excitotoxicity in ALS. *J. Neurochem.*, **85**, 680–689.

187 Kawahara, Y., Ito, K., Sun, H., Aizawa, H., Kanazawa, I., and Kwak, S. (2004) Glutamate receptors: RNA editing and death of motor neurons. *Nature*, **427**, 801.

188 Kazazian, H.H., Jr. (2004) Mobile elements: drivers of genome evolution. *Science*, **303**, 1626–1632.

189 Eisenberg, E., Nemzer, S., Kinar, Y., Sorek, R., Rechavi, G., and Levanon, E.Y. (2005) Is abundant A-to-I RNA editing primate-specific? *Trends Genet.*, **21**, 77–81.

190 Neeman, Y., Levanon, E.Y., Jantsch, M.F., and Eisenberg, E. (2006) RNA editing level in the mouse is determined by the genomic repeat repertoire. *RNA*, **12**, 1802–1809.

191 Neeman, Y., Dahary, D., Levanon, E.Y., Sorek, R., and Eisenberg, E. (2005) Is there any sense in antisense editing? *Trends Genet.*, **21**, 544–547.

192 Kawahara, Y. and Nishikura, K. (2006) Extensive adenosine-to-inosine editing detected in Alu repeats of antisense RNAs reveals scarcity of sense-antisense duplex formation. *FEBS Lett.*, **580**, 2301–2305.

193 Sorek, R., Ast, G., and Graur, D. (2002) Alu-containing exons are alternatively spliced. *Genome Res.*, **12**, 1060–1067.

194 Lev-Maor, G., Sorek, R., Levanon, E.Y., Paz, N., Eisenberg, E., and Ast, G. (2007) RNA-editing-mediated exon evolution. *Genome Biol.*, **8**, R29.

195 Zhang, Z. and Carmichael, G.G. (2001) The fate of dsRNA in the nucleus: a p54(nrb)-containing complex mediates the nuclear retention of promiscuously A-to-I edited RNAs. *Cell*, **106**, 465–475.

196 Schramke, V. and Allshire, R. (2003) Hairpin RNAs and retrotransposon LTRs effect RNAi and chromatin-based gene silencing. *Science*, **301**, 1069–1074.

197 Schramke, V. and Allshire, R. (2004) Those interfering little RNAs! Silencing and eliminating chromatin. *Curr. Opin. Genet. Dev.*, **14**, 174–180.

198 Perepelitsa-Belancio, V. and Deininger, P. (2003) RNA truncation by premature polyadenylation attenuates human mobile element activity. *Nat. Genet.*, **35**, 363–366.

199 Han, J.S., Szak, S.T., and Boeke, J.D. (2004) Transcriptional disruption by the L1 retrotransposon and implications for mammalian transcriptomes. *Nature*, **429**, 268–274.

200 Paz, N., Levanon, E.Y., Amariglio, N., Heimberger, A.B., Ram, Z., Constantini, S., Barbash, Z.S., Adamsky, K., *et al.* (2008) Altered adenosine-to-inosine RNA editing in human cancer. *Genome Res.*, **17**, 1586–1595.

201 Bartel, D.P. (2004) MicroRNAs: genomics, biogenesis, mechanism, and function. *Cell*, **116**, 281–297.

202 Hammond, S.M. (2005) Dicing and slicing: the core machinery of the RNA interference pathway. *FEBS Lett.*, **579**, 5822–5829.

203 Fire, A., Xu, S., Montgomery, M.K., Kostas, S.A., Driver, S.E., and Mello, C.C. (1998) Potent and specific genetic interference by double-stranded RNA in *Caenorhabditis elegans*. *Nature*, **391**, 806–811.

204 Scadden, A.D. and Smith, C.W. (2001) RNAi is antagonized by A-->I hyper-editing. *EMBO Rep.*, **2**, 1107–1111.

205 Tonkin, L.A. and Bass, B.L. (2003) Mutations in RNAi rescue aberrant chemotaxis of ADAR mutants. *Science*, **302**, 1725.

206 Griffiths-Jones, S., Saini, H.K., van Dongen, S., and Enright, A.J. (2008) miRBase: tools for microRNA genomics. *Nucleic Acids Res.*, **36**, D154–D158.

207 Du, T. and Zamore, P.D. (2005) microPrimer: the biogenesis and function of microRNA. *Development*, **132**, 4645–4652.

208 Pfeffer, S., Sewer, A., Lagos-Quintana, M., Sheridan, R., Sander, C., Grasser, F.A., van Dyk, L.F., Ho, C.K., *et al.* (2005) Identification of microRNAs of the herpesvirus family. *Nat. Methods*, **2**, 269–276.

209 Ahmed, M., Taylor, W., Smith, P.R., and Becker, M.A. (1999) Accelerated transcription of PRPS1 in X-linked overactivity of normal human phosphoribosylpyrophosphate synthetase. *J. Biol. Chem.*, **274**, 7482–7488.

210 Spitsin, S. and Koprowski, H. (2008) Role of uric acid in multiple sclerosis. *Curr. Top. Microbiol. Immunol.*, **318**, 325–342.

211 Liang, H. and Landweber, L.F. (2007) Hypothesis: RNA editing of microRNA target sites in humans? *RNA*, **13**, 463–467.

212 Care, A., Catalucci, D., Felicetti, F., Bonci, D., Addario, A., Gallo, P., Bang, M.L., Segnalini, P., *et al.* (2007) MicroRNA-133 controls cardiac hypertrophy. *Nat. Med.*, **13**, 613–618.

213 Garzon, R., Fabbri, M., Cimmino, A., Calin, G.A., and Croce, C.M. (2006) MicroRNA expression and function in cancer. *Trends Mol. Med*, **12**, 580–587.

214 He, L., He, X., Lowe, S.W., and Hannon, G.J. (2007) microRNAs join the p53 network–another piece in the tumour-suppression puzzle. *Nat. Rev. Cancer*, **7**, 819–822.

215 Casey, J.L. (2006) RNA editing in hepatitis delta virus. *Curr. Top. Microbiol. Immunol.*, **307**, 67–89.

216 Cattaneo, R., Schmid, A., Eschle, D., Baczko, K., ter Meulen, V., and Billeter, M.A. (1988) Biased hypermutation and other genetic changes in defective measles viruses in human brain infections. *Cell*, **55**, 255–265.

217 Taylor, D.R., Puig, M., Darnell, M.E., Mihalik, K., and Feinstone, S.M. (2005) New antiviral pathway that mediates hepatitis C virus replicon interferon sensitivity through ADAR1. *J. Virol.*, **79**, 6291–6298.

218 Scadden, A.D. and Smith, C.W. (1997) A ribonuclease specific for inosine-containing RNA: a potential role in antiviral defence? *EMBO J.*, **16**, 2140–2149.

219 Patterson, J.B., Thomis, D.C., Hans, S.L., and Samuel, C.E. (1995) Mechanism of interferon action: double-stranded RNA-specific adenosine deaminase from human cells is inducible by alpha and gamma interferons. *Virology*, **210**, 508–511.

220 Yang, J.H., Luo, X., Nie, Y., Su, Y., Zhao, Q., Kabir, K., Zhang, D., and Rabinovici, R. (2003) Widespread inosine-containing mRNA in lymphocytes regulated by ADAR1 in response to inflammation. *Immunology*, **109**, 15–23.

221 Tenoever, B.R., Ng, S.L., Chua, M.A., McWhirter, S.M., Garcia-Sastre, A., and Maniatis, T. (2007) Multiple functions of the IKK-related kinase IKKepsilon in interferon-mediated antiviral immunity. *Science*, **315**, 1274–1278.

222 Qi, P., Han, J.X., Lu, Y.Q., Wang, C., and Bu, F.F. (2006) Virus-encoded microRNAs: future therapeutic targets? *Cell. Mol. Immunol.*, **3**, 411–419.

223 Pfeffer, S., Zavolan, M., Grasser, F.A., Chien, M., Russo, J.J., Ju, J., John, B., Enright, A.J., *et al.* (2004) Identification of virus-encoded microRNAs. *Science*, **304**, 734–736.

224 Gandy, S.Z., Linnstaedt, S.D., Muralidhar, S., Cashman, K.A., Rosenthal, L.J., and Casey, J.L. (2007) RNA editing of the human herpesvirus 8 kaposin transcript eliminates its transforming activity and is induced during lytic replication. *J. Virol.*, **81**, 13544–13551.

225 Franca, R., Spadari, S., and Maga, G. (2006) APOBEC deaminases as cellular antiviral factors: a novel natural host defense mechanism. *Med. Sci. Monit.*, **12**, RA92–RA98.

226 Nie, Y., Hammond, G.L., and Yang, J.H. (2007) Double-stranded RNA deaminase ADAR1 increases host susceptibility to virus infection. *J. Virol.*, **81**, 917–923.

227 Wang, Z., Choi, M.K., Ban, T., Yanai, H., Negishi, H., Lu, Y., Tamura, T., Takaoka, A., *et al.* (2008) Regulation of innate immune responses by DAI (DLM-1/ZBP1) and other DNA-sensing molecules. *Proc. Natl. Acad. Sci. U.S.A.*, **105**, 5477–5482.

4 Posttranslational Modification of Sm Proteins: Diverse Roles in snRNP Assembly and Germ Line Specification

Graydon B. Gonsalvez and A. Gregory Matera

4.1 Introduction

Sm proteins and their paralogs are ubiquitous, highly expressed proteins found in all three domains of life. These proteins are involved in a diverse array of processes, usually involving RNA binding and/or stabilization. Although a number of functions have been attributed to Sm paralogs in eukaryotes, they are best known for their essential function in pre-mRNA splicing. Splicing is the process whereby intervening sequences (introns) are removed from pre-messenger (m) RNAs prior to their export from the nucleus, and eventual translation in the cytoplasm [1]. Splicing is now widely recognized as an effective mechanism of post-transcriptional gene regulation.

Pre-mRNAs are spliced by small nuclear ribonucleoproteins (snRNPs). The U1, U2, U4, U5, and U6 snRNPs are components of the major spliceosome and are responsible for splicing the majority of introns [1]. The minor spliceosome includes U11, U12, $U4_{atac}$, and $U6_{atac}$, which replace U1, U2, U4, and U6, respectively [2–4]. Both classes of spliceosome share the U5 snRNP, although variant U5 small non-coding RNAs (snRNAs) exist [5, 6] and might partition between the two classes. The Sm proteins and a closely related set of Sm-like proteins (Lsm) are the central protein components of spliceosomal snRNPs. In addition to the Sm proteins, each snRNP contains a unique snRNA and several snRNP-specific proteins (reviewed in [7]). Base pairing interactions between the snRNA and the pre-mRNA contribute to the specificity of splicing [8]. In addition, the splicing reaction requires intricate and dynamic base pairing interactions between the snRNAs themselves [9]. The Sm and Lsm proteins are thought to play roles in stabilizing the associated snRNAs and in facilitating these important interactions [10–13].

In the late 1970s, Lerner and Steitz [14] found that serum from patients with autoimmune diseases such as systemic lupus erythematosus (SLE) contained antibodies that strongly reacted with small nuclear proteins. Further studies demonstrated that these proteins were associated with snRNAs [14]. The complexes were therefore called snRNPs. The crucial function of these RNPs in splicing was

Posttranscriptional Gene Regulation: RNA Processing in Eukaryotes, First Edition. Edited by Jane Wu.

hypothesized shortly thereafter [15]. In the intervening years since this seminal discovery, a great deal has been learned about snRNPs and splicing. However, a very important question still remained. What was the epitope recognized by the SLE autoantibodies? The primary amino acid sequences of the Sm proteins recognized by the autoantibodies failed to reveal extended stretches of identity. For many years, it was therefore thought that the epitope would be structural in nature. However, when the human proteins were produced in *Escherichia coli*, they reacted relatively weakly with the autoantibodies [16–18]. This finding suggested that posttranslational modification of Sm proteins may be required for efficient antibody binding. An answer to this question was finally provided in 2000, when Lührmann and colleagues demonstrated that C-terminal arginine residues on a subset of Sm proteins were symmetrically dimethylated [19]. Interestingly, the posttranslationally modified Sm proteins were the same ones recognized by the autoantibodies. The investigators went on to prove that the methyl modifications present on the Sm proteins were the primary epitopes recognized by the SLE autoantibodies [19]. This finding framed the next important question for the snRNP field: What is the function of Sm protein methylation? Discussion of this central question will be the theme of this chapter.

4.2
Protein Methylation – Flavors and Functions

Although proteins have been known to contain methylated arginine residues since 1967 [20], this field of study has largely been overshadowed by studies of protein phosphorylation and the various biological processes regulated by this modification. With the advent of new tools such as methylation-specific antibodies, and the ability to efficiently detect and analyze methyl modifications using mass spectrometry, the field of protein methylation is experiencing a rebirth. A great deal has been learned in recent years about methylation of lysine and arginine residues present on histones. Certain methyl marks are typically correlated with transcriptional activation, whereas others are found in inactive or silent genes [21, 22]. This has led to the "histone code" hypothesis, which states that the transcriptional machinery is able to interpret the various modifications present on histones to determine the transcriptional state of the corresponding gene [23]. Undoubtedly, in the years to come, a great deal more will be learned about this exciting mechanism of transcriptional regulation. In addition to regulating transcription, arginine methylation has been shown to play essential roles in protein localization, ribosome processing, signal transduction, and DNA repair [24–26].

Arginine residues can contain either a single (monomethylarginine (MMA)) or a dimethyl modification [26]. Dimethylarginine residues can be further subdivided into two categories: asymmetric dimethylarginine (aDMA) or symmetric dimethylarginine (sDMA); see Figure 4.1. The most common dimethyl modification is

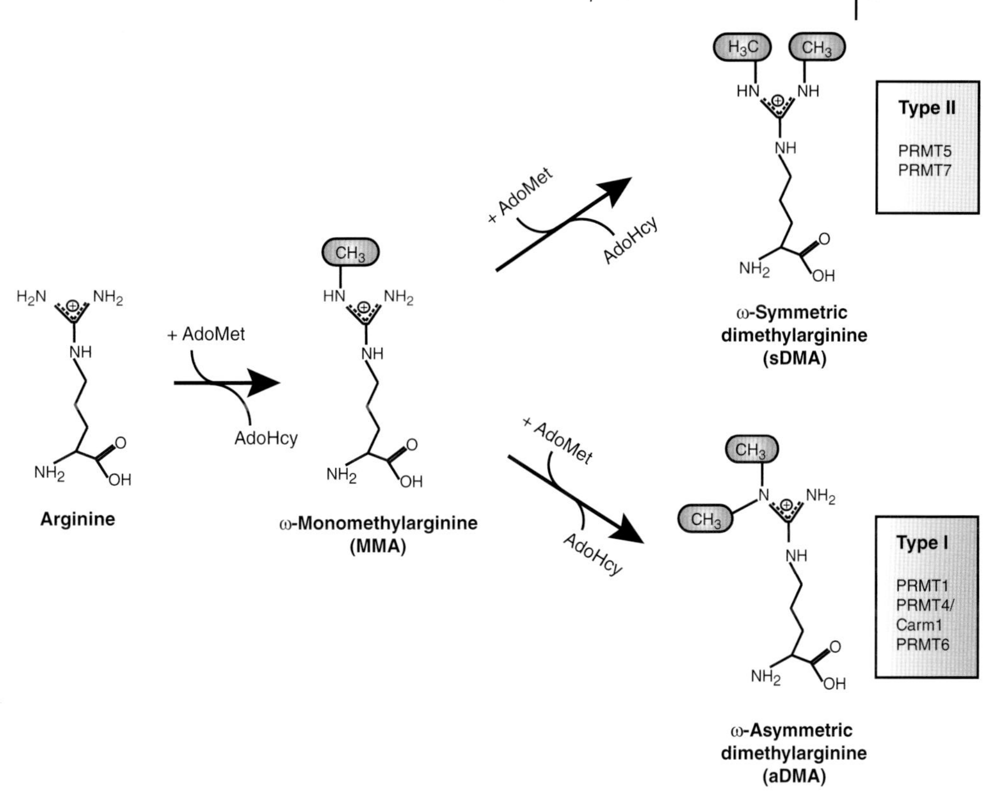

Figure 4.1 Arginine residues can contain three different types of methyl modifications. Type I and type II enzymes can both catalyze the addition of a single methyl group to the guanidinium group of arginine, thereby generating monomethylarginine (MMA). This is generally thought to represent an intermediate step *in vivo* prior to dimethylation. Type I enzyme catalyzes the formation of asymmetric dimethylarginine (aDMA) in which both methyl groups are found on the same nitrogen atom. PRMT1, 4, and 6 have been shown to possess type I activity. PRMTs of the type II category catalyze the formation of symmetric dimethylarginine (sDMA) in which the methyl groups are symmetrically distributed between both nitrogen atoms of the side chain. PRMT5 and 7 have been conclusively shown to possess type II activity. PRMT9 may be a type II enzyme, but further work is needed before a firm conclusion can be drawn.

aDMA, which is catalyzed by protein arginine methyltransferases (PRMTs) of the type I family. In contrast, PRMTs of the type II family catalyze the less common sDMA modification. Both types of enzymes are capable of monomethylating their substrates, and this is thought to represent an intermediate step in dimethylation. To date, nine PRMTs (PRMT1–9) have been identified in mammals. Of these, PRMT5, PRMT7, and PRMT9/FBXO11 have been shown to possess type II activity [27–30]. No enzymatic activity has yet been demonstrated for PRMT2 and PRMT8.

Type I enzymes are the remainder, with PRMT1 being the most abundant arginine methyltransferase *in vivo*.

4.3 Sm Proteins Contain sDMA- and aDMA-Modified Arginines

In order to elucidate the epitope recognized by SLE autoantibodies, Brahms *et al.* [19] isolated nuclei from human cells and purified snRNPs from these fractions. Using a combination of mass spectrometry and protein sequencing, they found that numerous C-terminal arginine residues in SmB, SmD1, and SmD3 were sDMA modified [19, 31]. They also found that Lsm4, which is a component of the U6 snRNP, contained sDMA residues [31]. However, no evidence of MMA or aDMA residues was detected on purified Sm or Lsm proteins. The authors went on to prove that the sDMA modification present on these proteins was the major epitope recognized by the patient autoantibodies [19]. Consistent with this view, peptides derived from the Sm proteins that contained unmethylated or aDMA-modified residues were not recognized by the autoantibodies [19].

Several years later, Miranda *et al.* [32] showed that Sm proteins also contained aDMA residues. Interestingly, they found this modification solely on nuclear Sm proteins; cytoplasmic Sm proteins contained only sDMA residues [32]. These findings were validated in a recent study by Bedford and coworkers [33]. These authors demonstrated that SmB was a substrate for the type I enzyme PRMT4/CARM1. Although PRMT4 knockout mice die shortly after birth, mouse embryonic fibroblast (MEF) cell lines could be established from homozygous animals [34]. In comparison to control cell lines, SmB present in the knockout MEFs was still sDMA modified but lacked aDMA [33].

Although several reports have independently verified the sDMA modification status of Sm proteins [19, 31, 35–39], several questions remain with regard to Sm protein aDMA modification. Which sites are aDMA modified in these proteins? Although a candidate region has been identified for SmB, the precise location is not known. The C-terminus of SmB contains a "PMG" domain, and such regions are typically rich in proline, methionine, and glycine residues. One or more arginine residues within the SmB PMG domain are aDMA modified [33]. In contrast to SmB, SmD1 and SmD3 do not contain a canonical PMG domain. Thus, it is currently unknown what residues in SmD1 and SmD3 are aDMA modified or even what enzymes are responsible for this modification. Finally, why were aDMA residues not observed in the snRNPs purified and analyzed by Brahms *et al.* [19]? One possible difference between these reports is that Brahms *et al.* [19] used human cells to purify snRNPs, whereas Miranda *et al.* [32] and Cheng *et al.* [33] used mouse cell lines for their studies. It is therefore possible that aDMA modification of Sm proteins is murine specific. A more likely explanation is that the purification protocols used by the two groups isolated different pools of snRNPs, some of which contained aDMA-modified Sm proteins, and others that were devoid of this modification. Further experimentation is needed to resolve these questions. The balance of this chapter will, therefore, focus on the well-documented Sm protein sDMA modifications.

4.4 SnRNP Assembly, the Survival Motor Neuron (SMN) Complex, and Spinal Muscular Atrophy (SMA)

The splicing of pre-mRNA is carried out by snRNPs of the Sm and Sm-like (Lsm) family. The U1, U2, U4, $U4_{atac}$, U5, U11, and U12 snRNPs each contain a common core of seven Sm proteins bound to a unique snRNA. The U6 and $U6_{atac}$ snRNPs contain a related set of seven Lsm proteins also bound to a unique snRNA [3, 7]; the evolutionary origin of these fascinating proteins is covered in Section 4.15. Relatively little is known about the assembly of U6 and $U6_{atac}$, but their biogenesis is thought to occur solely within the nucleus [40–44]. Much more is known about the biogenesis of the Sm-class snRNPs. Their assembly is complex and highly regulated, and takes place in the nucleus and cytoplasm of the cell (Figure 4.2) [7]. The biogenesis of Sm-class snRNPs will be the main focus of this chapter.

The Sm-class snRNPs are transcribed by RNA polymerase II and subsequently transported to the cytoplasm via the export factors PHAX, CRM1, and the cap-binding complex [45, 46]. In the cytoplasm, the Sm proteins are assembled onto their respective snRNAs by the survival motor neurons (SMNs) complex (reviewed in [47, 48]). The SMN complex is a large, biochemically stable, oligomeric complex containing SMNs, several associated proteins termed Gemins (Gemin 2–8), and UNRIP/STRAP (reviewed in [49]). The Sm proteins by themselves have relatively low RNA-binding specificity. The SMN protein in conjunction with Gemin5 ensures that the Sm proteins are correctly loaded onto the proper snRNAs [50, 51]. Once the Sm proteins have been assembled onto the snRNA, the 7-methylguanosine cap of the snRNA is hypermethylated to a 2,2,7-trimethylguanosine (TMG) cap by the enzyme TGS1 [52, 53]. The SMN complex is thought to aid in recruiting TGS1 to the snRNP [52]. This modified cap structure serves as the docking site for the import adapter Snurportin1 [54]. Subsequently, Snurportin1, the SMN complex, and the import receptor Importin-β mediate import of the snRNP into the nucleus [55–58]. Further rounds of modification and assembly are thought to take place in Cajal bodies (CBs) [7, 59, 60] before the snRNP is finally ready to function in pre-mRNA splicing. Importantly, the SMN complex localizes within CBs and appears to play a role in targeting the newly imported snRNPs to this nuclear body [61–64].

The SMN complex has received a great deal of attention in recent years. Not only does this complex play a prominent role at multiple steps in the snRNP assembly pathway (Figure 4.2), but it is also interesting from a medical standpoint. In 1995, it was discovered that mutations that reduce SMN protein levels result in the inherited neuromuscular disorder spinal muscular atrophy (SMA [65]. The biological function attributed to the SMN complex suggests that SMA may be an snRNP-related disease. While there are considerable data supporting this

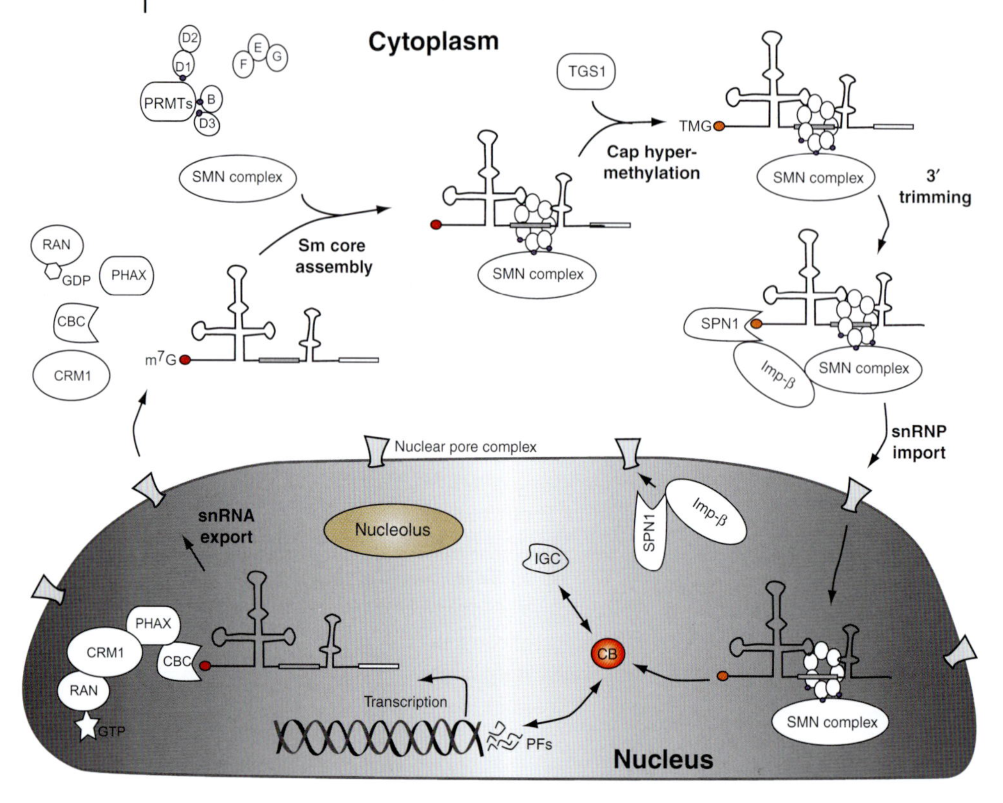

Figure 4.2 The biogenesis of Sm-class snRNPs occurs in a stepwise fashion and involves maturation events in both the nucleus and cytoplasm of the cell. The snRNA is transcribed by RNA polymerase II and is subsequently exported to the cytoplasm in a complex containing CRM1, Ran-GTP, PHAX, and the cap-binding complex (CBC). This complex is disassembled upon export to the cytoplasm and is bound by the survival motor neuron (SMN) complex. The SMN complex plays a pivotal role in the assembly of Sm-class snRNPs by specifically loading Sm proteins onto the newly exported snRNA. Subsequent to and dependent on Sm-core assembly, the cap structure of the snRNA is hypermethylated by the enzyme TGS1. The cap is converted from a 7-methylguanosine (m7G) cap to a 2,2,7-trimethylguanosine (TMG) cap. The SMN complex is thought to recruit TGS1 to the snRNP. Next, sequences at the 3′ end of the snRNA are trimmed. The exonuclease that performs this digestion has not yet been identified. A specific import adapter called Snurportin1 (SPN1) is then recruited to the maturing snRNP. SPN1 specifically recognizes the TMG cap structure of the snRNP. An import complex containing SPN1, the import receptor importin-β (Imp-β), and SMN then mediate the import of the snRNP into the nucleus. Once in the nucleus, the import complex is disassembled, and SPN1 and Imp-β are shuttled back to the cytoplasm. The SMN complex, however, remains associated with the snRNP and targets it to the Cajal body (CB). Further rounds of maturation occur within this nuclear body including the association of numerous snRNP-specific proteins and modification of the snRNA. The assembled snRNP is then targeted to perichromatin fibrils (PFs) to participate in pre-mRNA splicing or to interchromatin granule clusters (IGCs) for storage.

hypothesis, non-snRNP-related functions have also been attributed to SMN. Thus, current animal models implicate snRNP-dependent and -independent functions of SMNs as etiological factors in SMA [66–71].

How does the SMN complex ensure correct loading of Sm proteins onto snRNA? Two requirements for this assembly step are that the complex members efficiently and specifically recognize both Sm proteins and snRNA. The Sm site of snRNAs appears to be recognized specifically by Gemin5 [50]. A number of SMN complex members, including SMN itself, have been shown to specifically bind Sm proteins [72–80]. While the SMN complex members appear to bind the common N-terminal Sm domain found in all Sm and Sm-like proteins, SMN is thought to preferentially interact with the methylated C-termini of SmB, SmD1, and SmD3. Several reports have lent credence to this view. SMN binds with a much higher affinity to C-terminal peptides from SmD1 and SmD3 if they contain sDMA residues [81]. Interestingly, in contrast to sDMA, aDMA or MMA residues do not increase the SMN–Sm association [81]. Using co-immunoprecipitation experiments, it was shown that the SMN–Sm association was disrupted upon treating cells with general methyltransferase inhibitors [31, 81]. Lastly, SMN contains a Tudor domain, mutations in which disrupt Sm protein association [82]. This domain has recently been shown to be a methyl-binding module [83]. Therefore, there is ample evidence to indicate that the SMN–Sm association is mediated, at least in part, by sDMA modification.

4.5 PRMT5 and PRMT7–The Sm Protein Methyltransferases

At the time of its identification, PRMT5 was the only known type II methyltransferase. Using *in vitro* experiments, PRMT5 was shown to sDMA modify residues in Sm proteins, myelin basic protein (MBP), and histones [27, 28]. In cytoplasmic lysates, PRMT5 is found in a complex along with Sm proteins as well as cofactors, pICln and MEP50/WD45 [37, 84]. Interestingly, both pICln and MEP50/WD45 are capable of directly binding to Sm proteins via the N-terminal Sm domain [37, 84, 85]. They may therefore act as specificity factors for Sm protein methylation *in vivo*, targeting PRMT5 to the appropriate substrates. In addition, they may serve to regulate the activity of PRMT5. Consistent with this view, PRMT5 has a much lower enzymatic activity when purified from bacteria as opposed to mammalian cell lines [37, 86]. Furthermore, PRMT5 complexes containing pICln and MEP50/WD45 purified from native lysates display a very specific substrate range and do not methylate histones [84]. It is not known, however, if the activity and specificity of recombinant PRMT5 can be increased by the addition of exogenous pICln and MEP50/WD45. It is also possible that PRMT5 is posttranslationally modified itself and that this modification is required for optimal enzymatic activity. In this vein, it is interesting to note that PRMT5 was originally identified in a yeast two-hybrid screen using Janus kinase 2 as bait (PRMT5 was initially termed Janus kinase-binding protein 1 (JBP1) [28]).

PRMT7 was recently identified and shown to methylate Sm proteins *in vitro* [29]. Like PRMT5, PRMT7 purified from human cell lysates has a much higher enzymatic activity than recombinant enzyme [29]. The substrate specificity of purified PRMT7 is, however, much broader than that of PRMT5. PRMT7 was shown to methylate a variety of substrates including histones, MBP, GAR, and Sm proteins [29]. In fact, the purified protein could even methylate a peptide composed solely of GR repeats [29]. The interacting partners of PRMT7 are not known, but presumably, such factors exist to confer substrate specificity to the enzyme *in vivo*.

In order to determine which enzymes were responsible for methylating Sm proteins *in vivo*, Gonsalvez *et al.* [39], specifically depleted HeLa cells of PRMT5 or PRMT7 using RNAi. Interestingly, both enzymes were found to be required for this modification–sDMA modification of Sm proteins was reduced in both depleted lysates [39]. This effect is likely to be direct since depleting one enzyme did not codeplete the other [39]. In addition, both PRMT5 and PRMT7 were able to bind to recombinant Sm proteins, and a previous report showed that the two enzymes did not associate with each other [29, 39]. PRMT5 and PRMT7 do not always share the same substrate however. Coilin, a CB marker protein requires PRMT5 but not PRMT7 for its sDMA modification (G.B.G. and A.G.M., unpublished results). In addition, several uncharacterized proteins were reduced in sDMA modification in PRMT5 depleted lysates but not in PRMT7-depleted lysates [39].

On a mechanistic level, PRMT5 and PRMT7 appear to act nonredundantly in Sm protein methylation, as overexpression of one enzyme could not functionally compensate for the loss of the other [39]. Based on these observations, we envision two scenarios by which PRMT5 and PRMT7 function to methylate Sm proteins (Figure 4.3). In the first model, PRMT5 and PRMT7 independently associate with Sm proteins and catalyze their methylation. It is possible that each enzyme methylates a distinct set of residues on the Sm proteins. An initial methylation event by one enzyme could also serve as a recognition site for the other. Consistent with this idea, PRMT7, but not PRMT5, was able to bind the C-terminus of SmD3 in the presence of a methyltransferase inhibitor [39]. This suggests that in order for PRMT5 to bind and methylate Sm proteins, it requires a PRMT7-derived methylation mark. A similar scenario has been described for histone posttranslational modification. Methylation of histone H4 arginine 3 by PRMT1 is required for subsequent acetylation and/or methylation of other H4 residues [87]. Thus, this initial methylation event by PRMT1 serves as a signal for subsequent modification. An alternate model suggests that PRMT7 somehow regulates the activity of PRMT5. In this scenario, PRMT7 activates PRMT5, thus enabling it to sDMA modify Sm proteins. Once modified, the Sm proteins are handed off to the SMN complex. Because PRMT5 and PRMT7 do not associate *in vivo* [29], this regulation is likely to be indirect. For instance, PRMT7 might methylate an unknown protein that in turn regulates the activity of PRMT5 (Figure 4.3). However, given that PRMT5 and PRMT7 do not always methylate the same targets [39], any regulation of PRMT5 activity by PRMT7 would have to be substrate specific.

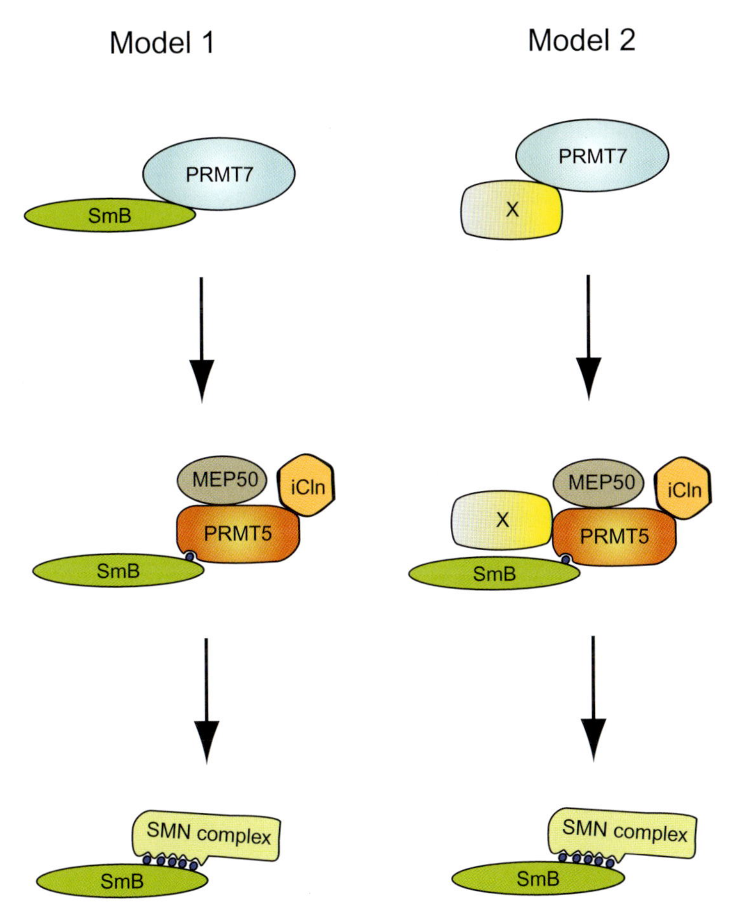

Figure 4.3 Two models can be envisioned by which PRMT5 and PRMT7 function in Sm protein methylation. Model 1 suggests that subsequent to translation, Sm proteins first encounter PRMT7. PRMT7 monomethylates or symmetrically dimethylates the Sm proteins. This in turn serves as a mark for PRMT5 to complete the symmetric dimethylation of the Sm proteins. Once methylated by PRMT5, the Sm proteins are transferred to the SMN complex for loading onto the snRNA. An alternate model (model 2) suggests that PRMT7 regulates the activity of an unknown protein (X) that in turn regulates the activity of PRMT5. In this model, PRMT5 is solely responsible for symmetrically dimethylating the Sm protein. As with model 1, the Sm proteins once methylated are handed off to the SMN complex for Sm core assembly.

In both models, PRMT5 has been placed downstream of PRMT7 in the snRNP assembly pathway. Several lines of evidence argue for this placement. First, the activity of the SMN complex is stimulated in an ATP-dependent manner by the PRMT5 complex [88]. Second, Sm proteins can only be transferred to the SMN complex from the PRMT5 complex [37]. This suggests the existence of functional cross-talk between the two complexes *in vivo*. Lastly, using co-immunoprecipitation experiments, a stable *in vivo* association could easily be detected between PRMT5 and Sm proteins, but not between PRMT7 and Sm proteins (G.B.G. and A.G.M., unpublished results). Thus, binding and methylation by PRMT7 is likely to be rapid and transient. Collectively, these observations argue that PRMT5 methylation is the rate-limiting step before Sm proteins are transferred to the SMN complex for assembly onto snRNA.

Recently, two groups have identified a potential PRMT9 using bioinformatics techniques. Importantly, the two identified genes are not identical and hence are referred to in the literature as PRMT9 4q31 [89] and PRMT9 2p16.3 [30]. The latter gene is also referred to as FBOX011 [30]. No enzymatic activity has yet been demonstrated for PRMT9 4q31. In contrast, Cook *et al.* [30] have shown that FLAG-tagged human PRMT9 2p16.3/FBOX011 purified from HeLa cells possesses type II activity. GST-tagged PRMT9 2p16.3/FBOX011 purified from bacteria also appears to sDMA modify a variety of substrates [30]. A few caveats, however, warrant further investigation of the *in vivo* enzymatic activity of PRMT9 2p16.3/FBOX011. Several reports have now shown that anti-FLAG M2 agarose beads nonspecifically precipitates MEP50/WD45 and PRMT5 [90–93]. Thus, proteins purified using FLAG affinity columns are contaminated by a type II methyltransferase–PRMT5. Additionally, it is not currently known if the substrates methylated by PRMT9 2p16.3/FBOX011 *in vitro* are also methylated by PRMT9 *in vivo*. Functional *in vivo* studies are therefore required to resolve these questions.

4.6 Sm Protein Methylation is Required for sn/RNP Assembly in Mammals

Although it had been demonstrated that SMN bound with a higher affinity to sDMA-modified Sm proteins in comparison to unmodified proteins, it was never actually determined whether these methyl modifications were required for snRNP assembly *in vivo*. In order to test this, PRMT5 or PRMT7 was depleted from cells and snRNP assembly was examined [39]. Specifically, two consecutive steps in the snRNP biogenesis pathway–Sm core assembly and TMG capping–were monitored. Both steps were disrupted upon depletion of either PRMT5 or PRMT7 [39]. These defects were not as severe, however, as when SMN itself was depleted [39]. In contrast to Sm core assembly and TMG capping, nuclear import of snRNPs was not affected by reduced methylation. The residual snRNPs that were assembled following PRMT5 or PRMT7 depletion were still imported into the nucleus [39]. Once in the nucleus, the snRNPs are normally targeted to CBs for further processing [7, 59, 60]. In fact, these nuclear bodies appear to be sensitive to the level of ongoing snRNP biogenesis. Depleting core snRNP assembly factors such as SMN, PHAX, TGS1, and SPN1 correlate with a breakdown of CBs [62–64]. Conversely, stimulating snRNP assembly by overexpressing Sm proteins causes an increase in CB number [94]. In contrast to SMN depletion, loss of PRMT5 or PRMT7 caused a much milder CB disruption phenotype [39].

Based on these observations, Gonsalvez *et al.* [39] hypothesized that PRMT5 and PRMT7 function in the snRNP assembly pathway by sDMA modifying Sm proteins and increasing their affinity for the SMN complex. In the absence of this modification, snRNP assembly likely occurs at a reduced rate. Therefore, phenotypes observed upon depletion of PRMT5 or PRMT7 are not as severe as those observed upon SMN depletion. It is possible that, under these conditions, backup mechanisms exist to ensure at least a modest level of snRNP assembly, because treating cells with saturating amounts of methyltransferase inhibitors did not completely inhibit snRNP biogenesis [39]. In support of this scenario, several SMN complex members have been shown to bind to Sm proteins irrespective of their methylation status [72–80]. Furthermore, SMN itself can also associate with Sm proteins via the N-terminal Sm motif present in all Sm proteins [95]. Therefore, it seems likely that the primary purpose of Sm protein methylation is to increase the efficiency and perhaps the kinetics of the snRNP assembly pathway.

A study by Khusial *et al.* [96] also examined the requirement for Sm protein methylation in snRNP assembly and arrived at a different conclusion. These authors analyzed the localization and incorporation of a mutant FLAG-SmD3 construct into snRNPs [96]. The four arginine residues normally methylated in wild-type SmD3 were mutated to leucine. Therefore, this mutant SmD3 construct cannot be sDMA modified. Khusial *et al.* [96] found that the mutant behaved similarly to wild-type FLAG-SmD3. The mutant was primarily localized within the nucleus, it sedimented in a similar pattern as wild-type FLAG-SmD3 on sucrose gradients, and the mutant was associated with other Sm proteins, snRNAs, PRMT5, iCln, and SMN [96]. Based on these findings, the authors concluded that methylation of SmD3 was not required for snRNP assembly [96].

A few caveats in the experimental design, however, make this interpretation somewhat premature. The experiments were done in a background that contained endogenous wild-type SmD3. The co-immunoprecipitated proteins could therefore have come from mixed snRNP complexes (the U4/U5/U6 tri-snRNP for instance), some that contained the FLAG-SmD3 mutant, and others that contained endogenous wild-type SmD3. In order to effectively interpret these results, the level of endogenous wild-type SmD3 needs to be compared with the level of exogenous mutant FLAG-SmD3. A better experiment would be to deplete cells of endogenous SmD3 using RNAi and to transfect into these cells a mutant FLAG-SmD3 construct that is resistant to degradation by the siRNAs. Using this strategy, the incorporation of the mutant SmD3 construct into snRNPs could be examined under conditions of greatly reduced wild-type SmD3. In fact, if the cells survive this treatment, it would indicate that the mutant construct is able to incorporate into snRNPs and to function properly in splicing. Furthermore, conclusions based on this study have to be limited to the SmD3 protein. The authors [96] do not suggest that methylation of Sm proteins is dispensable for snRNP assembly since SmB, SmD1, and endogenous SmD3 were all methylated under the experimental conditions used. Therefore, these methylated Sm proteins could potentially compensate for the presence of the exogenously expressed mutant FLAG-SmD3.

4.7
Sm Protein Methylation in *Drosophila*

In contrast to humans, where Sm protein methylation appears to be required for efficient snRNP assembly, the situation in *Drosophila* is a bit more complex. In an initial attempt at answering this question, Gonsalvez *et al.* [38] focused on a loss of function mutant in the *dart5/capsuléen* gene (*dart5-1*), the fly ortholog of human *PRMT5*. The mutant contains a transposon inserted within the coding sequence of *dart5*, thereby resulting in the production of a truncated protein that lacks the C-terminal methyltransferases domain. As expected, lysates prepared for homozygous *dart5-1* flies indicated the absence of full-length Dart5 protein [38]. The truncated product is likely to be unstable because it could not be reliably detected on Western blots. Also consistent with *dart5* being the *Drosophila* ortholog of PRMT5, Sm proteins in the mutant lysates could not be detected using two methylation-specific antibodies, Y12 and SYM10 [38]. Thus, the Sm proteins were not sDMA modified in the mutant lysates. Surprisingly however, the homozygous *dart5* mutants were completely viable [38]. Furthermore, unlike hypomorphic mutations in *Smn* [70], the *dart5-1* homozygotes did not display SMA-like phenotypes [38]. The observation that the mutants were viable despite expressing Sm proteins that were not methylated suggests that snRNP assembly is unperturbed. SnRNP assembly assays monitoring Sm core assembly, TMG capping, and nuclear import confirmed this hypothesis; *dart5-1* homozygous mutants expressed equivalent levels of snRNP as wild-type flies [38].

Characterization of the *dart5* mutant phenotype was performed prior to identification of human PRMT7 as a type II methyltransferase. Like humans, *Drosophila* express an ortholog of PRMT7, called Dart7 (CG9882 [97]). Using RNAi in *Drosophila* cell culture, Gonsalvez *et al.* [98] demonstrated that like Dart5, Dart7 was also required for Sm protein methylation. Furthermore, as observed in human cells, the two enzymes functioned independently of each other. Dart5 overexpression was unable to compensate for the depletion of Dart7 with respect to Sm protein methylation and vice versa [98]. Interestingly, unlike *dart5*, which was not required for organismal viability, *dart7* is an essential gene. Dart7 depletion results in early pupal lethality [98]. It does not appear, however, that the Dart7-associated lethality is caused by snRNP assembly defects. Lysates prepared from these depleted animals contained residually methylated Sm proteins, as judged by Y12 and SYM10 Western blotting [98]. In contrast, fully viable *dart5* mutants express Sm proteins that are completely Y12 and SYM10 negative [38]. A more feasible explanation is that the Dart7-depleted animals are dying due to aberrant methylation of a non-Sm target. It will be interesting to determine the identity of these *in vivo* targets. Currently, the Sm proteins are the only identified *in vivo* targets of Dart7 and PRMT7.

A caveat associated with the examination of *dart5* and *dart7* phenotypes and drawing conclusions related to snRNP assembly is the possibility that these enzymes could methylate a variety of substrates *in vivo* in addition to Sm proteins.

In the case of PRMT5, certain histone residues have been identified as *in vivo* targets (see Section 4.13). Thus, care has to be taken with correlating a particular organismal phenotype with defects observed on a molecular level. In order to circumvent this limitation, Gonsalvez *et al.* [98] employed a similar strategy to the one used by Khusial *et al.* [96]. A series of SmD1 constructs were constructed that were tagged with the Venus fluorescent protein (VFP). In addition to a wild-type control, two mutant constructs were also made, one in which the entire C-terminal RG tail was deleted (SmD1ΔRG) and another in which the arginine residues within the tail were changed to lysine (SmD1(R-K)). Unlike arginine, Dart5 and Dart7 are unable to methylate lysine residues, and since it contains a primary amino group, it cannot be symmetrically dimethylated by any enzyme. Unlike the previous study by Khusial *et al.* [96], Gonsalvez *et al.* [98] were able to use the power of *Drosophila* genetics to examine the transgenic SmD1 constructs in the absence of any endogenous SmD1 protein. SmD1 was chosen as the focus of their study for three main reasons. First, SmD1, unlike SmB and SmD3, does not form part of the U7 snRNP [99]. As such, the phenotypes observed can be attributed specifically to spliceosomal snRNPs. Second, prior to the formation of the Sm core, SmD1 exists in a subcomplex with SmD2, whereas SmB and SmD3 are part of the same subcomplex [100]. Loss of SmD1 methylation could potentially inhibit the incorporation of the entire D1/D2 subcomplex into the Sm core, thus making the phenotype more severe. In contrast, if methylation mutants in either SmB or SmD3 were made, the wild-type methylation status of the other could ameliorate the severity of the snRNP phenotype. Lastly, as observed by Zhang *et al.* [101], truncation of the C-terminal tail of *Saccharomyces cerevisiae* SmD1 produced a more severe phenotype than truncation of the C-termini of either SmB or SmD3. These data suggest that the RG tail of SmD1 plays a critical role in metazoan snRNP assembly.

The three SmD1 constructs were expressed on a lethal SmD1 mutant background. Consistent with the situation observed in *S. cerevisiae*, the VFP-SmD1ΔRG construct was unable to complement the lethal phenotype of the mutant SmD1 allele [98]. In contrast, both WT VFP-SmD1 and VFP-SmD1(R-K) were able to produce viable offspring [98]. Thus, both proteins can functionally substitute for endogenous SmD1. Importantly, Sm core assembly, TMG capping, and nuclear import of snRNPs were all found to be unaffected in the VFP-SmD1(R-K) strain [98]. Based on these results, and the results obtained using *dart5* [38], it appears that symmetric arginine dimethylation is completely dispensable for snRNP assembly in *Drosophila*.

4.8 Unresolved Questions: Sm Protein Methylation and snRNP Assembly

A possible caveat of the recent experiments by Gonsalvez *et al.* [98] is that SmB and SmD3 are still methylated in the same strain. Thus, in order to conclusively show that methylation is not required for snRNP assembly in *Drosophila,* three mutant alleles (loss of function mutants of SmB, SmD1, and SmD3) and three transgenes (methylation negative mutants of SmB, SmD1, and SmD3) would all need to be expressed on the same genetic background. Even with a genetically tractable organism such as *Drosophila,* this experiment poses many challenges and would require a considerable time investment. However, in the quest to determine whether Sm protein methylation is truly required for snRNP assembly, this experiment would represent the *coup de grâce.*

It is interesting to note that the viable SmD1(R-K) construct displayed a reduced affinity for binding SMN [98]. In fact, it bound SMN to the same level as the nonviable SmD1ΔRG construct [98]. A similar situation was observed upon depletion of Dart5; the SmB–SMN association was greatly reduced [38]. Yet, the *dart5* mutants were also clearly viable [38]. In displaying this preference for methylated Sm protein, *Drosophila* SMN behaves exactly like its human counterpart. Like human SMN, *Drosophila* SMN also contains a highly conserved Tudor domain. Why then, are these mutants viable? Clearly, SMN is required for efficient snRNP assembly in *Drosophila* [70]. Therefore, it is possible that upon disruption of efficient SMN–Sm binding, backup mechanisms ensure proper and continued snRNP assembly. It would be quite interesting therefore to determine the nature of these backup mechanisms. Genetic screens to identify synthetic lethal targets of the *dart5* mutant should shed light on this topic.

4.9
Conclusion – The Evolution of snRNP Assembly

Over the past years, studies from several laboratories have demonstrated that a few differences exist in the mechanism of snRNP assembly between flies and mammals. Although Sm proteins in *Drosophila* are methylated by two distinct enzymes, much like they are in mammals, this modification is less important for snRNP assembly in fruit flies. Other differences exist as well. For instance, the human SMN protein is phosphorylated on N-terminal serine residues, and this modification is required for optimal SMN activity [102]. Importantly, the N-terminus of the human protein that contains the phosphorylated residues is not conserved in invertebrates. Additionally, in humans, the SMN protein is present in a large, biochemically stable, oligomeric complex. Unambiguous orthologs of most of the SMN complex members have yet to be identified in *Drosophila.*

In order to assimilate and understand these apparent differences, we take a Darwinian approach. We hypothesize that during the course of evolution, layers of regulation have been added to the snRNP assembly pathway. The SMN protein, for instance, is present in fission yeast genomes but is completely absent from budding yeast [103, 104]. To take this argument a step further, Sm proteins in *S. cerevisiae* are not sDMA modified, as they lack appropriate RG residues in their C-terminal tails [11, 12]. In this view, *Drosophila* represents an intermediate between *S. cerevisiae* and humans. In *Drosophila*, the Sm proteins are clearly methylated, and SMN is certainly required for snRNP assembly. However, the known posttranslational modifications of SMN and Sm proteins appear to be dispensable for snRNP assembly. In light of these findings, it is intriguing that evolution has favored retention of this methyl modification in *Drosophila*. Why would these methyl marks be present on Sm proteins, if not for snRNP assembly? Could these marks represent a more ancestral function of Sm proteins that is unrelated to snRNP assembly and splicing? This brings us to our next topic–the requirement for Sm protein methylation in specification of the germ line.

4.10
Sm Proteins Are Required for Germ Cell Specification

Although the *dart5* mutants were viable, possessed wild-type levels of snRNPs, and did not display any SMA-like phenotypes, they were not without a phenotype. The mutants were "grandchild-less" [38, 105]. In other words, progenies from mutant *dart5* females were viable, but irrespective of their genotype, they were sterile. Further examination revealed that germ cell specification was defective in these mutants [38, 105]. The reason for this rather peculiar phenotype has to do with the manner in which germ cells are specified in *Drosophila*. Unlike mammals, where germ cell specification begins rather late in embryogenesis, in *Drosophila*, this process is initiated during oogenesis and is completed very early in embryogenesis. Disruption of this process often results in viable offspring that are agametic and therefore sterile–hence the moniker "grandchild-less."

A detailed description of *Drosophila* oogenesis and germ cell specification is beyond the scope of this chapter. Several excellent reviews have been published on this topic, and for those interested in learning the intricacies of *Drosophila* oogenesis, we refer you to these publications [106–111]. For the purposes of this chapter, we will summarize decades of painstaking scientific endeavor in a few sentences.

For most of oogenesis, the egg chamber contains 16 cells. One of these cells, the posterior-most cell, is the oocyte (Figure 4.4). The other 15 cells are called nurse cells because they supply essential nutrients to the growing oocyte. As the egg develops from a germ line stem cell, factors responsible for establishing germ line fate are produced by the nurse cells, transported, and localized to the posterior of the growing oocyte (Figure 4.4, red gradient). These factors, functioning in a hierarchical manner, mediate the assembly of a specialized cytoplasm at the posterior of the egg known as the pole plasm (Figures 4.4 and 4.5). Proper specification of the pole plasm is required for the establishment of the germ cell fate. At the top of this cascade is a protein named Oskar [112, 113]. During oogenesis, *oskar* mRNA produced by the nurse cells is specifically transported and localized to the posterior [114]. Consequently, Oskar protein is only translated in this region of the oocyte. A complex system of translational regulation further ensures that *oskar* mRNA is not precociously translated prior to posterior localization [115–118].

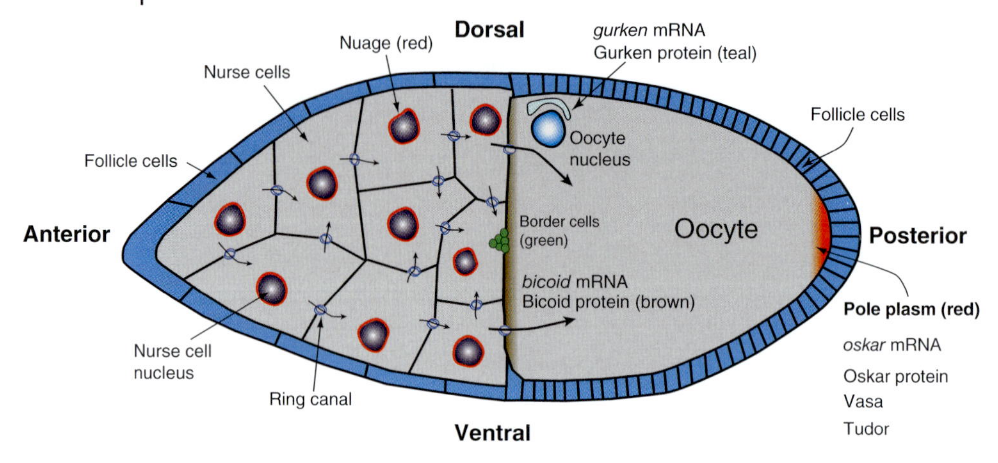

Figure 4.4 A stage 10 *Drosophila* egg chamber is illustrated. The four axis of the egg chamber (anterior, posterior, dorsal, and ventral) are shown. The egg chamber consists of 15 nurse cells, 1 oocyte, and many follicle cells of somatic origin. Unlike the oocyte, the nurse cells are transcriptionally active and provide the bulk of mRNAs and proteins required by the growing oocyte. The nurse cells share a common cytoplasm and are connected to each other by ring canals. The arrows indicate the direction of cytoplasmic flow. The oocyte in a stage 10 egg chamber is the largest cells and is positioned at the posterior. The axis of the egg chamber is established at earlier developmental time points by the localized transport of mRNAs encoding fate determinants. These factors are produced by the nurse cells and are subsequently transported and localized within the oocyte. The localization of *oskar* mRNA (red gradient) at the posterior establishes the formation of a specialized cytoplasm known as pole plasm. Proper formation of the pole plasm is essential for germ cell specification and abdominal patterning. The localization of the *oskar* mRNA also establishes the anterior–posterior axis. The localization of *gurken* mRNA to the dorsal surface of the oocyte nucleus is essential for proper specification of the dorsal–ventral axis. Finally, the localization of *bicoid* mRNA at the anterior (brown gradient) is responsible for specifying anterior structures during embryogenesis.

Oskar protein mediates the recruitment of Vasa to the posterior, and together, both proteins function to recruit Tudor [119–122]. The activities of all three proteins are essential for pole plasm specification [106].

Recently, a gene called *valois,* which shares a similar grandchild-less phenotype to *dart5,* was cloned and characterized [123–125]. Interestingly, *valois* was found to be the *Drosophila* ortholog of human MEP50/WD45 [123, 124], a protein that is part of the human PRMT5 complex [84, 126]. Anne and Mechler [124] confirmed an *in vivo* interaction between Valois and Dart5/Capsuléen. Furthermore, consistent with the observations in human cells [39], Dart5 levels were drastically reduced in *valois* mutants [38]. Thus, complex formation between Dart5 and Valois may be required for their mutual stability. Further characterization of the mutant phenotypes revealed that Oskar and Vasa were properly localized in both *dart5* and *valois* mutants [38, 105, 124]. However, in the case of both mutants, Tudor was mislocal-

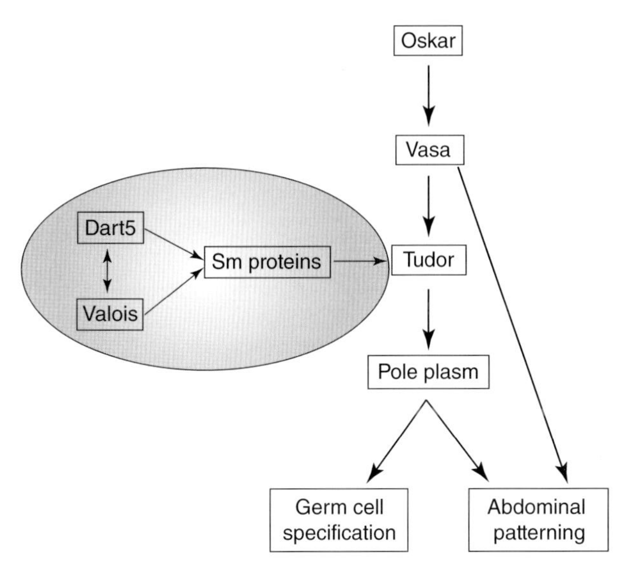

Figure 4.5 This schematic represents a simplified version of the germ cell specification pathway. The Oskar protein functions at the top of the pathway and along with Vasa and Tudor, is required for the specification of the pole plasm. These proteins also function in patterning the abdomen of the embryo. Mutants in *dart5* and *valois* do not affect the localization of either Oskar or Vasa in the oocyte. Thus, Dart5 and Valois function downstream of Oskar and Vasa in the germ cell specification pathway. However, both mutants affect Tudor localization within the egg chamber. The only known *in vivo* targets of Dart5 and Valois are the Sm proteins. Based on this and other observations, we propose that the main function of Dart5 and Valois in the germ cell specification pathway is to methylate Sm proteins. The Sm proteins in turn anchor Tudor at the posterior where it is required for the proper specification of the pole plasm.

ized within the egg chamber [38, 105, 124], suggesting that Dart5 and Valois function via Tudor to establish germ cell fate.

Since *dart5* and *valois* mutants do not affect Oskar or Vasa function, but do affect the localization of Tudor, does this mean they can be placed directly upstream of Tudor in the germ cell specification pathway? The answer does not appear to be as simple as that since *dart5* and *valois* mutants have additional phenotypes not shared by *tudor* mutants. Mutant *dart5* and *valois* males are sterile whereas loss-of-function *tudor* males are fertile [38, 127]. Thus, Dart5 and Valois play additional roles in the male germ line. In order for Dart5 to act directly upstream of Tudor, the *dart5* phenotype should be at least as strong as that of *tudor*. Only about 15% of the progenies from *tudor* null females survive and hatch into larvae [127]. In contrast, 65% of the progenies from *dart5-1* females hatch [38]. Thus, in this regard, the *tudor* phenotype is stronger. Furthermore, Dart5 does not appear to function downstream of Vasa in a classical sense since *vasa* mutants do not affect Dart5 activity [38].

4.11
Dart5, Valois, Sm Proteins, and Tudor Anchoring

How then do Dart5 and Valois function to specify the germ line? Since mutations in both genes affect Tudor localization, it is worth considering the Tudor protein in a bit more detail. Within the egg chamber, Tudor localizes to the pole plasm and also to a perinuclear region called the nuage [120] (Figure 4.4). The nuage is an RNP-rich, electron-dense structure found on the cytoplasmic face of the nurse cell nuclei. The specific function of the nuage is not known, but many factors required for germ cell specification localize to the nuage in addition to the pole plasm. Based on these findings, it has been hypothesized that the nuage serves as an RNP assembly zone, where factors required for pole plasm transport are recruited and loaded onto localized mRNAs. Interestingly, like Tudor, Valois also localizes to the nuage and pole plasm [124]. Dart5 on the other hand dot not display any specific localization pattern [105].

The Tudor protein is the founding member of a family of proteins that share a common structural motif called Tudor domains. As described previously, the SMN protein that plays a central role in the snRNP assembly pathway contains a Tudor domain. SMN contains a single Tudor domain, whereas Tudor itself contains eleven such domains. Several reports have demonstrated that SMN binds with a very high affinity to sDMA-modified proteins [31, 38, 39, 81]. These findings were validated by the recent discovery that Tudor domains are methyl-binding modules [83]. Thus, it is quite interesting that the disruption of a methyltransferase complex in *Drosophila* results in Tudor delocalization. Based on these findings, we hypothesize that Dart5-mediated methylation of Sm proteins is required for Tudor anchoring (Figure 4.5). Since Valois and Tudor share a similar localization pattern, it is likely that once methylated, Sm proteins are transferred to Tudor via Valois. Consistent with this hypothesis, Tudor has been shown to directly associate with both Sm proteins and Valois [105, 124]. Importantly, the association between Tudor and Sm proteins was methylation dependent [105].

The crux of this hypothesis is that Sm proteins play a critical role in the germ line by anchoring Tudor at the pole plasm. A caveat of this interpretation is that, in the absence of Dart5 and Valois, a failure to methylate a non-Sm target could result in the observed phenotypes. While this is certainly possible, multiple lines of evidence argue for a prominent role for Sm proteins in the germ line. In *C. elegans*, Sm proteins are important in maintaining the integrity of P-granules [128], structures that are functionally similar to the *Drosophila* nuage and pole plasm. Interestingly, this phenotype was strictly Sm dependent. Depletion of core splicing proteins did not result in P-granule disruption [128]. The orthologous structure to the nuage and pole plasm in *Xenopus* oocytes is the so-called "mitochondrial cloud." Using immunoelectron microscopy, Sm proteins have been shown to be components of the mitochondrial cloud [129]. Conservation of this critical Sm function in the germ line can even be observed in mammals. In spermatocytes, the chromatoid body is the functional equivalent of the nuage and pole plasm. Not only are Sm proteins found within the chromatoid body, but an *in vivo* interaction was also detected between Sm proteins and another chromatoid body component, mouse tudor repeat protein 1 (MTR1) [130]. Thus, there is ample evidence to indicate that Sm proteins play a vital role in germ cell specification. The available data suggest that this function is likely to be mediated via a Tudor domain containing protein. In *Drosophila*, this protein is most likely Tudor itself.

The *dart5* mutant will be a valuable tool in understanding the mechanism by which Sm proteins function in the germ line. Although Sm protein depletion in *C. elegans* resulted in P-granule disruption, the treatment also resulted in early embryonic lethality [128]. The lethality is likely to result from the disruption of pre-mRNA splicing as a similar phenotype was also observed when core splicing factors were depleted [128, 131]. In contrast, the splicing and germ cell specification functions of Sm proteins have been separated in the *dart5* mutants. Homozygous *dart5* mutants are viable, and snRNP assembly is not affected [38].

4.12
Unresolved Questions: Sm Proteins and Germ Cell Specification

Several questions remain to be answered. Gonsalvez *et al.* [38, 98] have shown that Sm proteins are methylated by two enzymes – Dart5 and Dart7. Is Dart7 also required for germ cell specification? If our hypothesis is correct, and Sm protein methylation is a critical determinant for germ line specification, then Dart7 will also be required for this process. However, Dart7-depleted animals die very early during development [98], thus making this question difficult to answer. In order to answer this question, a complete loss-of-function *dart7* allele is needed. Using a germ line clone strategy, the mutant isoform of Dart7 could then be expressed exclusively in the ovary. Every other tissue would express wild-type Dart7, thus enabling the organism to survive. Attempts to deplete Dart7 specifically in the ovary using RNAi have not been successful (G.B.G. and A.G.M., unpublished results). During oogenesis, the ovary appears to be somewhat resistant to dsRNA-mediated gene silencing [132].

In order to truly probe the mechanism of Sm protein-mediated germ cell specification, functional studies are needed. The core of our hypothesis is that Tudor binds to Sm proteins via their methyl modifications and is consequently anchored at the pole plasm. A necessary requirement of this model, therefore, is that the Sm proteins themselves be present at the pole plasm. Studies aimed at answering this question are under way, but preliminary indications are that Sm proteins are present at the pole plasm at the appropriate stage required for Tudor anchoring (G.B.G. and A.G.M., unpublished results). To fully test this hypothesis, however, Sm mutants that are abolished or reduced in binding to Tudor, yet fully capable of assembling into snRNPs and functioning in pre-mRNA splicing, will be needed. In other words, viable Sm mutants that are capable of sustaining life but unable to specify germ cell fate are needed to answer this question. Since this experiment will be done in a strain expressing wild-type Dart5, the phenotype can be correlated specifically to the mutant Sm protein.

4.13 The Transcriptional Functions of PRMT5

For most of this chapter, we have focused on Sm proteins as the main *in vivo* substrates for PRMT5. However, it is becoming increasingly clear that PRMT5, along with other PRMTs, also methylates arginine residues present in histones. Initial studies using chromosome immunoprecipitation (ChIP) experiments indicated that PRMT5 is present at the promoter of cell cycle-regulated genes such as *cyclin E1* and the myc target gene, *cad* [133, 134]. Overexpression of PRMT5 resulted in repression of these genes, suggesting a negative role for PRMT5 in transcriptional regulation [133, 134]. Subsequent studies demonstrated that PRMT5 dimethylated arginine 8 in histone H3 (H3R8) and arginine 3 in histone H4 (H4R3) [135]. These modifications were associated with the repression of *suppressor of tumorigenicity 7* (*ST7*) and *nonmetastatic 23* (*NM23*) [135]. PRMT5 also functions in association with chromatin remodeling complexes and by directly methylating SPT5, a transcription elongation factor [134, 136].

The observation that PRMT5 is involved in repressing several tumor suppressor genes has led to the notion that PRMT5 expression levels may be correlated with the occurrence of certain types of cancer. Consistent with this belief, PRMT5 overexpression was shown to result in increased growth of NIH 3T3 cells, whereas PRMT5 depletion caused the opposite phenotype [135]. A recent study validated the physiological significance of this hypothesis. PRMT5 levels were found to be significantly elevated in a number of lymphoid cancer cell lines as well as in clinical samples from mantle cell lymphomas [137]. Concomitant with increased PRMT5 levels, dimethylation of H3R8 and H4R3 was also increased [137]. Interestingly, the expression level of PRMT5 in these cell lines appears to be controlled at the level of translation. The cancer cell lines were found to be deficient in two micro (mi)RNAs – miR-92b and miR-96 [137]. Reintroduction of these miRNAs into the cancer cell lines reduced the level of PRMT5 [137]. Further work will be needed to establish whether reduction of PRMT5 levels can be used as an effective therapy to slow the progression of cancer.

The involvement of PRMT5 in histone modification extends beyond the realm of cancer; it also appears to be involved in mammalian germ cell specification. In a recent study, PRMT5 was shown to be expressed in primordial germ cells (PGCs) of the mouse at embryonic day (E7.5), a time point critical in their development and specification [138]. Importantly, PRMT5 was found in a complex along with Blimp1, a transcriptional repressor protein previously shown to be involved in germ cell specification [138–140]. Interestingly, PRMT5 and Blimp1 are present in the nuclei of PGCs at E8.5, and this correlates with high levels of histone H4R3 dimethylation [138]. Given the previously documented functions of PRMT5, it is likely that the Blimp1/PRMT5 complex represses the expression of a set of target genes. The inappropriate expression of these genes might disrupt the progression of PGCs toward a germ cell lineage. A single candidate gene, termed *Dhx1*, was identified as being regulated by the Blimp1/PRMT5 complex. The expression of *Dhx1* was reduced at E8.5 when Blimp1 and PRMT5 were nuclear, and was induced at E11.5 when both proteins translocated to the cytoplasm [138]. Thus, the Blimp1/PRMT5 complex performs an essential regulatory function between E7.5 and E11.5 that is required for proper establishment of the germ cell fate.

In considering the function of PRMT5 in murine germ cell specification, it is worth noting that *dart5* mutants are male sterile. Unlike the female germ line, the function of *Dart5* in the male germ line in *Drosophila* is not likely to involve Tudor. Homozygous male *dart5* mutants are sterile and have defects in spermatocyte maturation, whereas *tudor* males are fertile [38, 127]. It is therefore possible that much like PRMT5, Dart5 functions in a transcriptional capacity in the male germ line. Chromosome immunoprecipitation followed by microarray analysis (ChIP on Chip) should help elucidate this question. In this regard, it is interesting to note that *Drosophila* expresses a Blimp1 homolog – dBlimp1 [141, 142]. It is not currently known, however, if dBlimp1 and Dart5 associate *in vivo*.

The function of PRMT5 in methylating histone residues is extremely well conserved. The *Arabidopsis* ortholog of PRMT5 (*AtPRMT5*) was recently identified and shown to dimethylate H4R3 [143, 144]. Remarkably, this methyl mark was correlated with reduced expression of *Flowering Locus C* (*FLC*) [144]. Mutants of *AtPRMT5* displayed increased *FLC* expression, delayed flowering time, and a slow-growth phenotype [143, 144]. Thus, *AtPRMT5* appears to play a role in the transition from vegetative to reproductive development in *Arabidopsis*. It is therefore quite interesting that PRMT5 functions in germ cell specification in mice, fruit flies, and in a conceptually similar process in plants.

4.14
Arginine Methylation – No Longer in the Shadow of Phosphorylation

A great deal has been learned in the past few years about the various biological processes controlled by arginine methylation. Yet these advances are likely to represent the tip of the iceberg. With the advent of new technologies for analyzing this important posttranslational modification, significant discoveries are being made at a very rapid pace. For the better part of this decade, histone methylation was though to be an irreversible process, a static epigenetic mark. As a consequence of these methyl marks, certain genes are maintained in a transcriptionally active state, whereas other are repressed [145]. These fates were believed to be programmed and set early on in development, thus enabling differentiation along a variety of lineages. Consequently, the transcriptome of a muscle cell does not completely overlap with that of a germ cell. This static epigenetic hypothesis breaks down, however, when considering embryonic stem (ES) cells. Many genes in ES cells contain both active and repressive methylation marks on the same gene [146]. During the course of differentiation, many of these resolve into exclusively active or repressed states, suggesting that a certain level of plasticity exists *in vivo* [146]. Clearly, in order for this to occur, there must exist one or more enzymes capable of removing these methylation marks. A flurry of recent papers has proven the existence of the long sought-after demethylases.

Among the first demethylases to be identified were those that catalyze the removal of methyl groups from lysine residues. Lysine residues on histones are found in one of three modification states: mono-, di-, or trimethylated. The first demethylase to be identified was LSD1, and it was shown to be capable of removing mono- and dimethyl marks [147, 148]. Interestingly, LSD1 is specific for the lysine 4 residue in histone H3 (H3K4), a modification often found at the promoter region of actively transcribed genes [147]. Subsequently, enzymes containing Jumonji-C (JmjC) domains were identified and also shown to remove mono- and dimethyl marks from lysine residues [149, 150]. Like LSD1, the JmjC domain-containing proteins displayed specificity for certain lysine residues. Recently, two JmjC domain-containing proteins, JHDM3 and JARID1, were identified as the enzymes responsible for demethylating trimethyl lysine residues [151–157].

Until recently, a similar enzymatic activity had not been described for methylated arginine residues. PAD4, an enzyme belonging to the peptidylarginine deiminase (PADI) family, was shown to be capable of removing methyl groups from arginine residues [158, 159]. However, it is not a demethylase in the true sense of the word since it converts methyl arginine to citrulline, thereby permanently changing the primary amino acid sequence of the histone. Recently, JMJD6, a JmjC domain-containing protein, was identified as a true arginine demethylase [160]. JMJD6 is capable of removing mono- and dimethyl marks from select arginine residues and has no activity toward dimethylated lysine residues [160]. It is possible that other arginine demethylases exist as well because H3R17 and H3R26 are apparently not substrates for JMJD6 [160].

These recent discoveries will undoubtedly be pivotal in our understanding of methylation as a key regulator of numerous biological processes. Many nonhistone proteins, including Sm proteins and several RNA-binding proteins, are also methylated on arginine residues. Is JMJD6 capable of removing these methyl marks as well? If so, processes such as snRNP assembly, and nucleocytoplasmic transport of proteins and RNPs could be controlled in response to environmental stimuli. The dogma that the transcriptome of adult differentiated cells is fixed by epigenetic programs set during development has been challenged. In a groundbreaking discovery, two groups recently demonstrated that differentiated adult cells could be reprogrammed back to a stem cell fate [161, 162]. Continued investigation of methyltransferases and demethylases will ultimately help unlock the secret pathways by which cells differentiate, and will provide us with useful insights into specific mechanisms that could be used to control and even reverse this process.

4.15
Sm Proteins – Doughnut-Shaped Relics of Our RNA Past

As mentioned in the introduction of this chapter, paralogs of the Sm proteins can be found in all three domains of life – bacteria, archaea, and eukarya. Hfq, the bacterial ortholog of Sm proteins, is found in many bacterial species and appears to function in numerous RNA-related processes. For instance, Hfq has been shown to target several mRNAs for degradation, to regulate the translation of a subset of mRNAs, and to mediate certain RNA–RNA interactions [163–166]. Consequently, Hfq is thought to function as a general RNA chaperone. Consistent with this notion, mutation of Hfq results in pleiotropic phenotypes such as decreased growth rate, UV sensitivity, and increased cell length [167, 168]. Interestingly, only after the three-dimensional structure of Hfq was solved was it realized that Hfq is, in fact, an Sm paralog [164, 166]. Analogous to eukaryotic Sm and Lsm protein, Hfq oligomerizes into a six-membered ring and binds RNA within a central basic pore [166].

In archaea, several members of the Sm superfamily have been identified. Much like their bacterial and eukaryotic counterparts, the archaeal Sm proteins also oligomerize into doughnut-shaped structures. Each archaeal species expresses up to three Sm proteins – Sm1, Sm2, or Sm3, with Sm1 being the most common archaeal Sm protein. In contrast to eukaryotic Sm proteins, which form seven-membered hetero-oligomeric rings, the archaeal Sm proteins homo-oligomerize [169–172]. In this respect, they are more similar to Hfq. In fact, an archaeal Hfq homolog was recently identified in *Methanococcus jannaschii* [173, 174]. Interestingly, unlike other archaeal species, *M. jannaschii* does not express a canonical Sm protein. Therefore, in *M. jannaschii*, the Hfq-like protein is a likely Sm substitute. In a striking demonstration of functional conservation, the *M. jannaschii* Hfq-like protein could partially compensate for loss of function *hfq* mutants in *E. coli* [175]. The physiological functions of the archaeal Sm proteins are unknown, but they have been shown to bind U-rich RNA sequences that are similar to the Sm site of snRNAs [169, 172]. The only demonstrated *in vivo* substrate of archaeal Sm proteins is pre-RNAse P RNA [172]. Therefore, archaeal Sm proteins may function in processing transfer (t)RNA. Additionally, archaeal Sm proteins are often found next to genes encoding ribosome proteins, in what appears to be an operon [170]. Therefore, it is also possible that archeal Sm proteins function in processing ribosomal RNA.

The three-dimensional structures of bacterial, archaeal, and eukaryotic Sm proteins are remarkably similar. In fact, many of these Sm structures can be directly superimposed [171, 172]. It is therefore widely believed that current-day Sm proteins have evolved from a common ancestral gene. Bacteria express a single Sm-like protein, archaea express up to 3, and eukaryotes express upward of 16 Sm proteins. Therefore, gene duplication, followed by specialization of function is likely to have occurred. Consequently, bacterial Hfq has evolved into the many Sm and Lsm proteins found in modern-day eukaryotes. Through the course of evolution, the Sm and Lsm proteins have acquired eukaryotic specific functions such as splicing, telomere maintenance, and histone pre-mRNA processing. Consistent with this gain of function hypothesis, Hfq as well as most archaeal Sm proteins does not have N- or C-terminal extensions. In contrast, many eukaryotic Sm and Lsm proteins have relatively long C-terminal extensions. Importantly, in the case of SmB, SmD1, and SmD3, the C-terminal extensions also contain methylated arginine residues, suggesting further specialization of function.

How do Sm proteins function in the germ line? Although their identification in the germ line specification pathway is more recent, it may represent a more ancestral function for these proteins than pre-mRNA splicing. Germ line specification in *Drosophila* involves the transport and localization of a number of mRNAs that encode cell fate determining factors. Consistent with their ancestral role, Sm proteins may function to regulate these mRNAs. For instance, *oskar* mRNA localization to the posterior of the oocyte is critical for germ cell specification [114]. The *oskar* mRNPs that are *en route* to the pole plasm must be translationally repressed [115–118]. Analogous to the function of Hfq, Sm proteins may regulate the translation of *oskar* mRNA. It has recently been shown that *oskar* mRNA forms oligomers *in vivo* and that the formation of these large higher-order *oskar* mRNPs is required for translational repression [176]. Sm proteins may function as an RNA chaperone to facilitate intermolecular interactions between *oskar* mRNAs, thus enabling the formation of a large translationally quiescent *oskar* mRNP. While it is not currently known precisely how Sm proteins function in the germ line, we believe that they do more than just anchor Tudor at the pole plasm. Whatever their germ line function, it will likely involve one or more RNAs. Time will reveal what secrets these ancient RNA-binding proteins have yet to tell.

References

1 Sanford, J.R. and Caceres, J.F. (2004) Pre-mRNA splicing: life at the centre of the central dogma. *J. Cell Sci.*, **117**, 6261–6263.

2 Tarn, W.Y. and Steitz, J.A. (1996) A novel spliceosome containing U11, U12, and U5 snRNPs excises a minor class (AT-AC) intron *in vitro*. *Cell*, **84**, 801–811.

3 Will, C.L. and Luhrmann, R. (2005) Splicing of a rare class of introns by the U12-dependent spliceosome. *Biol. Chem.*, **386**, 713–724.

4 Tarn, W.Y. and Steitz, J.A. (1996) Highly diverged U4 and U6 small nuclear RNAs required for splicing rare AT-AC introns. *Science*, **273**, 1824–1832.

5 Chen, L., Lullo, D.J., Ma, E., Celniker, S.E., Rio, D.C., and Doudna, J.A. (2005) Identification and analysis of U5 snRNA variants in *Drosophila*. *RNA*, **11**, 1473–1477.

6 Sontheimer, E.J. and Steitz, J.A. (1992) Three novel functional variants of human U5 small nuclear RNA. *Mol. Cell. Biol.*, **12**, 734–746.

7 Matera, A.G., Terns, R.M., and Terns, M.P. (2007) Non-coding RNAs: lessons from the small nuclear and small nucleolar RNAs. *Nat. Rev. Mol. Cell Biol.*, **8**, 209–220.

8 Reed, R. (2000) Mechanisms of fidelity in pre-mRNA splicing. *Curr. Opin. Cell Biol.*, **12**, 340–345.

9 Valadkhan, S. (2007) The spliceosome: caught in a web of shifting interactions. *Curr. Opin. Struct. Biol.*, **17**, 310–315.

10 Bordonne, R. and Tarassov, I. (1996) The yeast SME1 gene encodes the homologue of the human E core protein. *Gene*, **176**, 111–117.

11 Roy, J., Zheng, B., Rymond, B.C., and Woolford, J.L., Jr. (1995) Structurally related but functionally distinct yeast Sm D core small nuclear ribonucleoprotein particle proteins. *Mol. Cell. Biol.*, **15**, 445–455.

12 Rymond, B.C. (1993) Convergent transcripts of the yeast PRP38-SMD1 locus encode two essential splicing factors, including the D1 core polypeptide of small nuclear

ribonucleoprotein particles. *Proc. Natl. Acad. Sci. U.S.A.*, **90**, 848–852.

13 Mayes, A.E., Verdone, L., Legrain, P., and Beggs, J.D. (1999) Characterization of Sm-like proteins in yeast and their association with U6 snRNA. *EMBO J.*, **18**, 4321–4331.

14 Lerner, M.R. and Steitz, J.A. (1979) Antibodies to small nuclear RNAs complexed with proteins are produced by patients with systemic lupus erythematosus. *Proc. Natl. Acad. Sci. U.S.A.*, **76**, 5495–5499.

15 Lerner, M.R., Boyle, J.A., Mount, S.M., Wolin, S.L., and Steitz, J.A. (1980) Are snRNPs involved in splicing? *Nature*, **283**, 220–224.

16 Ou, Y., Sun, D., Sharp, G.C., and Hoch, S.O. (1997) Screening of SLE sera using purified recombinant Sm-D1 protein from a baculovirus expression system. *Clin. Immunol. Immunopathol.*, **83**, 310–317.

17 Rokeach, L.A., Haselby, J.A., and Hoch, S.O. (1992) Overproduction of a human snRNP-associated Sm-D autoantigen in *Escherichia coli* and *Saccharomyces cerevisiae*. *Gene*, **118**, 247–253.

18 Wagatsuma, M., Asami, N., Miyachi, J., Uchida, S., Watanabe, H., and Amann, E. (1993) Antibody recognition of the recombinant human nuclear antigens RNP 70 kD, SS-A, SS-B, Sm-B, and Sm-D by autoimmune sera. *Mol. Immunol.*, **30**, 1491–1498.

19 Brahms, H., Raymackers, J., Union, A., de Keyser, F., Meheus, L., and Lührmann, R. (2000) The C-terminal RG dipeptide repeats of the spliceosomal Sm proteins D1 and D3 contain symmetrical dimethylarginines, which form a major B-cell epitope for anti-Sm autoantibodies. *J. Biol. Chem.*, **275**, 17122–17129.

20 Paik, W.K. and Kim, S. (1967) Enzymatic methylation of protein fractions from calf thymus nuclei. *Biochem. Biophys. Res. Commun.*, **29**, 14–20.

21 Martin, C. and Zhang, Y. (2005) The diverse functions of histone lysine methylation. *Nat. Rev. Mol. Cell Biol.*, **6**, 838–849.

22 Wysocka, J., Allis, C.D., and Coonrod, S. (2006) Histone arginine methylation and its dynamic regulation. *Front Biosci.*, **11**, 344–355.

23 Strahl, B.D. and Allis, C.D. (2000) The language of covalent histone modifications. *Nature*, **403**, 41–45.

24 Bachand, F. (2007) Protein arginine methyltransferases: from unicellular eukaryotes to humans. *Eukaryot. Cell*, **6**, 889–898.

25 Bedford, M.T. (2007) Arginine methylation at a glance. *J. Cell Sci.*, **120**, 4243–4246.

26 Bedford, M.T., and Richard, S. (2005) Arginine methylation an emerging regulator of protein function. *Mol. Cell*, **18**, 263–272.

27 Branscombe, T.L., Frankel, A., Lee, J.H., Cook, J.R., Yang, Z., Pestka, S., and Clarke, S. (2001) PRMT5 (Janus kinase-binding protein 1) catalyzes the formation of symmetric dimethylarginine residues in proteins. *J. Biol. Chem.*, **276**, 32971–32976.

28 Pollack, B.P., Kotenko, S.V., He, W., Izotova, L.S., Barnoski, B.L., and Pestka, S. (1999) The human homologue of the yeast proteins Skb1 and Hsl7p interacts with Jak kinases and contains protein methyltransferase activity. *J. Biol. Chem.*, **274**, 31531–31542.

29 Lee, J.H., Cook, J.R., Yang, Z.H., Mirochnitchenko, O., Gunderson, S.I., Felix, A.M., Herth, N., Hoffmann, R., and Pestka, S. (2005) PRMT7, a new protein arginine methyltransferase that synthesizes symmetric dimethylarginine. *J. Biol. Chem.*, **280**, 3656–3664.

30 Cook, J.R., Lee, J.H., Yang, Z.H., Krause, C.D., Herth, N., Hoffmann, R., and Pestka, S. (2006) FBXO11/PRMT9, a new protein arginine methyltransferase, symmetrically dimethylates arginine residues. *Biochem. Biophys. Res. Commun.*, **342**, 472–481.

31 Brahms, H., Meheus, L., de Brabandere, V., Fischer, U., and Lührmann, R. (2001) Symmetrical dimethylation of arginine residues in spliceosomal Sm protein B/B' and the Sm-like protein LSm4, and their interaction with the SMN protein. *RNA*, **7**, 1531–1542.

32 Miranda, T.B., Khusial, P., Cook, J.R., Lee, J.H., Gunderson, S.I., Pestka, S., Zieve, G.W., and Clarke, S. (2004) Spliceosome Sm proteins D1, D3, and B/B' are asymmetrically dimethylated at arginine residues in the nucleus. *Biochem. Biophys. Res. Commun.*, **323**, 382–387.

33 Cheng, D., Cote, J., Shaaban, S., and Bedford, M.T. (2007) The arginine methyltransferase CARM1 regulates the coupling of transcription and mRNA processing. *Mol. Cell*, **25**, 71–83.

34 Yadav, N., Lee, J., Kim, J., Shen, J., Hu, M.C., Aldaz, C.M., and Bedford, M.T. (2003) Specific protein methylation defects and gene expression perturbations in coactivator-associated arginine methyltransferase 1-deficient mice. *Proc. Natl. Acad. Sci. U.S.A.*, **100**, 6464–6468.

35 Boisvert, F.M., Cote, J., Boulanger, M.C., Cleroux, P., Bachand, F., Autexier, C., and Richard, S. (2002) Symmetrical dimethylarginine methylation is required for the localization of SMN in Cajal bodies and pre-mRNA splicing. *J. Cell Biol.*, **159**, 957–969.

36 Boisvert, F.M., Cote, J., Boulanger, M.C., and Richard, S. (2003) A proteomic analysis of arginine-methylated protein complexes. *Mol. Cell. Proteomics*, **2**, 1319–1330.

37 Friesen, W.J., Paushkin, S., Wyce, A., Massenet, S., Pesiridis, G.S., Van Duyne, G., Rappsilber, J., Mann, M., and Dreyfuss, G. (2001) The methylosome, a 20S complex containing JBP1 and pICln, produces dimethylarginine-modified Sm proteins. *Mol. Cell. Biol.*, **21**, 8289–8300.

38 Gonsalvez, G.B., Rajendra, T.K., Tian, L., and Matera, A.G. (2006) The Sm-protein methyltransferase, dart5, is essential for germ-cell specification and maintenance. *Curr. Biol.*, **16**, 1077–1089.

39 Gonsalvez, G.B., Tian, L., Ospina, J.K., Boisvert, F.M., Lamond, A.I., and Matera, A.G. (2007) Two distinct arginine methyltransferases are required for biogenesis of Sm-class ribonucleoproteins. *J. Cell Biol.*, **178**, 733–740.

40 Boelens, W.C., Palacios, I., and Mattaj, I.W. (1995) Nuclear retention of RNA as a mechanism for localization. *RNA*, **1**, 273–283.

41 Terns, M.P., Grimm, C., Lund, E., and Dahlberg, J.E. (1995) A common maturation pathway for small nucleolar RNAs. *EMBO J.*, **14**, 4860–4871.

42 Vankan, P., McGuigan, C., and Mattaj, I.W. (1990) Domains of U4 and U6 snRNAs required for snRNP assembly and splicing complementation in *Xenopus* oocytes. *EMBO J.*, **9**, 3397–3404.

43 Pante, N., Jarmolowski, A., Izaurralde, E., Sauder, U., Baschong, W., and Mattaj, I.W. (1997) Visualizing nuclear export of different classes of RNA by electron microscopy. *RNA*, **3**, 498–513.

44 Spiller, M.P., Boon, K.L., Reijns, M.A., and Beggs, J.D. (2007) The Lsm2-8 complex determines nuclear localization of the spliceosomal U6 snRNA. *Nucleic Acids Res.*, **35**, 923–929.

45 Izaurralde, E., Lewis, J., Gamberi, C., Jarmolowski, A., McGuigan, C., and Mattaj, I.W. (1995) A cap-binding protein complex mediating U snRNA export. *Nature*, **376**, 709–712.

46 Ohno, M., Segref, A., Bachi, A., Wilm, M., and Mattaj, I.W. (2000) PHAX, a mediator of U snRNA nuclear export whose activity is regulated by phosphorylation. *Cell*, **101**, 187–198.

47 Meister, G., Eggert, C., and Fischer, U. (2002) SMN-mediated assembly of RNPs: a complex story. *Trends Cell Biol.*, **12**, 472–478.

48 Paushkin, S., Gubitz, A.K., Massenet, S., and Dreyfuss, G. (2002) The SMN complex, an assemblyosome of ribonucleoproteins. *Curr. Opin. Cell Biol.*, **14**, 305–312.

49 Pellizzoni, L. (2007) Chaperoning ribonucleoprotein biogenesis in health and disease. *EMBO Rep.*, **8**, 340–345.

50 Battle, D.J., Lau, C.K., Wan, L., Deng, H., Lotti, F., and Dreyfuss, G. (2006) The Gemin5 protein of the SMN complex identifies snRNAs. *Mol. Cell*, **23**, 273–279.

51 Pellizzoni, L., Yong, J., and Dreyfuss, G. (2002) Essential role for the SMN

complex in the specificity of snRNP assembly. *Science*, **298**, 1775–1779.

52 Mouaikel, J., Narayanan, U., Verheggen, C., Matera, A.G., Bertrand, E., Tazi, J., and Bordonne, R. (2003) Interaction between the small-nuclear-RNA cap hypermethylase and the spinal muscular atrophy protein, survival of motor neuron. *EMBO Rep.*, **4**, 616–622.

53 Mouaikel, J., Verheggen, C., Bertrand, E., Tazi, J., and Bordonne, R. (2002) Hypermethylation of the cap structure of both yeast snRNAs and snoRNAs requires a conserved methyltransferase that is localized to the nucleolus. *Mol. Cell*, **9**, 891–901.

54 Huber, J., Cronshagen, U., Kadokura, M., Marshallsay, C., Wada, T., Sekine, M., and Lührmann, R. (1998) Snurportin1, an m3G-cap-specific nuclear import receptor with a novel domain structure. *EMBO J.*, **17**, 4114–4126.

55 Narayanan, U., Achsel, T., Lührmann, R., and Matera, A.G. (2004) Coupled *in vitro* import of U snRNPs and SMN, the spinal muscular atrophy protein. *Mol. Cell*, **16**, 223–234.

56 Narayanan, U., Ospina, J.K., Frey, M.R., Hebert, M.D., and Matera, A.G. (2002) SMN, the spinal muscular atrophy protein, forms a pre-import snRNP complex with snurportin1 and importin beta. *Hum. Mol. Genet.*, **11**, 1785–1795.

57 Massenet, S., Pellizzoni, L., Paushkin, S., Mattaj, I.W., and Dreyfuss, G. (2002) The SMN complex is associated with snRNPs throughout their cytoplasmic assembly pathway. *Mol. Cell. Biol.*, **22**, 6533–6541.

58 Ospina, J.K., Gonsalvez, G.B., Bednenko, J., Darzynkiewicz, E., Gerace, L., and Matera, A.G. (2005) Cross-talk between snurportin1 subdomains. *Mol. Biol. Cell*, **16**, 4660–4671.

59 Jady, B.E., Darzacq, X., Tucker, K.E., Matera, A.G., Bertrand, E., and Kiss, T. (2003) Modification of Sm small nuclear RNAs occurs in the nucleoplasmic Cajal body following import from the cytoplasm. *EMBO J.*, **22**, 1878–1888.

60 Nesic, D., Tanackovic, G., and Kramer, A. (2004) A role for Cajal bodies in the final steps of U2 snRNP biogenesis. *J. Cell Sci.*, **117**, 4423–4433.

61 Carvalho, T., Almeida, F., Calapez, A., Lafarga, M., Berciano, M.T., and Carmo-Fonseca, M. (1999) The spinal muscular atrophy disease gene product, SMN: a link between snRNP biogenesis and the Cajal (coiled) body. *J. Cell Biol.*, **147**, 715–728.

62 Shpargel, K.B. and Matera, A.G. (2005) Gemin proteins are required for efficient assembly of Sm-class ribonucleoproteins. *Proc. Natl. Acad. Sci. U.S.A.*, **102**, 17372–17377.

63 Girard, C., Neel, H., Bertrand, E., and Bordonne, R. (2006) Depletion of SMN by RNA interference in HeLa cells induces defects in Cajal body formation. *Nucleic Acids Res.*, **34**, 2925–2932.

64 Lemm, I., Girard, C., Kuhn, A.N., Watkins, N.J., Schneider, M., Bordonne, R., and Luhrmann, R. (2006) Ongoing U snRNP biogenesis is required for the integrity of Cajal bodies. *Mol. Biol. Cell*, **17**, 3221–3231.

65 Lefebvre, S., Burglen, L., Reboullet, S., Clermont, O., Burlet, P., Viollet, L., Benichou, B., Cruaud, C., Millasseau, P., Zeviani, M., *et al.* (1995) Identification and characterization of a spinal muscular atrophy-determining gene. *Cell*, **80**, 155–165.

66 Winkler, C., Eggert, C., Gradl, D., Meister, G., Giegerich, M., Wedlich, D., Laggerbauer, B., and Fischer, U. (2005) Reduced U snRNP assembly causes motor axon degeneration in an animal model for spinal muscular atrophy. *Genes Dev.*, **19**, 2320–2330.

67 Carrel, T.L., McWhorter, M.L., Workman, E., Zhang, H., Wolstencroft, E.C., Lorson, C., Bassell, G.J., Burghes, A.H., and Beattie, C.E. (2006) Survival motor neuron function in motor axons is independent of functions required for small nuclear ribonucleoprotein biogenesis. *J. Neurosci.*, **26**, 11014–11022.

68 Avila, A.M., Burnett, B.G., Taye, A.A., Gabanella, F., Knight, M.A., Hartenstein, P., Cizman, Z., Di Prospero, N.A., Pellizzoni, L., Fischbeck, K.H., and Sumner, C.J. (2007) Trichostatin A increases SMN expression and survival in a mouse model of spinal muscular atrophy. *J. Clin. Invest.*, **117**, 659–671.

69 Gabanella, F., Butchbach, M.E., Saieva, L., Carissimi, C., Burghes, A.H., and Pellizzoni, L. (2007) Ribonucleoprotein assembly defects correlate with spinal muscular atrophy severity and preferentially affect a subset of spliceosomal snRNPs. *PLoS One*, **2**, e921.

70 Rajendra, T.K., Gonsalvez, G.B., Walker, M.P., Shpargel, K.B., Salz, H.K., and Matera, A.G. (2007) A *Drosophila melanogaster* model of spinal muscular atrophy reveals a function for SMN in striated muscle. *J. Cell Biol.*, **176**, 831–841.

71 McWhorter, M.L., Boon, K.L., Horan, E.S., Burghes, A.H., and Beattie, C.E. (2007) The SMN binding protein Gemin2 is not involved in motor axon outgrowth. *Dev. Neurobiol.*, **68** (2), 182–194.

72 Charroux, B., Pellizzoni, L., Perkinson, R.A., Shevchenko, A., Mann, M., and Dreyfuss, G. (1999) Gemin3: a novel DEAD box protein that interacts with SMN, the spinal muscular atrophy gene product, and is a component of gems. *J. Cell Biol.*, **147**, 1181–1194.

73 Charroux, B., Pellizzoni, L., Perkinson, R.A., Yong, J., Shevchenko, A., Mann, M., and Dreyfuss, G. (2000) Gemin4. A novel component of the SMN complex that is found in both gems and nucleoli. *J. Cell Biol.*, **148**, 1177–1186.

74 Friesen, W.J. and Dreyfuss, G. (2000) Specific sequences of the Sm and Sm-like (Lsm) proteins mediate their interaction with the spinal muscular atrophy disease gene product (SMN). *J. Biol. Chem.*, **275**, 26370–26375.

75 Gubitz, A.K., Mourelatos, Z., Abel, L., Rappsilber, J., Mann, M., and Dreyfuss, G. (2002) Gemin5, a novel WD repeat protein component of the SMN complex that binds Sm proteins. *J. Biol. Chem.*, **277**, 5631–5636.

76 Baccon, J., Pellizzoni, L., Rappsilber, J., Mann, M., and Dreyfuss, G. (2002) Identification and characterization of Gemin7, a novel component of the survival of motor neuron complex. *J. Biol. Chem.*, **277**, 31957–31962.

77 Pellizzoni, L., Baccon, J., Rappsilber, J., Mann, M., and Dreyfuss, G. (2002) Purification of native survival of motor neurons complexes and identification of Gemin6 as a novel component. *J. Biol. Chem.*, **277**, 7540–7545.

78 Carissimi, C., Saieva, L., Baccon, J., Chiarella, P., Maiolica, A., Sawyer, A., Rappsilber, J., and Pellizzoni, L. (2006) Gemin8 is a novel component of the survival motor neuron complex and functions in small nuclear ribonucleoprotein assembly. *J. Biol. Chem.*, **281**, 8126–8134.

79 Carissimi, C., Baccon, J., Straccia, M., Chiarella, P., Maiolica, A., Sawyer, A., Rappsilber, J., and Pellizzoni, L. (2005) Unrip is a component of SMN complexes active in snRNP assembly. *FEBS Lett.*, **579**, 2348–2354.

80 Otter, S., Grimmler, M., Neuenkirchen, N., Chari, A., Sickmann, A., and Fischer, U. (2007) A comprehensive interaction map of the human survival of motor neuron (SMN) complex. *J. Biol. Chem.*, **282**, 5825–5833.

81 Friesen, W.J., Massenet, S., Paushkin, S., Wyce, A., and Dreyfuss, G. (2001) SMN, the product of the spinal muscular atrophy gene, binds preferentially to dimethylarginine-containing protein targets. *Mol. Cell*, **7**, 1111–1117.

82 Buhler, D., Raker, V., Lührmann, R., and Fischer, U. (1999) Essential role for the tudor domain of SMN in spliceosomal U snRNP assembly: implications for spinal muscular atrophy. *Hum. Mol. Genet.*, **8**, 2351–2357.

83 Cote, J. and Richard, S. (2005) Tudor domains bind symmetrical dimethylated arginines. *J. Biol. Chem.*, **280**, 28476–28483.

84 Meister, G., Eggert, C., Buhler, D., Brahms, H., Kambach, C., and Fischer, U. (2001) Methylation of Sm proteins by a complex containing PRMT5 and the putative U snRNP assembly factor pICln. *Curr. Biol.*, **11**, 1990–1994.

85 Pu, W.T., Krapivinsky, G.B., Krapivinsky, L., and Clapham, D.E. (1999) pICln inhibits snRNP biogenesis by binding core spliceosomal proteins. *Mol. Cell. Biol.*, **19**, 4113–4120.

86 Rho, J., Choi, S., Seong, Y.R., Cho, W.K., Kim, S.H., and Im, D.S. (2001) Prmt5, which forms distinct homo-oligomers, is a member of the protein-arginine methyltransferase family. *J. Biol. Chem.*, **276**, 11393–11401.

87 Huang, S., Litt, M., and Felsenfeld, G. (2005) Methylation of histone H4 by arginine methyltransferase PRMT1 is essential *in vivo* for many subsequent histone modifications. *Genes Dev.*, **19**, 1885–1893.

88 Meister, G. and Fischer, U. (2002) Assisted RNP assembly: SMN and PRMT5 complexes cooperate in the formation of spliceosomal UsnRNPs. *EMBO J.*, **21**, 5853–5863.

89 Lee, J., Sayegh, J., Daniel, J., Clarke, S., and Bedford, M.T. (2005) PRMT8, a new membrane-bound tissue-specific member of the protein arginine methyltransferase family. *J. Biol. Chem.*, **280**, 32890–32896.

90 Baillat, D., Hakimi, M.A., Naar, A.M., Shilatifard, A., Cooch, N., and Shiekhattar, R. (2005) Integrator, a multiprotein mediator of small nuclear RNA processing, associates with the C-terminal repeat of RNA polymerase II. *Cell*, **123**, 265–276.

91 Lee, J.H. and Skalnik, D.G. (2005) CpG-binding protein (CXXC finger protein 1) is a component of the mammalian Set1 histone H3-Lys4 methyltransferase complex, the analogue of the yeast Set1/COMPASS complex. *J. Biol. Chem.*, **280**, 41725–41731.

92 Nishioka, K. and Reinberg, D. (2003) Methods and tips for the purification of human histone methyltransferases. *Methods*, **31**, 49–58.

93 Yanagida, M., Hayano, T., Yamauchi, Y., Shinkawa, T., Natsume, T., Isobe, T., and Takahashi, N. (2004) Human fibrillarin forms a sub-complex with splicing factor 2-associated p32, protein arginine methyltransferases, and tubulins alpha 3 and beta 1 that is independent of its association with preribosomal ribonucleoprotein complexes. *J. Biol. Chem.*, **279**, 1607–1614.

94 Sleeman, J.E., Ajuh, P., and Lamond, A.I. (2001) snRNP protein expression enhances the formation of Cajal bodies containing p80-coilin and SMN. *J. Cell Sci.*, **114**, 4407–4419.

95 Azzouz, T.N., Pillai, R.S., Dapp, C., Chari, A., Meister, G., Kambach, C., Fischer, U., and Schumperli, D. (2005) Toward an assembly line for U7 snRNPs: interactions of U7-specific Lsm proteins with PRMT5 and SMN complexes. *J. Biol. Chem.*, **280**, 34435–34440.

96 Khusial, P.R., Vaidya, K., and Zieve, G.W. (2005) The symmetrical dimethylarginine post-translational modification of the SmD3 protein is not required for snRNP assembly and nuclear transport. *Biochem. Biophys. Res. Commun.*, **337**, 1119–1124.

97 Boulanger, M.C., Miranda, T.B., Clarke, S., Di Fruscio, M., Suter, B., Lasko, P., and Richard, S. (2004) Characterization of the *Drosophila* protein arginine methyltransferases DART1 and DART4. *Biochem. J.*, **379**, 283–289.

98 Gonsalvez, G.B., Praveen, K., Hicks, H.J., Tian, L., and Matera, A.G. (2008) Sm protein methylation is dispensable for snRNP assembly in *Drosophila melanogaster. RNA*, **14** (5), 878–887.

99 Pillai, R.S., Will, C.L., Lührmann, R., Schumperli, D., and Muller, B. (2001) Purified U7 snRNPs lack the Sm proteins D1 and D2 but contain Lsm10, a new 14 kDa Sm D1-like protein. *EMBO J.*, **20**, 5470–5479.

100 Raker, V.A., Plessel, G., and Lührmann, R. (1996) The snRNP core assembly pathway: identification of stable core protein heteromeric complexes and an snRNP subcore particle *in vitro*. *EMBO J.*, **15**, 2256–2269.

101 Zhang, D., Abovich, N., and Rosbash, M. (2001) A biochemical function for the Sm complex. *Mol. Cell*, **7**, 319–329.

102 Grimmler, M., Bauer, L., Nousiainen, M., Korner, R., Meister, G., and Fischer, U. (2005) Phosphorylation regulates the activity of the SMN complex during assembly of spliceosomal U snRNPs. *EMBO Rep.*, **6**, 70–76.

103 Hannus, S., Buhler, D., Romano, M., Seraphin, B., and Fischer, U. (2000) The *Schizosaccharomyces pombe* protein Yab8p and a novel factor, Yip1p, share structural and functional similarity with the spinal muscular atrophy-associated proteins SMN and SIP1. *Hum. Mol. Genet.*, **9**, 663–674.

104 Liu, Q., Fischer, U., Wang, F., and Dreyfuss, G. (1997) The spinal muscular atrophy disease gene product, SMN, and its associated protein SIP1 are in a complex with spliceosomal snRNP proteins. *Cell*, **90**, 1013–1021.

105 Anne, J., Ollo, R., Ephrussi, A., and Mechler, B.M. (2007) Arginine methyltransferase Capsuleen is essential for methylation of spliceosomal Sm proteins and germ cell formation in *Drosophila*. *Development*, **134**, 137–146.

106 Mahowald, A.P. (2001) Assembly of the *Drosophila* germ plasm. *Int. Rev. Cytol.*, **203**, 187–213.

107 de Cuevas, M., Lilly, M.A., and Spradling, A.C. (1997) Germline cyst formation in *Drosophila*. *Annu. Rev. Genet.*, **31**, 405–428.

108 Santos, A.C. and Lehmann, R. (2004) Germ cell specification and migration in *Drosophila* and beyond. *Curr. Biol.*, **14**, R578–R589.

109 Johnstone, O. and Lasko, P. (2001) Translational regulation and RNA localization in *Drosophila* oocytes and embryos. *Annu. Rev. Genet.*, **35**, 365–406.

110 Riechmann, V. and Ephrussi, A. (2001) Axis formation during *Drosophila* oogenesis. *Curr. Opin. Genet. Dev.*, **11**, 374–383.

111 Huynh, J.R. and St Johnston, D. (2004) The origin of asymmetry: early polarisation of the *Drosophila* germline cyst and oocyte. *Curr. Biol.*, **14**, R438–R449.

112 Lehmann, R. and Nusslein-Volhard, C. (1986) Abdominal segmentation, pole cell formation, and embryonic polarity require the localized activity of oskar, a maternal gene in *Drosophila*. *Cell*, **47**, 141–152.

113 Ephrussi, A. and Lehmann, R. (1992) Induction of germ cell formation by oskar. *Nature*, **358**, 387–392.

114 Kim-Ha, J., Smith, J.L., and Macdonald, P.M. (1991) oskar mRNA is localized to the posterior pole of the *Drosophila* oocyte. *Cell*, **66**, 23–35.

115 Kim-Ha, J., Kerr, K., and Macdonald, P.M. (1995) Translational regulation of oskar mRNA by bruno, an ovarian RNA-binding protein, is essential. *Cell*, **81**, 403–412.

116 Markussen, F.H., Michon, A.M., Breitwieser, W., and Ephrussi, A. (1995) Translational control of oskar generates short OSK, the isoform that induces pole plasma assembly. *Development*, **121**, 3723–3732.

117 Rongo, C., Gavis, E.R., and Lehmann, R. (1995) Localization of oskar RNA regulates oskar translation and requires Oskar protein. *Development*, **121**, 2737–2746.

118 Vanzo, N.F. and Ephrussi, A. (2002) Oskar anchoring restricts pole plasm formation to the posterior of the *Drosophila* oocyte. *Development*, **129**, 3705–3714.

119 Breitwieser, W., Markussen, F.H., Horstmann, H., and Ephrussi, A. (1996) Oskar protein interaction with Vasa represents an essential step in polar granule assembly. *Genes Dev.*, **10**, 2179–2188.

120 Bardsley, A., McDonald, K., and Boswell, R.E. (1993) Distribution of tudor protein in the *Drosophila* embryo suggests separation of functions based on site of localization. *Development*, **119**, 207–219.

121 Hay, B., Jan, L.Y., and Jan, Y.N. (1990) Localization of vasa, a component of *Drosophila* polar granules, in maternal-effect mutants that alter embryonic anteroposterior polarity. *Development*, **109**, 425–433.

122 Lasko, P.F. and Ashburner, M. (1990) Posterior localization of vasa protein correlates with, but is not sufficient for, pole cell development. *Genes Dev.*, **4**, 905–921.

123 Cavey, M., Hijal, S., Zhang, X., and Suter, B. (2005) *Drosophila* valois encodes a divergent WD protein that is required for Vasa localization and Oskar protein accumulation. *Development*, **132**, 459–468.

124 Anne, J. and Mechler, B.M. (2005) Valois, a component of the nuage and pole plasm, is involved in assembly of these structures, and binds to Tudor and the methyltransferase Capsuleen. *Development*, **132**, 2167–2177.

125 Schüpbach, T. and Wieschaus, E. (1986) Germline autonomy of maternal-effect mutations altering the embryonic body pattern of *Drosophila*. *Dev. Biol.*, **113**, 443–448.

126 Friesen, W.J., Wyce, A., Paushkin, S., Abel, L., Rappsilber, J., Mann, M., and Dreyfuss, G. (2002) A novel WD repeat protein component of the methylosome binds Sm proteins. *J. Biol. Chem.*, **277**, 8243–8247.

127 Thomson, T. and Lasko, P. (2004) *Drosophila* tudor is essential for polar granule assembly and pole cell specification, but not for posterior patterning. *Genesis*, **40**, 164–170.

128 Barbee, S.A., Lublin, A.L., and Evans, T.C. (2002) A novel function for the Sm proteins in germ granule localization during *C. elegans* embryogenesis. *Curr. Biol.*, **12**, 1502–1506.

129 Bilinski, S.M., Jaglarz, M.K., Szymanska, B., Etkin, L.D., and Kloc, M. (2004) Sm proteins, the constituents of the spliceosome, are components of nuage and mitochondrial cement in *Xenopus* oocytes. *Exp. Cell Res.*, **299**, 171–178.

130 Chuma, S., Hiyoshi, M., Yamamoto, A., Hosokawa, M., Takamune, K., and Nakatsuji, N. (2003) Mouse Tudor Repeat-1 (MTR-1) is a novel component of chromatoid bodies/nuages in male germ cells and forms a complex with snRNPs. *Mech. Dev.*, **120**, 979–990.

131 Barbee, S.A. and Evans, T.C. (2006) The Sm proteins regulate germ cell specification during early *C. elegans* embryogenesis. *Dev. Biol.*, **291** (1), 132–143.

132 Kennerdell, J.R., Yamaguchi, S., and Carthew, R.W. (2002) RNAi is activated during *Drosophila* oocyte maturation in a manner dependent on aubergine and spindle-E. *Genes Dev.*, **16**, 1884–1889.

133 Fabbrizio, E., El Messaoudi, S., Polanowska, J., Paul, C., Cook, J.R., Lee, J.H., Negre, V., Rousset, M., Pestka, S., Le Cam, A., and Sardet, C. (2002) Negative regulation of transcription by the type II arginine methyltransferase PRMT5. *EMBO Rep.*, **3**, 641–645.

134 Pal, S., Yun, R., Datta, A., Lacomis, L., Erdjument-Bromage, H., Kumar, J., Tempst, P., and Sif, S. (2003) mSin3A/histone deacetylase 2- and PRMT5-containing Brg1 complex is involved in transcriptional repression of the Myc target gene cad. *Mol. Cell. Biol.*, **23**, 7475–7487.

135 Pal, S., Vishwanath, S.N., Erdjument-Bromage, H., Tempst, P., and Sif, S. (2004) Human SWI/SNF-associated PRMT5 methylates histone H3 arginine 8 and negatively regulates expression of ST7 and NM23 tumor suppressor genes. *Mol. Cell. Biol.*, **24**, 9630–9645.

136 Kwak, Y.T., Guo, J., Prajapati, S., Park, K.J., Surabhi, R.M., Miller, B., Gehrig, P., and Gaynor, R.B. (2003) Methylation of SPT5 regulates its interaction with RNA polymerase II and transcriptional elongation properties. *Mol. Cell*, **11**, 1055–1066.

137 Pal, S., Baiocchi, R.A., Byrd, J.C., Grever, M.R., Jacob, S.T., and Sif, S. (2007) Low levels of miR-92b/96 induce PRMT5 translation and H3R8/H4R3 methylation in mantle cell lymphoma. *EMBO J.*, **26**, 3558–3569.

138 Ancelin, K., Lange, U.C., Hajkova, P., Schneider, R., Bannister, A.J., Kouzarides, T., and Surani, M.A. (2006) Blimp1 associates with Prmt5 and directs histone arginine methylation in mouse germ cells. *Nat. Cell Biol.*, **8**, 623–630.

139 Ohinata, Y., Payer, B., O'Carroll, D., Ancelin, K., Ono, Y., Sano, M., Barton, S.C., Obukhanych, T., Nussenzweig, M., Tarakhovsky, A., Saitou, M., and Surani, M.A. (2005) Blimp1 is a critical determinant of the germ cell lineage in mice. *Nature*, **436**, 207–213.

140 Vincent, S.D., Dunn, N.R., Sciammas, R., Shapiro-Shalef, M., Davis, M.M., Calame, K., Bikoff, E.K., and Robertson, E.J. (2005) The zinc finger transcriptional repressor Blimp1/Prdm1 is dispensable for early axis formation but is required for specification of primordial germ cells in the mouse. *Development*, **132**, 1315–1325.

141 Agawa, Y., Sarhan, M., Kageyama, Y., Akagi, K., Takai, M., Hashiyama, K., Wada, T., Handa, H., Iwamatsu, A., Hirose, S., and Ueda, H. (2007) *Drosophila* Blimp-1 is a transient transcriptional repressor that controls timing of the ecdysone-induced developmental pathway. *Mol. Cell. Biol.*, **27**, 8739–8747.

142 Ng, T., Yu, F., and Roy, S. (2006) A homologue of the vertebrate SET domain and zinc finger protein Blimp-1 regulates terminal differentiation of the tracheal system in the *Drosophila* embryo. *Dev. Genes Evol.*, **216**, 243–252.

143 Pei, Y., Niu, L., Lu, F., Liu, C., Zhai, J., Kong, X., and Cao, X. (2007) Mutations in the Type II protein arginine methyltransferase AtPRMT5 result in pleiotropic developmental defects in *Arabidopsis. Plant Physiol.*, **144**, 1913–1923.

144 Wang, X., Zhang, Y., Ma, Q., Zhang, Z., Xue, Y., Bao, S., and Chong, K. (2007) SKB1-mediated symmetric dimethylation of histone H4R3 controls flowering time in *Arabidopsis. EMBO J.*, **26**, 1934–1941.

145 Bernstein, B.E., Meissner, A., and Lander, E.S. (2007) The mammalian epigenome. *Cell*, **128**, 669–681.

146 Mikkelsen, T.S., Ku, M., Jaffe, D.B., Issac, B., Lieberman, E., Giannoukos, G., Alvarez, P., Brockman, W., Kim, T.K., Koche, R.P., Lee, W., Mendenhall, E., O'Donovan, A., Presser, A., Russ, C., Xie, X., Meissner, A., Wernig, M., Jaenisch, R., Nusbaum, C., Lander, E.S., and Bernstein, B.E. (2007) Genome-wide maps of chromatin state in pluripotent and lineage-committed cells. *Nature*, **448**, 553–560.

147 Shi, Y., Lan, F., Matson, C., Mulligan, P., Whetstine, J.R., Cole, P.A., Casero, R.A., and Shi, Y. (2004) Histone demethylation mediated by the nuclear amine oxidase homolog LSD1. *Cell*, **119**, 941–953.

148 Shi, Y.J., Matson, C., Lan, F., Iwase, S., Baba, T., and Shi, Y. (2005) Regulation of LSD1 histone demethylase activity by its associated factors. *Mol. Cell*, **19**, 857–864.

149 Tsukada, Y., Fang, J., Erdjument-Bromage, H., Warren, M.E., Borchers, C.H., Tempst, P., and Zhang, Y. (2006) Histone demethylation by a family of JmjC domain-containing proteins. *Nature*, **439**, 811–816.

150 Yamane, K., Toumazou, C., Tsukada, Y., Erdjument-Bromage, H., Tempst, P., Wong, J., and Zhang, Y. (2006) JHDM2A, a JmjC-containing H3K9 demethylase, facilitates transcription activation by androgen receptor. *Cell*, **125**, 483–495.

151 Cloos, P.A., Christensen, J., Agger, K., Maiolica, A., Rappsilber, J., Antal, T., Hansen, K.H., and Helin, K. (2006) The putative oncogene GASC1 demethylates tri- and dimethylated lysine 9 on histone H3. *Nature*, **442**, 307–311.

152 Fodor, B.D., Kubicek, S., Yonezawa, M., O'Sullivan, R.J., Sengupta, R., Perez-Burgos, L., Opravil, S., Mechtler, K., Schotta, G., and Jenuwein, T. (2006) Jmjd2b antagonizes H3K9 trimethylation at pericentric heterochromatin in mammalian cells. *Genes Dev.*, **20**, 1557–1562.

153 Klose, R.J., Yamane, K., Bae, Y., Zhang, D., Erdjument-Bromage, H., Tempst, P., Wong, J., and Zhang, Y. (2006) The transcriptional repressor JHDM3A demethylates trimethyl histone H3 lysine 9 and lysine 36. *Nature*, **442**, 312–316.

154 Whetstine, J.R., Nottke, A., Lan, F., Huarte, M., Smolikov, S., Chen, Z., Spooner, E., Li, E., Zhang, G., Colaiacovo, M., and Shi, Y. (2006) Reversal of histone lysine trimethylation by the JMJD2 family of histone demethylases. *Cell*, **125**, 467–481.

155 Klose, R.J., Yan, Q., Tothova, Z., Yamane, K., Erdjument-Bromage, H., Tempst, P., Gilliland, D.G., Zhang, Y., and Kaelin, W.G., Jr. (2007) The retinoblastoma binding protein RBP2 is an H3K4 demethylase. *Cell*, **128**, 889–900.

156 Lee, N., Zhang, J., Klose, R.J., Erdjument-Bromage, H., Tempst, P., Jones, R.S., and Zhang, Y. (2007) The trithorax-group protein Lid is a histone H3 trimethyl-Lys4 demethylase. *Nat. Struct. Mol. Biol.*, **14**, 341–343.

157 Liang, G., Klose, R.J., Gardner, K.E., and Zhang, Y. (2007) Yeast Jhd2p is a histone H3 Lys4 trimethyl demethylase. *Nat. Struct. Mol. Biol.*, **14**, 243–245.

158 Cuthbert, G.L., Daujat, S., Snowden, A.W., Erdjument-Bromage, H., Hagiwara, T., Yamada, M., Schneider, R., Gregory, P.D., Tempst, P., Bannister, A.J., and Kouzarides, T. (2004) Histone deimination antagonizes arginine methylation. *Cell*, **118**, 545–553.

159 Wang, Y., Wysocka, J., Sayegh, J., Lee, Y.H., Perlin, J.R., Leonelli, L., Sonbuchner, L.S., McDonald, C.H., Cook, R.G., Dou, Y., Roeder, R.G., Clarke, S., Stallcup, M.R., Allis, C.D., and Coonrod, S.A. (2004) Human PAD4 regulates histone arginine methylation levels via demethylimination. *Science*, **306**, 279–283.

160 Chang, B., Chen, Y., Zhao, Y., and Bruick, R.K. (2007) JMJD6 is a histone arginine demethylase. *Science*, **318**, 444–447.

161 Takahashi, K., Tanabe, K., Ohnuki, M., Narita, M., Ichisaka, T., Tomoda, K., and Yamanaka, S. (2007) Induction of pluripotent stem cells from adult human fibroblasts by defined factors. *Cell*, **131**, 861–872.

162 Yu, J., Vodyanik, M.A., Smuga-Otto, K., Antosiewicz-Bourget, J., Frane, J.L., Tian, S., Nie, J., Jonsdottir, G.A., Ruotti, V., Stewart, R., Slukvin, I.I., and Thomson, J.A. (2007) Induced pluripotent stem cell lines derived from human somatic cells. *Science*, **318** (5858), 1917–1920.

163 Hajnsdorf, E. and Regnier, P. (2000) Host factor Hfq of *Escherichia coli* stimulates elongation of poly(A) tails by poly(A) polymerase I. *Proc. Natl. Acad. Sci. U.S.A.*, **97**, 1501–1505.

164 Moller, T., Franch, T., Hojrup, P., Keene, D.R., Bachinger, H.P., Brennan, R.G., and Valentin-Hansen, P. (2002) Hfq: a bacterial Sm-like protein that mediates RNA-RNA interaction. *Mol. Cell*, **9**, 23–30.

165 Vytvytska, O., Moll, I., Kaberdin, V.R., von Gabain, A., and Blasi, U. (2000) Hfq (HF1) stimulates ompA mRNA decay by interfering with ribosome binding. *Genes Dev.*, **14**, 1109–1118.

166 Zhang, A., Wassarman, K.M., Ortega, J., Steven, A.C., and Storz, G. (2002) The Sm-like Hfq protein increases OxyS RNA interaction with target mRNAs. *Mol. Cell*, **9**, 11–22.

167 Muffler, A., Traulsen, D.D., Fischer, D., Lange, R., and Hengge-Aronis, R. (1997) The RNA-binding protein HF-I plays a global regulatory role which is largely, but not exclusively, due to its role in expression of the sigmaS subunit of RNA polymerase in *Escherichia coli*. *J. Bacteriol.*, **179**, 297–300.

168 Tsui, H.C., Leung, H.C., and Winkler, M.E. (1994) Characterization of broadly pleiotropic phenotypes caused by an hfq insertion mutation in *Escherichia coli* K-12. *Mol. Microbiol.*, **13**, 35–49.

169 Achsel, T., Stark, H., and Luhrmann, R. (2001) The Sm domain is an ancient RNA-binding motif with oligo(U) specificity. *Proc. Natl. Acad. Sci. U.S.A.*, **98**, 3685–3689.

170 Collins, B.M., Harrop, S.J., Kornfeld, G.D., Dawes, I.W., Curmi, P.M., and Mabbutt, B.C. (2001) Crystal structure of a heptameric Sm-like protein complex from archaea: implications for the structure and evolution of snRNPs. *J. Mol. Biol.*, **309**, 915–923.

171 Mura, C., Cascio, D., Sawaya, M.R., and Eisenberg, D.S. (2001) The crystal structure of a heptameric archaeal Sm protein: implications for the eukaryotic snRNP core. *Proc. Natl. Acad. Sci. U.S.A.*, **98**, 5532–5537.

172 Toro, I., Thore, S., Mayer, C., Basquin, J., Seraphin, B., and Suck, D. (2001) RNA binding in an Sm core domain: X-ray structure and functional analysis of an archaeal Sm protein complex. *EMBO J.*, **20**, 2293–2303.

173 Sauter, C., Basquin, J., and Suck, D. (2003) Sm-like proteins in Eubacteria: the crystal structure of the Hfq protein from *Escherichia coli*. *Nucleic Acids Res.*, **31**, 4091–4098.

174 Valentin-Hansen, P., Eriksen, M., and Udesen, C. (2004) The bacterial Sm-like protein Hfq: a key player in RNA transactions. *Mol. Microbiol.*, **51**, 1525–1533.

175 Nielsen, J.S., Boggild, A., Andersen, C.B., Nielsen, G., Boysen, A., Brodersen, D.E., and Valentin-Hansen, P. (2007) An Hfq-like protein in archaea: crystal structure and functional characterization of the Sm protein from *Methanococcus jannaschii*. *RNA*, **13**, 2213–2223.

176 Chekulaeva, M., Hentze, M.W., and Ephrussi, A. (2006) Bruno acts as a dual repressor of oskar translation, promoting mRNA oligomerization and formation of silencing particles. *Cell*, **124**, 521–533.

5
Structure, Function, and Biogenesis of Small Nucleolar Ribonucleoprotein Particles

Katherine S. Godin and Gabriele Varani

5.1
Introduction

Two families of small nucleolar ribonucleoprotein (snoRNP) particles are responsible for the posttranscriptional modification of RNA in eukaryotes and archea [1]. Each RNP is composed of a noncoding RNA (small nucleolar RNA (snoRNA)) and a set of associated proteins common to each of two functional and structural classes. Both the box H/ACA and box C/D snoRNPs localize predominately to the nucleolus, a subcompartment of the nucleus where ribosome assembly occurs. They contain an enzymatic activity responsible for the modification of specific nucleotides, predominately within ribosomal RNA precursors (pre-rRNA), as well as for pre-rRNA nucleolytic cleavage (for review, see [2]). The two most common posttranscriptional modifications (isomerization of uridine to pseudouridine, Ψ, and methylation of the ribose sugar 2′hydroxyl, Nm; Figure 5.1) are executed by the box H/ACA [3, 4] and box C/D snoRNPs [5–7], respectively.

Both modifications are widespread in the mature ribosomal RNAs of all living organisms. Furthermore, they tend to cluster in highly conserved regions of the ribosome, which are generally devoid of proteins in the assembled structure but near functionally important regions [8]. These modifications are assumed to be important for folding, stability, and ultimately enzymatic function, though the detailed enzymatic mechanism is often unclear [9]. NMR studies of small model rRNA hairpins containing Nm and ψ indicate that these modifications do not significantly alter the structure from their unmodified counterparts [10, 11]; they enhance the thermal stability possibly through additional hydrogen bonds or base stacking interactions. Isomerization of uridine to ψ provides an additional hydrogen bond donor that can stabilize RNA–RNA or RNA–protein interactions, establishes a local conformational preference for the C3′-endo conformation of the sugar, and restricts the base to the *anti* conformation [12, 13] protecting the N1–H bond from the solvent and rigidifying the phosphodiester backbone. NMR studies of Nm-modified nucleotides also reveal a conformational preference for the C3′-endo conformation and increased structural rigidity [10], while the added methyl group is predicted to promote base stacking interactions. Unfortunately,

Posttranscriptional Gene Regulation: RNA Processing in Eukaryotes, First Edition. Edited by Jane Wu.

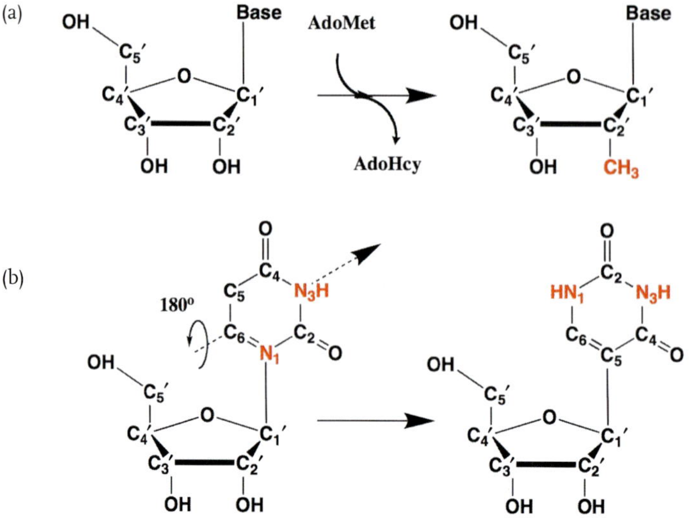

Figure 5.1 (a) Methylation of the target nucleotide 2′OH ribose is assisted by the conversion of S-adenosymethionine (AdoMet) into S-adenosylhomocysteine (AdoHcy). (b) Isomerization of uridine into pseudouridine (Ψ) requires 180° rotation around the N3-C6 axis.

this information comes from a limited number of theoretical predictions and small model oligonucleotides; the global effect of these modifications on the structure and function of the ribosome remains to be fully established.

In yeast, defects in pseudouridylation or ribose methylation alter growth rate, the processing rate of pre-rRNA, and the activity of the ribosome [8, 9]. Although the loss of individual modifications generally has no noticeable effect on growth [14], mutations that deactivate the snoRNP catalytic site and therefore generate a global loss of Ψ or Nm modifications lead to near lethal growth phenotypes [15]. The loss of multiple modified nucleosides, especially within the same functionally significant region, impairs growth and affects the ribosome in different combinatorial ways, suggesting that the more subtle effects of the individual modifications have a synergistic impact when combined [16].

A distinct subset of snoRNPs modifies small nuclear RNPs (snRNPs) instead of rRNA, within functionally significant regions as well. This specialized group of RNPs localizes within Cajal bodies, a nuclear compartment that is the site for snRNP maturation [17]. Small Cajal body RNPs (scaRNPs) are differentiated from snoRNPs by a unique Cajal body-specific localization sequence within the guide scaRNA, known as the CAB box [18]. The scaRNAs subclass includes box C/D scaRNAs, box H/ACA scaRNAs, chimeric C/D-H/ACA scaRNAs, and human telomerase (TR), a chimeric enzyme H/ACA scaRNA. Human TR contains an H/ACA-like domain at its 3′ end that recruits two sets of the standard H/ACA snoRNP proteins [19]. However, the TR scaRNA has no known function or pseudouridylation target [20] and is incapable of enzymatic modification (as discussed in [21]). *In vitro*, the domain is dispensable for TR activity, yet it is required *in vivo* for the processing, stability, and nuclear localization of TR [19, 22]. Thus, this domain is likely to be responsible for the subnuclear localization of TR [22].

Ribosomal RNA modification by Ψ and Nm occurs in all three phylogenic kingdoms, but cellular compartmentalization of snoRNPs is unique to eukaryotes. In archaea, modifications of both rRNA and transfer RNA (tRNA) is executed in the cytoplasm by RNP complexes containing C/D or H/ACA small RNA (sRNA) [23]. Eukaryotic and archaea sno/small RNP (sRNP) proteins share striking sequence homology, suggesting that these complexes are ancient and evolve from a common ancestor, and that their function has been conserved throughout evolution [23]. Archaeal sRNA and their protein components are typically smaller than their eukaryotic snoRNP, perhaps defining the minimal characteristics necessary for both classes to function. In eubacteria, Ψ and Nm are generated by stand-alone enzymes that lack an RNA component (see the review in [24, 25]), but retain strong structural homology to the sno/sRNA enzyme. These stand-alone enzymes account for only ~14 rRNA modifications in *Escherichia coli*, while the RNA-guided snoRNPs produce ~100 modifications in *Saccharomyces cerevisiae* and ~200 in humans. Clearly, the addition of the RNA component has allowed for an expansion of the specificity of these particles.

In mammals, this unique mode of specificity has allowed for new regulatory roles as evident in an intriguing subset of brain-specific snoRNAs discovered through high-throughput screens for noncoding RNAs in mice [26]. Unlike the canonical snoRNAs, brain-specific snoRNAs lack complementarities to pre-rRNA or pre-snRNAs, thus the “orphan snoRNA” designation. Insight into the biological functions is limited but, surprisingly, one orphan C/D snoRNA, MBII-52, targets the serotonin receptor 2C pre-mRNA and assembles into a bona fide snoRNP [27, 28] that inhibits nucleolar ADAR2 A-to-I RNA editing upon 2′O-methylation of its substrate [28].

Numerous studies over the past decade have aimed to elucidate the molecular basis for the functions of both sno/sRNPs classes and of their protein and RNA constituents. For a detailed discussion of the sno/sRNP components and of past structural accomplishment, refer to the review in [29]. Within the last two years, these early piecemeal achievements have come together in the most complete structures of box C/D and box H/ACA sRNP particles [30–32], allowing for an understanding of each component's role in enzymatic activity.

5.2
The Guide RNA

The specificity of both the C/D and H/ACA classes is determined by sequence complimentary between the target substrate and the snoRNA component (Figure 5.2). The presence of these complementary sequences identifies the target and orients it within the active site. In this way, these RNPs can then be specific for many different target nucleotides by using a common set of enzymes. In addition

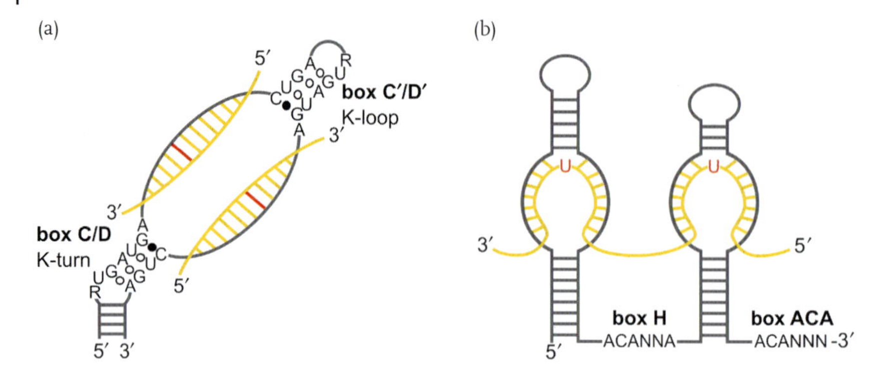

Figure 5.2 Secondary structure of the box C/D and box H/ACA snoRNA and their target RNAs. (a) The box C/D snoRNA (gray) contains two C/D boxes found at the termini and in the center of the snoRNA sequence. The guide sequence pairs with target (yellow) positions the target (red bond) five nucleotides from either D box. (b) The box H/ACA snoRNA (gray) contains the characteristic ACA sequence located three nucleotides from the 3′-terminus and the box H in the hinge between the adjacent hairpins. The target uridine (U) is positioned 14–16 nucleotides from each sequence by the pseudouridylation pocket.

to the sequences that specify the target, both classes of snoRNAs contain unique sequence elements that are highly conserved between archaea and eukaryotes and are necessary for protein recognition and assembly of the mature RNP [29]. Box C/D sno/sRNAs are defined by two sets of conserved sequence elements known as box C and box D that are joined by two guide sequences and fold into asymmetric K-turn motifs: a canonical K-turn containing the terminal stem-loop designated box C/D and a noncanonical K-loop located in the middle of the sno/sRNA known as box C′/D′ (Figure 5.2a). The box H/ACA snoRNA consists of two conserved hairpin motifs each encoding a guide sequence and two similar sequence elements, box ACA at the 3′ end and box H linking the two hairpins; archaeal box H/ACA sRNAs contain a single hairpin with a 3′ end box ACA motif (Figure 5.2b).

5.3
The Core sno/sRNP Proteins

Both classes of sno/sRNAs serve as platforms for the assembly of a group of highly conserved class-specific proteins that recognize the sno/sRNA's unique sequences and architectures (Figures 5.2a and 5.3a). Each set of class-specific proteins includes an enzyme and auxiliary proteins that participate in the stabilization of the snoRNA and facilitate the sno/sRNA enzyme interaction. The initial identification of the C/D and H/ACA sno/sRNP enzymatic components, fibrillarin and Cbf5, respectively, was based on sequence homology with known stand-alone methylases [24] and Ψ synthases [33] and verified by mutations in yeast that pro-

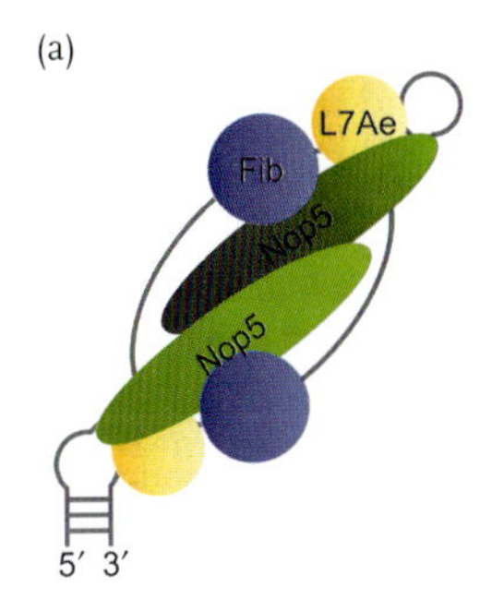

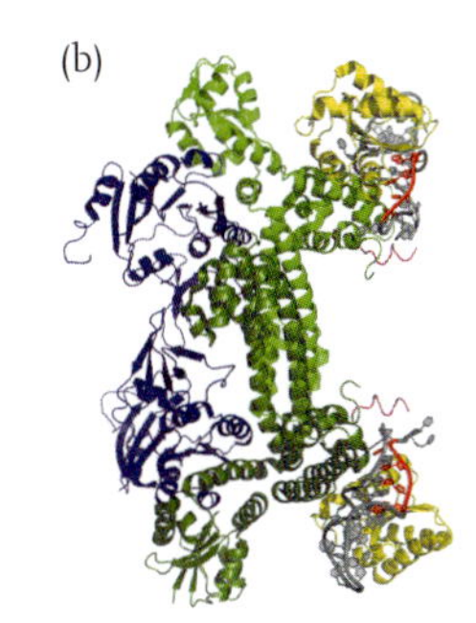

Figure 5.3 Architecture of class C/D sno/sRNPs. (a) In the bipartite model, archaeal C/D snoRNP proteins are symmetrically bound at each C/D element. L7Ae (yellow) associates with the Nop5 (varied green for clarity) C-terminus and recognizes the classical K-turns at both C/D boxes. Fibrillarin (blue) binds the Nop5 N-terminus, while the Nop5 coiled-coil domain separates the protein-bound C/D boxes. (b) Structure of the half-mer bound sRNP (3ID5) resembles the bipartite model. A new α-helical GAEK segment (hot pink) protrudes from the Nop5 C-terminal domain separating the two guide strands highlighted in orange. (c) In the di-sRNP model, four sets of proteins share two sRNAs and assume a eukaryotic-like asymmetric arrangement where L7Ae dissociates from one C/D element to allow for productive Nm.

duced defects in snoRNP processing [15, 34]. Although fibrillarin and Cbf5 are the only catalytic components, snoRNP activity, stability, and nucleolar localization require all the class-specific snoRNP proteins (reviewed in [35]).

Eukaryotic C/D snoRNAs associate with the methylase (fibrillarin), with a K-turn binding protein 15.5-kDa protein (Snu13 in yeast), and with two other proteins known as Nop56 and Nop58 that interact with fibrillarin and facilitate recognition of the C/D and C′/D′ boxes. Snu13 and 15.5-kDa proteins share a common ancestor (archaeal protein L7Ae), a K-turn binding component of both C/D and H/ACA sRNPs, the spliceosome, and the ribosome [36]. Eukaryotic Nop56 and Nop58 are very similar proteins and appear to have diverged from a common archaeal ancestor (Nop5, also known as Nop56/58) [37]. Nop56 is the only component that is dispensable for activity *in vitro* [38]; every other protein is required [39, 40]. Both Nop56 and Nop58 proteins are essential in yeast, indicating that their functions are not redundant despite their similarity.

Box H/ACA snoRNAs associate with four unique proteins as well: Cbf5, Nop10, Nhp2, and Gar1. All are essential for viability in yeast, and all but Gar1 are necessary for RNP stability *in vivo* [41–47]. Nop10, the smallest of the four proteins, has no significant sequence or structural homology to other known proteins; in combination with Cbf5, it forms the minimal snoRNP necessary for pseudouridylation *in vitro* [48]. Nhp2 is homologous to the C/D protein 15.5 kDa/Snu13 but lacks specificity for K-turns [49] and in archaea, Nhp2 is replaced by ribosomal protein L7Ae (as in C/D snoRNPs).

5.4
Assembly and Structural Organization of sno/sRNPs

The catalytic mechanism of both fibrillarin and Cbf5 must be analogous to those of the corresponding stand-alone bacterial homologs, and indeed, the archaeal Cbf5 can isomerize U55 of tRNA, like TruB, as a stand-alone enzyme. However, Cbf5 and fibrillarin are generally unable to modify a substrate on their own even when the guide snoRNA is present [50, 51]. Structures of the complete substrate-bound H/ACA sRNP and analysis of the catalytic activity of these particles suggest strong mechanistic similarities to TruB and are the first to describe the contribution of each protein to H/ACA sno/sRNP activity. Our understanding of box C/D sno/sRNPs structure and function is more limited but rapidly progressing along a similar trajectory from biochemical assembly studies to recent electron microscopy (EM) [31, 52] and X-ray crystallography structures [32, 53].

5.5
Asymmetric Assembly, Structure, and Activity of the Box C/D

5.5.1
In vitro Reconstitution of C/D sno/sRNPs

The methylation guide sno/sRNPs in archaea and eukaryotes are fundamentally similar in structure and function. *In vitro*, reconstitution of active archaeal C/D sRNPs follows an ordered stepwise process initiated by the L7Ae recognition of the distinct K-turn within the C/D motif [36, 51] and followed by the cooperative binding of L7Ae to the C′/D′ domain [54]. Nop5 and fibrillarin assemble on the sRNA following these initial interactions [55]; since fibrillarin is incapable of independent association with RNA, it is likely introduced to the RNP as a heterodimer with Nop5 *in vivo* [56–58]. As a result, archaeal sRNPs are assumed to be organized symmetrically with all three core proteins located at each C/D element (Figure 5.3a). However, binding of L7Ae at only one C/D unit is sufficient for activity [41], suggesting that box C/D sRNPs could be functionally asymmetric.

In eukaryotic C/D snoRNPs, the four core proteins are asymmetrically distributed at each C/D element (Figure 5.3a). This conclusion was reached from *in vivo* [40, 42, 55] and *in vitro* [43] reconstitution experiments, suggesting that Nop58, Nop56, and 15.5 kDa/Snu13 have exclusive interactions at either one or the other C/D unit. Initiation of snoRNP assembly begins with the L7Ae homolog, 15.5 kDa, association with the conserved C/D K-turn; recruitment of a second 15.5-kDa protein to the C′/D′ site may occur instead through interactions with Nop56 or fibrillarin. The position of Nop58 and Nop56 in the particle were mapped by UV cross-linking of U25 snoRNPs reconstituted in *Xenopus* oocytes [42], showing that Nop58 associates with the C box and Nop56 with the C′ box. On the other hand, fibrillarin associates with both D boxes that lie near the methylation site, consistent with the requirement for one enzyme per target substrate [55].

5.5.2 The Initial Bipartite C/D sRNP Model

At first, archaeal box C/D sRNPs were thought to contain two sets of core proteins, L7Ae, Nop5, and fibrillarin per one bipartite guide sRNA (Figure 5.3a). One set of core proteins associated with each of the C/D and C′/D′ boxes and the two protein assembly sites were connected by homodimerization of the Nop5 coiled-coil domain. However, inconsistencies of this model were apparent in the first crystal structures of Nop5/fibrillarin heterotetramers [56, 57]. Nop5 is predominately α-helical and displays three functionally distinct domains: the N-terminus, C-terminus, and coiled-coiled domains. Fibrillarin recognition occurs through the Nop5 N-terminal domain (NTD), but the crystal structures from *Archaeoglobus fulgidus* (*Af*) and *Pyrococcus furiosus* (*Pf*) show that the orientation of this fibrillarin-bound domain varies. The strong structural similarities and molecular dynamics analysis suggest that this module is inherently flexible [56, 57]. The coiled-coil domains self-associate forming a four-helix bundle that links the heterodimers with the fibrillarin-bound Nop5 N-terminus, capping the bundled ends; the Nop5 C-terminus is unoccupied but known to mediate recruitment to the L7Ae-bound guide RNA and participate in catalysis by facilitating Ado-Met binding [56]. These initial structures revealed two inconsistencies. First, the sRNA C/D elements were separated by ~25–35A, a distance longer than the predicted 10–12 base paired A-form guide:substrate duplex. Second, the catalytic sites, separated by ~80–90A in the RNA-free Nop5-fibrillarin crystal structures, are too far from the expected site of modification [58].

The first crystal structure of a complete *Sulfolobus solfataricus* (*SS*) C/D sRNP bound to a half-mer C/D sRNA, a single C/D box, and a guide sequence, was consistent with the bipartite layout, but it was also enzymatically inactive (Figure 5.3b) [53]. The Nop5 dimer interface and orientation of the C-terminal domain (CTD) were superimposable with the previous Nop5 structures, except for the variably oriented fibrillarin-bound Nop5 NTD. As expected, L7Ae recognized the K-turn in the same way as the known L7Ae-bound K-turn RNA crystal structures [44, 45]. Nop5 recognition of the K-turn may be mediated by two highly conserved CTD residues (Q296 and R339 in *SS*) that specifically recognize the box C and box D sequences. The fact that the Nop5 CTD alone does not bind either L7Ae or the K-turn suggests the L7Ae interaction surface is incomplete and that L7Ae likely induces conformational changes [44], perhaps aligning the box C and box D for Nop5 recognition. In the presence of the bound half-mer sRNA, a new Nop5 CTD α-helical motif containing a GAEK sequence separates the two guide strands emerging from the sRNA K-turn (Figure 5.3b). This tetramer motif is conserved from Archaea to Eukarya, essential for activity, and UV cross-links without sequence specificity to the single-stranded sRNA guide sequence downstream of the box C and C′. From these observations, it seems likely that Nop5 positions the guide sequences for target RNA hybridization and subsequent modification. Furthermore the guide sequence downstream of the box C and C′ are necessary for sRNP assembly [46] but insufficient for enzymatic activity with half-mer sRNPs [53].

5.5.3
The di-C/D sRNP Model and Asymmetric Activity

Recently, a 27-Å EM structure of a fully active reconstituted archaeal C/D sRNP presented a solution to the sRNA placement problem and challenged the bipartite model by providing evidence of an unexpected and conserved di-sRNP particle (Figure 5.3c) [31, 46, 52]. Docking of the Nop5/fibrillarin and L7Ae crystal structures showed that the proteins assembled similarly to the bipartite model, but four sets of core proteins were found with two sRNAs. Although RNA placement could not be definitively determined, the EM density not occupied by protein suggests a surprising sRNA orientation. As in the bipartite model, each sRNA C/D motif associates with one set of core proteins; however, the sRNA is not parallel with the Nop5 coiled-coil domain bound within the same Nop5/fibrillarin heterotetramer, but it is instead perpendicular and bound by a second heterotetramer, while the remaining Nop5/fibrillarin halves are equally occupied by an additional sRNA (Figure 5.3c).

Additional EM studies and complementary activity assays identified structural distortions arising from half-mer sRNAs and deletions of the Nop5 coiled-coil domain that correlate with impaired Nm activity, suggesting that both Nop5 and the full sRNA optimally position fibrillarin and the target nucleotide [52]. However, even with a complete C/D sRNA in place, fibrillarin remains far from the guide sequence and from the expected site of modification. Therefore, flexibility of the filbrillarin-Nop5 NTD, as supported by previous crystal structures [56, 57], likely drives the correct placement of the enzyme at the modification site.

The most recent crystal structure and biochemical studies of a substrate-bound half-mer sRNP demonstrate the importance of fibrillarin conformational changes and present a provocative model for enzymatic activity [32]. The substrate-bound half-mer C/D sRNP resembles the substrate-free structure and further shows the GAEK α-helical motif wedged between the guide and substrate RNA directing the guide–substrate RNA duplex along the neighboring protein surface. This mode of substrate binding fits within the di-sRNP EM density showing two substrate-bound half-mer sRNPs complementing each other as predicted. Validation of the di-sRNP model also indicates that the position of the target site is more easily accessed by fibrillarin in *trans*; this is inconsistent with the bipartite model, where the box C/D target nucleotide would be modified by fibrillarin associated in *cis* with the box C/D bound Nop5/L7Ae. Furthermore, each guide sRNA binds one substrate while the *trans* L7Ae releases its K-turn feature, resulting in an asymmetric holoenzyme with remarkable resemblance to the predicted architecture of eukaryotic C/D snoRNP particles [42, 55] (Figure 5.3c). These hypotheses present an exciting new interpretation of the box C/D sno/sRNP activity; however, a structure of the complete particle is still needed to fully understand how the individual pieces combine together to activate sno/sRNA-guided methylation.

5.6 The Box H/ACA RNP Structure and Assembly of the Eukaryotic RNP

5.6.1 The Complete Substrate-Bound Box H/ACA

Recent crystal structures from Liang *et al.* [47] and Duan *et al.* [30] of the nearly complete and then of the complete substrate-bound H/ACA sRNP were remarkable achievements finally reconciling the target U placement within the active site. The organization of all four core proteins around the guide–substrate RNA duplex is consistent with previous structures of RNA-free and -bound complexes (Figure 5.4a,b) [45, 48, 49, 59–62]. Guide RNA recognition involves a composite surface formed by Cbf5, Nop10, and L7Ae; the ACA motif is positioned at the pseudouridine synthase and archaeosine tRNA-guanine transglycosylase (PUA) domain of Cbf5, and the apical stem loop is recognized by Nop10 and L7Ae. The recognition of the ACA signature sequence and lower stem by the PUA domain and the upper stem by L7Ae and Nop10 orients the pseudouridylation pocket across the catalytic face of Cbf5. Upon substrate binding, the pseudouridylation pocket of the guide RNA becomes ordered and makes additional contacts with conserved residues of Cbf5. Mutation of certain residues within this region is associated with dyskeratosis congenita and supports a link between this disease and ribosome modification [59, 63]. L7Ae not only binds at the predicted K-loop found at the apical

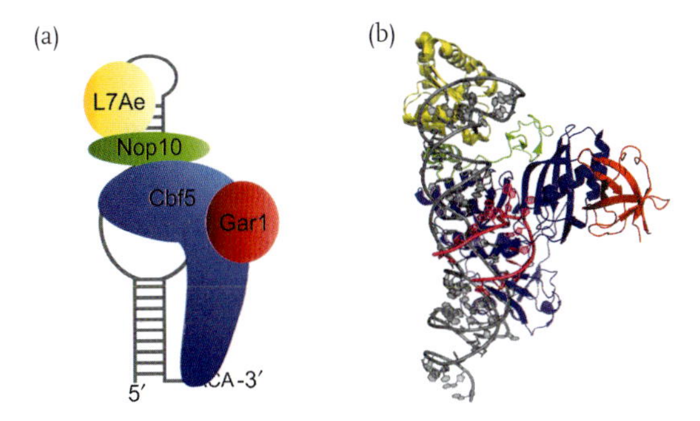

Figure 5.4 Architecture of box H/ACA sno/sRNPs. (a) Cartoon model of the assembled sRNP. Cbf5 (blue) recognizes box ACA and with Nop10 (green) secures the Ψ pocket of the guide sRNA (gray). L7Ae (yellow) contacts Nop10 and recognizes the K-turn. Gar1 (red) is far from the RNA and interacts solely with Cbf5. (b) Structure of the complete substrate-bound H/ACA sRNP (3HAY). The substrate RNA is colored in magenta, while the other components are colored as in (a).

stem loop of the sRNAs but also interacts with the Nop10 C-terminus. The highly conserved linker region of Nop10 makes intimate contacts with Cbf5 and with the apical stem of the sRNA, as originally proposed based on the structure of the Cbf5–Nop10 heterodimer [43]. In the substrate-bound full complex, Gar1 interacts with Cbf5 at the same position as in the other known Gar1-containing complexes [48, 59] and does not interact with either the guide or substrate RNA, nor does it appear to directly contribute to an RNA structural rearrangement upon substrate loading.

The substrate-bound guide sRNA shares similar features with structures of the protein-free and -bound RNA complexes [61, 64, 65], but the conformation, substrate docking, and H/ACA sRNP functionality vary depending on the bound proteins. In general, sequence-specific recognition constrains the substrate to adopt a U-shaped conformation and to form a unique three-helical junction with the guide sRNA where the two pseudocontinuous helices of the duplexed pseudouridylation pocket coaxially stack with the guide upper and lower stems [58]. The impact of the core proteins on the structural arrangement of the substrate sRNA has been widely inferred from biochemical assays reporting on ψ activity [50], and fluorescence anisotropy measuring conformational changes experienced by the target U [62]. Predictions from these earlier studies that L7Ae rearranges the guide sRNA to properly direct the substrate were recently confirmed with L7Ae-free and -bound Cbf5–Nop10 sRNP crystal structures [47]. In these enzymatically active H/ACA sRNPs, L7Ae binds to and anchors the upper stem causing a rotation of the guide–substrate helix that productively places the substrate deep in the active site. Additionally, L7Ae and substrate binding allowed for the flipping of the substrate base, as observed in the stand-alone enzyme TruB, via highly conserved histidine residues and substrate recognition by the Cbf5 thumb loop.

Cbf5 is closely related to TruB, a "stand-alone" *E. coli* ψ synthase that isomerizes U55 within the T-loop of all elongator tRNAs [49, 60, 66]. Structure determination of *E. coli* TruB bound to the TψC stem loop of the tRNA revealed that this ψ synthase positions its substrate uridine in its active site by flipping the base and inserting a histidine residue into the space vacated by it [66]. This histidine is conserved in Cbf5 (H80 in *Pf* Cbf5), and mutational analysis indicates that it is functionally important. Comparison of box H/ACA structures indicates that Cbf5 employs another histidine that is spatially close to His80 and His63 (which is conserved in Cbf5 orthologs but not in TruB orthologs), and that both histidines play an important role in substrate recognition and positioning in the active site. A second Cbf5 feature shared with TruB, the so-called thumb loop, additionally secures the substrate RNA on the face opposite to residues His80 and His63. No interactions occur between the thumb and the guide RNA, suggesting the thumb is dedicated to substrate recruitment. The interactions between the thumb and substrate RNA are largely nonspecific and therefore consistent with the need to accommodate a diverse family of substrate RNA sequences recognized by each guide ψ pocket.

With the guide and substrate RNAs in place, it is apparent that Gar1 does not participate in direct recognition of either RNA; instead, Gar1 is predicted to modulate the interaction between substrate RNA and the thumb motif of Cbf5. The residues within this loop of Cbf5 are disordered in the structures of Cbf5 and of the Nop10/Cbf5 heterodimer, but form additional interactions with a hydrophobic patch on Gar1 [59, 61]. Much like the substrate RNA, the thumb loop conformation depends on the sRNP constituents. Crystal structures capturing different thumb loop positions may define its range of motions or at least its accessible conformational space, from open in the Gar1/Nop10/Cbf5 substrate-bound RNP [61] to closed in the complete and Gar1-free sRNP [30]. A clearer role for Gar1 is emerging from biochemical assays, mutagenesis, and enzyme kinetic analysis of single-site Gar1 mutants that appear normal in single-turnover reactions but show a reduced turnover rate in multiple-turnover reactions, strongly suggesting that Gar1 facilitates product release. It is likely that Gar1 also contributes to substrate loading and that the interaction of the substrate with Gar1 may be critical for modulating target RNA loading and release.

5.6.2 Structure of Eukaryotic H/ACA snoRNPs

H/ACA sRNAs (and snoRNAs) that contain multiple stem-loop domains often have pseudouridylation guide sequences within each domain, suggesting that each stem loop is catalytically active and associates with a full complement of H/ACA sno/sRNP proteins. The multipartite architecture of these H/ACA sno/sRNPs was first observed from low-resolution EM images of eukaryotic particles [67] and was confirmed by RNA protection patterns consistent with the core interactions observed in archaea [68]. However, there are additional nucleotides 5′ to the stem loops that appear to interact with eukaryotic Cbf5 and imply that unique specificities and/or additional interparticle interactions may distinguish the eukaryotic snoRNP proteins from their archaeal counterparts.

5.6.3 Assembly of Eukaryotic H/ACA snoRNPs

In yeast and mammals, H/ACA snoRNP biogenesis is assisted by two assembly factors, Shq1 and Naf1 [69]. Shq1 associates with Cbf5 early during snoRNA transcription, possibly as a Cbf5-specific chaperone [70], but is displaced upon recruitment of Nap57 (Cbf5 in rat), Nhp2, Nop10, and Naf1 (a Gar1 structural homolog [71]) to the transcription site. Association of Nhp2 with Cbf5 requires the prior formation of the Nap57–Nop10 heterodimer [72], indicating that Nop10 mediates the association of Nhp2 with Nap57/Cbf5 analogous to the Nop10–L7Ae interaction observed in the structure of the archaeal H/ACA sRNP. The differences between archaea and eukaryotes are also apparent in the *in vitro* assembly of H/ACA snoRNPs that require the formation of the heterotrimeric complex of Nap57, Nop10, and Nhp2 (excluding Gar1) [72]. These three proteins are homologous to the archaeal proteins that contact the guide RNA in the crystal structure, suggesting that snoRNA recognition may require all three proteins. Alternatively, the complex may induce the appropriate kinked RNA structure on the H/ACA snoRNAs, or conformationally alter Nhp2 so that it can recognize the atypical K-turn found in eukaryotic snoRNAs.

5.7 Summary

Our current understanding of the structure and function relationship of box C/D and H/ACA snoRNPs has greatly benefited from genetic and biochemical studies delineating the interactions between the sno/sRNP components. Most significant have been the tremendous structural advances describing each individual protein component in atomic detail and their organization within the assembled snoRNP particle. However, outstanding questions remain. First, the structure of the enzymatically active C/D snoRNP remains unknown; second, the catalytic mechanism of each enzyme remains to be fully delineated; third, the structure and assembly pathway of the eukaryotic snoRNPs remain unclear. An even greater challenge is integrating the description of snoRNP maturation into known cellular processes, which still requires the identification of the factors that distinguish snoRNAs from other RNA polymerase II transcripts and an understanding of what these factors do during snoRNP processing, assembly, and nuclear trafficking. Last but not least, the impact of both Ψ and Nm modifications on the function of the ribosome and spliceosome remains far from clear. Thus, the sno/sRNP structural dissection presented here marks the tremendous achievement defining these two new classes of noncoding RNAs to date; however, to fully understand their biological function, much remains to be done.

References

1 Balakin, A.G., Smith, L., and Fournier, M.J. (1996) The RNA world of the nucleolus: two major families of small RNAs defined by different box elements with related functions. *Cell*, **86**, 823–834.

2 Bachellerie, J.P., Cavaille, J., and Huttenhofer, A. (2002) The expanding snoRNA world. *Biochimie*, **84**, 775–790.

3 Ni, J., Tien, A.L., and Fournier, M.J. (1997) Small nucleolar RNAs direct site-specific synthesis of pseudouridine in ribosomal RNA. *Cell*, **89**, 565–573.

4 Ganot, P., Bortolin, M.L., and Kiss, T. (1997) Site-specific pseudouridine formation in preribosomal RNA is guided by small nucleolar RNAs. *Cell*, **89**, 799–809.

5 Kiss-Laszlo, Z., Henry, Y., Bachellerie, J.P., Caizergues-Ferrer, M., and Kiss, T. (1996) Site-specific ribose methylation of preribosomal RNA: a novel function for small nucleolar RNAs. *Cell*, **85**, 1077–1088.

6 Nicoloso, M., Qu, L.H., Michot, B., and Bachellerie, J.P. (1996) Intron-encoded, antisense small nucleolar RNAs: the characterization of nine novel species points to their direct role as guides for the 2'-O-ribose methylation of rRNAs. *J. Mol. Biol.*, **260**, 178–195.

7 Tycowski, K.T., Shu, M.D., and Steitz, J.A. (1993) A small nucleolar RNA is processed from an intron of the human gene encoding ribosomal protein S3. *Genes Dev.*, **7**, 1176–1190.

8 Decatur, W.A., Liang, X.H., Piekna-Przybylska, D., and Fournier, M.J. (2007) Identifying effects of snoRNA-guided modifications on the synthesis and function of the yeast ribosome. *Methods Enzymol.*, **425**, 283–316.
9 King, T.H., Liu, B., McCully, R.R., and Fournier, M.J. (2003) Ribosome structure and activity are altered in cells lacking snoRNPs that form pseudouridines in the peptidyl transferase center. *Mol. Cell*, **11**, 425–435.
10 Kawai, G., Yamamoto, Y., Kamimura, T., Masegi, T., Sekine, M., Hata, T., Iimori, T., Watanabe, T., Miyazawa, T., and Yokoyama, S. (1992) Conformational rigidity of specific pyrimidine residues in tRNA arises from posttranscriptional modifications that enhance steric interaction between the base and the 2'-hydroxyl group. *Biochemistry*, **31**, 1040–1046.
11 Kim, N.K., Theimer, C.A., Mitchell, J.R., Collins, K., and Feigon, J. (2010) Effect of pseudouridylation on the structure and activity of the catalytically essential P6.1 hairpin in human telomerase RNA. *Nucleic Acids Res.*, **38**, 6746–6756.
12 Arnez, J.G. and Steitz, T.A. (1994) Crystal structure of unmodified tRNA(Gln) complexed with glutaminyl-tRNA synthetase and ATP suggests a possible role for pseudo-uridines in stabilization of RNA structure. *Biochemistry*, **33**, 7560–7567.
13 Yarian, C.S., Basti, M.M., Cain, R.J., Ansari, G., Guenther, R.H., Sochacka, E., Czerwinska, G., Malkiewicz, A., and Agris, P.F. (1999) Structural and functional roles of the N1- and N3-protons of psi at tRNA's position 39. *Nucleic Acids Res.*, **27**, 3543–3549.
14 Lowe, T.M. and Eddy, S.R. (1999) A computational screen for methylation guide snoRNAs in yeast. *Science*, **283**, 1168–1171.
15 Zebarjadian, Y., King, T., Fournier, M.J., Clarke, L., and Carbon, J. (1999) Point mutations in yeast CBF5 can abolish *in vivo* pseudouridylation of rRNA. *Mol. Cell Biol.*, **19**, 7461–7472.
16 Liang, X.H., Liu, Q., and Fournier, M.J. (2007) rRNA modifications in an intersubunit bridge of the ribosome strongly affect both ribosome biogenesis and activity. *Mol. Cell*, **28**, 965–977.
17 Verheggen, C., Lafontaine, D.L., Samarsky, D., Mouaikel, J., Blanchard, J.M., Bordonne, R., and Bertrand, E. (2002) Mammalian and yeast U3 snoRNPs are matured in specific and related nuclear compartments. *EMBO J.*, **21**, 2736–2745.
18 Richard, P., Darzacq, X., Bertrand, E., Jady, B.E., Verheggen, C., and Kiss, T. (2003) A common sequence motif determines the Cajal body-specific localization of box H/ACA scaRNAs. *EMBO J.*, **22**, 4283–4293.
19 Mitchell, J.R., Cheng, J., and Collins, K. (1999) A box H/ACA small nucleolar RNA-like domain at the human telomerase RNA 3' end. *Mol. Cell Biol.*, **19**, 567–576.
20 Chen, J.L., Blasco, M.A., and Greider, C.W. (2000) Secondary structure of vertebrate telomerase RNA. *Cell*, **100**, 503–514.
21 Meier, U.T. (2005) The many facets of H/ACA ribonucleoproteins. *Chromosoma*, **114**, 1–14.
22 Lukowiak, A.A., Narayanan, A., Li, Z.H., Terns, R.M., and Terns, M.P. (2001) The snoRNA domain of vertebrate telomerase RNA functions to localize the RNA within the nucleus. *RNA*, **7**, 1833–1844.
23 Omer, A.D., Ziesche, S., Decatur, W.A., Fournier, M.J., and Dennis, P.P. (2003) RNA-modifying machines in archaea. *Mol. Microbiol.*, **48**, 617–629.
24 Cheng, X. and Roberts, R.J. (2001) AdoMet-dependent methylation, DNA methyltransferases and base flipping. *Nucleic Acids Res.*, **29**, 3784–3795.
25 Hamma, T. and Ferre-D'Amare, A.R. (2006) Pseudouridine synthases. *Chem. Biol.*, **13**, 1125–1135.
26 Cavaillé, J., Buiting, K., Kiefmann, M., Lalande, M., Brannan, C.I., Horsthemke, B., Bachellerie, J.P., Brosius, J., and Hüttenhofer, A. (2000) Identification of brain-specific and imprinted small nucleolar RNA genes exhibiting an unusual genomic organization. *Proc. Natl. Acad. Sci. U.S.A.*, **97**, 14311–14316.
27 Soeno, Y., Taya, Y., Stasyk, T., Huber, L.A., Aoba, T., and Huttenhofer, A. (2010) Identification of novel ribonucleo-

protein complexes from the brain-specific snoRNA MBII-52. *RNA*, **16**, 1293–1300.

28 Vitali, P., Basyuk, E., Le Meur, E., Bertrand, E., Muscatelli, F., Cavaille, J., and Huttenhofer, A. (2005) ADAR2-mediated editing of RNA substrates in the nucleolus is inhibited by C/D small nucleolar RNAs. *J. Cell Biol.*, **169**, 745–753.

29 Reichow, S.L., Hamma, T., Ferre-D'Amare, A.R., and Varani, G. (2007) The structure and function of small nucleolar ribonucleoproteins. *Nucleic Acids Res.*, **35**, 1452–1464.

30 Duan, J., Li, L., Lu, J., Wang, W., and Ye, K. (2009) Structural mechanism of substrate RNA recruitment in H/ACA RNA-guided pseudouridine synthase. *Mol. Cell*, **34**, 427–439.

31 Bleichert, F., Gagnon, K.T., Brown, B.A., Maxwell, E.S., Leschziner, A.E., Unger, V.M., and Baserga, S.J. (2009). A dimeric structure for archaeal box C/D small ribonucleoproteins. *Science*, **325**, 1384–1387.

32 Xue, S., Wang, R., Yang, F., Terns, R., Terns, M., Zhang, X., Maxwell, E., and Li, H. (2010) Structural basis for substrate placement by an archaeal box C/D ribonucleoprotein particle. *Mol. Cell*, **39**, 939–949.

33 Koonin, E.V. (1996) Pseudouridine synthases: four families of enzymes containing a putative uridine-binding motif also conserved in dUTPases and dCTP deaminases. *Nucleic Acids Res.*, **24**, 2411–2415.

34 Tollervey, D., Lehtonen, H., Jansen, R., Kern, H., and Hurt, E.C. (1993) Temperature-sensitive mutations demonstrate roles for yeast fibrillarin in pre-rRNA processing, pre-rRNA methylation, and ribosome assembly. *Cell*, **72**, 443–457.

35 Matera, A.G., Terns, R.M., and Terns, M.P. (2007) Non-coding RNAs: lessons from the small nuclear and small nucleolar RNAs. *Nat. Rev. Mol. Cell Biol.*, **8**, 209–220.

36 Kuhn, J.F., Tran, E.J., and Maxwell, E.S. (2002) Archaeal ribosomal protein L7 is a functional homolog of the eukaryotic 15.5 kD/Snu13p snoRNP core protein. *Nucleic Acids Res.*, **30**, 931–941.

37 Newman, D.R., Kuhn, J.F., Shanab, G.M., and Maxwell, E.S. (2000) Box C/D snoRNA-associated proteins: two pairs of evolutionarily ancient proteins and possible links to replication and transcription. *RNA*, **6**, 861–879.

38 Lafontaine, D.L. and Tollervey, D. (1999) Nop58p is a common component of the box C+D snoRNPs that is required for snoRNA stability. *RNA*, **5**, 455–467.

39 Verheggen, C., Mouaikel, J., Thiry, M., Blanchard, J.M., Tollervey, D., Bordonne, R., Lafontaine, D.L., and Bertrand, E. (2001) Box C/D small nucleolar RNA trafficking involves small nucleolar RNP proteins, nucleolar factors and a novel nuclear domain. *EMBO J.*, **20**, 5480–5490.

40 Watkins, N.J., Dickmanns, A., and Luhrmann, R. (2002) Conserved stem II of the box C/D motif is essential for nucleolar localization and is required, along with the 15.5K protein, for the hierarchical assembly of the box C/D snoRNP. *Mol. Cell Biol.*, **22**, 8342–8352.

41 Omer, A.D., Zago, M., Chang, A., and Dennis, P.P. (2006) Probing the structure and function of an archaeal C/D-box methylation guide sRNA. *RNA*, **12**, 1708–1720.

42 Szewczak, L.B., DeGregorio, S.J., Strobel, S.A., and Steitz, J.A. (2002) Exclusive interaction of the 15.5 kD protein with the terminal box C/D motif of a methylation guide snoRNP. *Chem. Biol.*, **9**, 1095–1107.

43 Watkins, N.J., Newman, D.R., Kuhn, J.F., and Maxwell, E.S. (1998) *In vitro* assembly of the mouse U14 snoRNP core complex and identification of a 65-kDa box C/D-binding protein. *RNA*, **4**, 582–593.

44 Suryadi, J., Tran, E.J., Maxwell, E.S., and Brown, B.A., 2nd (2005) The crystal structure of the *Methanocaldococcus jannaschii* multifunctional L7Ae RNA-binding protein reveals an induced-fit interaction with the box C/D RNAs. *Biochemistry*, **44**, 9657–9672.

45 Hamma, T. and Ferre-D'Amare, A.R. (2004) Structure of protein L7Ae bound to a K-turn derived from an archaeal box H/ACA sRNA at 1.8 A resolution. *Structure*, **12**, 893–903.

46 Ghalei, H., Hsiao, H., Urlaub, H., Wahl, M., and Watkins, N. (2010) A novel Nop5-sRNA interaction that is required for efficient archaeal box C/D sRNP formation. *RNA*, **16**, 2341–2348. Epub ahead of print.

47 Liang, B., Zhou, J., Kahen, E.J., Terns, R., Terns, M.P., and Li, H. (2009) Structure of a functional ribonucleoprotein pseudouridine synthase bound to a substrate RNA. *Nat. Struct. Mol. Biol.*, **16**, 740–746.

48 Li, L. and Ye, K. (2006) Crystal structure of an H/ACA box ribonucleoprotein particle. *Nature*, **443**, 302–307.

49 Hamma, T., Reichow, S.L., Varani, G., and Ferre-D'Amare, A.R. (2005) The Cbf5-Nop10 complex is a molecular bracket that organizes box H/ACA RNPs. *Nat. Struct. Mol. Biol.*, **12**, 1101–1107.

50 Charpentier, B., Muller, S., and Branlant, C. (2005) Reconstitution of archaeal H/ACA small ribonucleoprotein complexes active in pseudouridylation. *Nucleic Acids Res.*, **33**, 3133–3144.

51 Omer, A.D., Ziesche, S., Ebhardt, H., and Dennis, P.P. (2002) *In vitro* reconstitution and activity of a C/D box methylation guide ribonucleoprotein complex. *Proc. Natl. Acad. Sci. U.S.A.*, **99**, 5289–5294.

52 Bleichert, F. and Baserga, S. (2010) Dissecting the role of conserved box C/D sRNA sequences in di-sRNP assembly and function. *Nucleic Acids Res.*, **38**, 8295–8305. Epub ahead of print.

53 Ye, K., Jia, R., Lin, J., Ju, M., Peng, J., Xu, A., and Zhang, L. (2009) Structural organization of box C/D RNA-guided RNA methyltransferase. *Proc. Natl. Acad. Sci. U.S.A.*, **106**, 13808–13813.

54 Rashid, R., Aittaleb, M., Chen, Q., Spiegel, K., Demeler, B., and Li, H. (2003) Functional requirement for symmetric assembly of archaeal box C/D small ribonucleoprotein particles. *J. Mol. Biol.*, **333**, 295–306.

55 Cahill, N.M., Friend, K., Speckmann, W., Li, Z.H., Terns, R.M., Terns, M.P., and Steitz, J.A. (2002) Site-specific cross-linking analyses reveal an asymmetric protein distribution for a box C/D snoRNP. *EMBO J.*, **21**, 3816–3828.

56 Aittaleb, M., Rashid, R., Chen, Q., Palmer, J.R., Daniels, C.J., and Li, H. (2003) Structure and function of archaeal box C/D sRNP core proteins. *Nat. Struct. Biol.*, **10**, 256–263.

57 Oruganti, S., Zhang, Y., Li, H., Robinson, H., Terns, M.P., Terns, R.M., Yang, W., and Li, H. (2007) Alternative conformations of the archaeal Nop56/58-fibrillarin complex imply flexibility in box C/D RNPs. *J. Mol. Biol.*, **371**, 1141–1150.

58 Zhang, X., Champion, E.A., Tran, E.J., Brown, B.A., 2nd, Baserga, S.J., and Maxwell, E.S. (2006) The coiled-coil domain of the Nop56/58 core protein is dispensable for sRNP assembly but is critical for archaeal box C/D sRNP-guided nucleotide methylation. *RNA*, **12**, 1092–1103.

59 Rashid, R., Liang, B., Baker, D.L., Youssef, O.A., He, Y., Phipps, K., Terns, R.M., Terns, M.P., and Li, H. (2006) Crystal structure of a Cbf5-Nop10-Gar1 complex and implications in RNA-guided pseudouridylation and dyskeratosis congenita. *Mol. Cell*, **21**, 249–260.

60 Manival, X., Charron, C., Fourmann, J.B., Godard, F., Charpentier, B., and Branlant, C. (2006) Crystal structure determination and site-directed mutagenesis of the *Pyrococcus abyssi* aCBF5-aNOP10 complex reveal crucial roles of the C-terminal domains of both proteins in H/ACA sRNP activity. *Nucleic Acids Res.*, **34**, 826–839.

61 Liang, B., Xue, S., Terns, R.M., Terns, M.P., and Li, H. (2007) Substrate RNA positioning in the archaeal H/ACA ribonucleoprotein complex. *Nat. Struct. Mol. Biol.*, **14**, 1189–1195.

62 Liang, B., Kahen, E.J., Calvin, K., Zhou, J., Blanco, M., and Li, H. (2008) Long-distance placement of substrate RNA by H/ACA proteins. *RNA*, **14**, 2086–2094.

63 Marrone, A. and Mason, P.J. (2003) Dyskeratosis congenita. *Cell. Mol. Life Sci.*, **60**, 507–517.

64 Jin, H., Loria, J.P., and Moore, P.B. (2007) Solution structure of an rRNA substrate bound to the pseudouridylation pocket of a box H/ACA snoRNA. *Mol. Cell*, **26**, 205–215.

65 Wu, H. and Feigon, J. (2007) H/ACA small nucleolar RNA pseudouridylation

pockets bind substrate RNA to form three-way junctions that position the target U for modification. *Proc. Natl. Acad. Sci. U.S.A.*, **104**, 6655–6660.

66 Hoang, C. and Ferre-D'Amare, A.R. (2001) Cocrystal structure of a tRNA Psi55 pseudouridine synthase: nucleotide flipping by an RNA-modifying enzyme. *Cell*, **107**, 929–939.

67 Watkins, N.J., Gottschalk, A., Neubauer, G., Kastner, B., Fabrizio, P., Mann, M., and Luhrmann, R. (1998) Cbf5p, a potential pseudouridine synthase, and Nhp2p, a putative RNA-binding protein, are present together with Gar1p in all H BOX/ACA-motif snoRNPs and constitute a common bipartite structure. *RNA*, **4**, 1549–1568.

68 Normand, C., Capeyrou, R., Quevillon-Cheruel, S., Mougin, A., Henry, Y., and Caizergues-Ferrer, M. (2006) Analysis of the binding of the N-terminal conserved domain of yeast Cbf5p to a box H/ACA snoRNA. *RNA*, **12**, 1868–1882.

69 Yang, P.K., Rotondo, G., Porras, T., Legrain, P., and Chanfreau, G. (2002) The Shq1p.Naf1p complex is required for box H/ACA small nucleolar ribonucleoprotein particle biogenesis. *J. Biol. Chem.*, **277**, 45235–45242.

70 Godin, K.S., Walbott, H., Leulliot, N., van Tilbeurgh, H., and Varani, G. (2009) The box H/ACA snoRNP assembly factor Shq1p is a chaperone protein homologous to Hsp90 cochaperones that binds to the Cbf5p enzyme. *J. Mol. Biol.*, **390**, 231–244.

71 Leulliot, N., Godin, K.S., Hoareau-Aveilla, C., Quevillon-Cheruel, S., Varani, G., Henry, Y., and Van Tilbeurgh, H. (2007) The box H/ACA RNP assembly factor Naf1p contains a domain homologous to Gar1p mediating its interaction with Cbf5p. *J. Mol. Biol.*, **371**, 1338–1353.

72 Wang, C. and Meier, U.T. (2004) Architecture and assembly of mammalian H/ACA small nucleolar and telomerase ribonucleoproteins. *EMBO J.*, **23**, 1857–1867.

6
Mechanistic Insights into Mammalian Pre-mRNA Splicing

Sebastian M. Fica, Eliza C. Small, Melissa Mefford, and Jonathan P. Staley

6.1
Introduction

Transcripts from most eukaryotic genes are synthesized as precursor mRNAs (pre-mRNAs), which contain noncoding introns that are subsequently removed in the nucleus before the mRNA is exported to the cytoplasm and translated into protein. Introns provide eukaryotic cells with a tremendous potential for expanding their proteomic diversity by allowing a single gene to encode multiple protein isoforms with distinct activities [1]. Indeed, up to 94% of human genes undergo alternative splicing [2]. Importantly, the relative proportions of distinct isoforms can be modulated to regulate expression of a gene [3]. Besides its role in diversifying gene expression, splicing is also important for efficient transcription [4], for mRNA export [5], for RNA localization [6], for transcript stability, and for efficient translation [7]. In addition, introns have been implicated in the biogenesis of noncoding RNAs (ncRNAs), including microRNAs (miRNAs; reviewed in [8]).

Introns are removed from pre-mRNA by the spliceosome, an incredibly dynamic multi-megadalton ribonucleoprotein (RNP) machine, whose core components are highly conserved across eukaryotes – from mammals to yeast [9]. Our understanding of the composition and workings of this unique cellular machine is informed primarily by work in HeLa cells and in the budding yeast *Saccharomyces cerevisiae*. The mechanism by which the spliceosome excises introns efficiently and accurately will be the focus of this chapter. For a broader view of alternative splicing, including its regulation, evolutionary implications, and pathogenic consequences, the reader can refer to other chapters of this book and to a series of excellent reviews [3, 10–12].

6.2
Chemistry of Splicing

Pre-mRNA splicing occurs by two sequential S_N2 transesterification reactions [13] (Figure 6.1a). In the first step, termed 5′ splice site (5′SS) cleavage, the 2′ hydroxyl

Posttranscriptional Gene Regulation: RNA Processing in Eukaryotes, First Edition. Edited by Jane Wu.

(a)

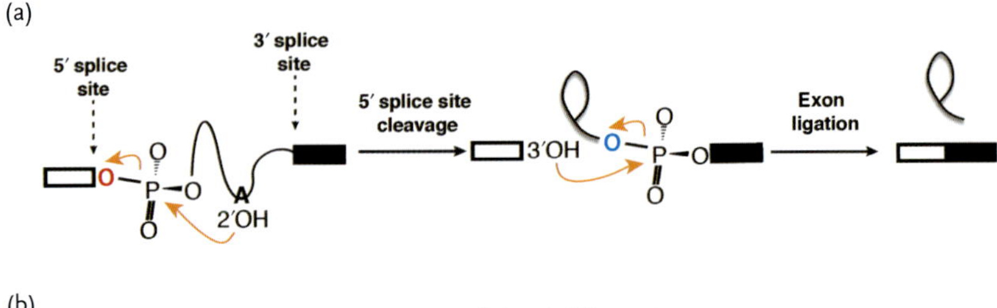

(b)

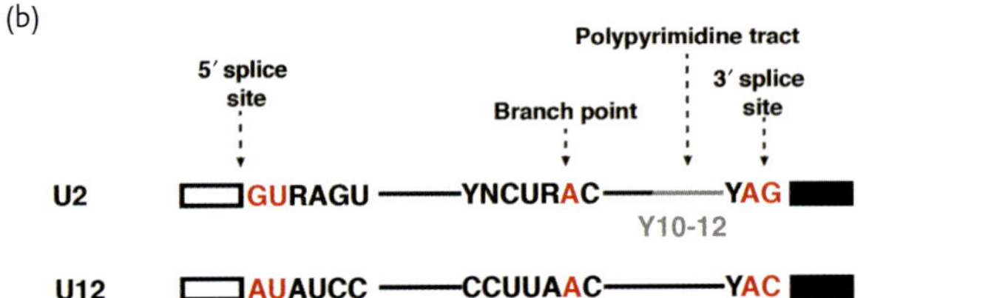

Figure 6.1 General mechanism and splice site consensus sequences for pre-mRNA splicing. (a) The two sequential phosphotransesterifications that occur during pre-mRNA splicing involve nucleophilic attack by hydroxyl groups at conserved junctions between the intron (black line) and 5′ exon (white box) and 3′ exon (black box). The direction of electron flow is indicated by orange arrows and the leaving group oxygens are colored red for 5′ splice site cleavage and blue for exon ligation. (b) The strongly conserved sequences for the major- and minor-class introns are indicated in red in the context of the full consensus. The polypyrimidine tract is depicted as a gray line. Notice that the branch point sequence shows much greater variation than the splice site sequences.

of a conserved adenosine (called the branch point or BP) acts as a nucleophile that attacks a phosphate at the 5′SS, generating a 5′ exon, with a free terminal 3′ hydroxyl, and a branched lariat intermediate. In the second step, termed exon ligation, the free 3′ hydroxyl of the 5′ exon attacks a phosphate at the 3′ splice site (3′SS) in the lariat intermediate, forming the mature mRNA product and excising the branched lariat intron. These reactions, catalyzed by the spliceosome, are indistinguishable from those catalyzed by self-splicing group II introns [14]. Because the group II reactions are necessarily catalyzed by RNA, it was proposed long ago that the spliceosome catalyzes pre-mRNA splicing through its RNA components [14–16]. We will return to this hypothesis when we review our understanding of the spliceosomal active site.

The reactive sites of the substrates are defined by consensus sequences (Figure 6.1b), although these are weaker in mammals than in lower eukaryotes, such as budding yeast. Furthermore, in humans and other metazoans, introns fall into two classes – a common, major class and a less-common, minor class [17]. The major class resembles introns from non-metazoans, whereas the minor class is unique to metazoans. The two classes of introns are excised by two distinct spliceosomes – the major and minor spliceosomes.

6.3 Composition and Assembly of the Spliceosome

Unlike other RNP machines, such as ribosomes, spliceosomes do not form a preassembled complex. Rather, they assemble *de novo* on every substrate from small nuclear ribonucleoprotein complexes (snRNPs), each consisting of one small RNA and associated proteins, and a number of other trans-acting factors that associate with the snRNPs during assembly (reviewed in [18]). The major spliceosome is comprised of the snRNPs U1, U2, U4, U5, and U6, while the minor spliceosome [19, 20] is comprised of the snRNPs U11, U12, $U4_{atac}$, U5, and $U6_{atac}$. While the U5 snRNP is the only common snRNP, the other snRNPs share similar small nuclear RNA (snRNA) sequences and secondary structures, protein compositions, and functions (see below, this section). Common transitions observed in both spliceosomes underscore fundamental requirements for their functions [17, 21]. Of the trans-acting factors, an important class includes DExD/H-box ATPases [22], which mediate important structural dynamics during the splicing cycle, as we shall discuss in Section 6.5.

In total, about 170 proteins associate with the mammalian spliceosome [9], although cryo-electron microscopy (cryo-EM) and mass spectrometry studies suggest individual complexes along the splicing pathway contain no more than 72 proteins at stoichiometric levels [23]. This discrepancy reflects the dynamic association and dissociation of splicing factors and the stringency of purification methods, as well as the involvement of some splicing factors in coupling splicing to other nuclear processes (reviewed in [7]). Roughly 80 of the proteins associated with mammalian splicing complexes are conserved all the way down to yeast [24], indicating that these factors function in the most basic mechanism of splicing.

In the classic view of spliceosome assembly, the spliceosome forms *de novo* on individual pre-mRNA substrates by sequential and coordinated addition of snRNPs [9] (Figure 6.2a). Then, the spliceosome is remodeled by the activity of trans-acting factors during the catalytic stages of splicing. Finally, upon formation of the mature mRNA, the spliceosome dissociates and recycles for a new round of splicing.

During assembly, spliceosomal snRNAs recognize the consensus sequences at the 5′SS and the BP. In the first assembly step for the major spliceosome, the U1 snRNP binds the pre-mRNA in an ATP-independent manner through specific base-paring interactions between the 5′ end of U1 snRNA and the 5′SS [26–28] (Figure 6.2a). In the case of the minor spliceosome, the U11 snRNP analogously utilizes its 5′ end to recognize the 5′SS of a minor-class intron [17]. Recent structural and genetic evidence suggests that the binding register of U1 in humans is somewhat flexible, which may explain how the U1–5′SS interaction can accommodate significant deviation of many natural 5′SS sequences from the consensus [29, 30]. This interaction is stabilized by serine-arginine (SR)-rich proteins [31], which *in vitro* can even bypass the requirement for U1 in the splicing cycle when present at high concentrations [32]. The SR proteins include an RNA recognition motif (RRM) for direct recognition of a substrate and an RS domain rich in

(a)

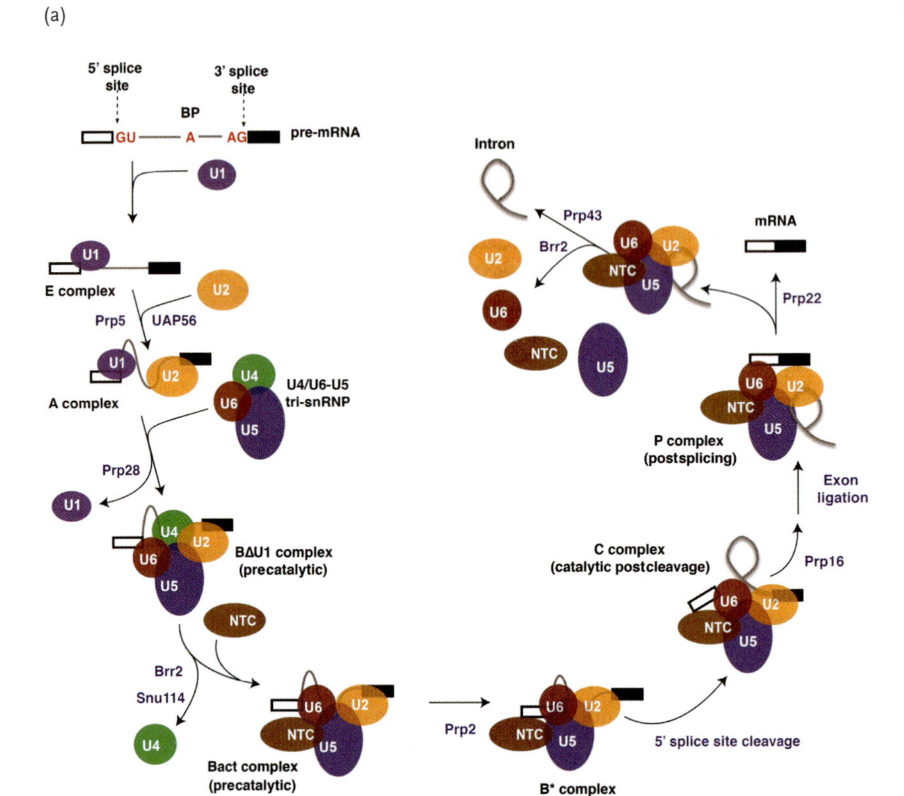

(b)

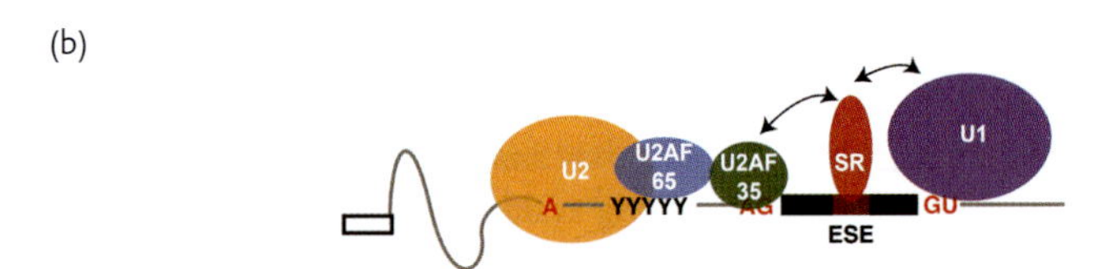

Figure 6.2 The splicing pathway. (a) The general assembly, rearrangements, and disassembly pathway during pre-mRNA splicing is cartooned. The intron is depicted as a gray line, with the assembling snRNPs and protein complexes shown as colored balls. Trans-acting factors that mediate each step are indicated in blue text; NTC denotes the Prp19-associated complex. (b) Interactions that occur between the U1 and U2 snRNPs during exon definition are mediated by SR proteins (red oval), which bind to defined sequences, such as ESEs (exonic splicing enhancers). Binding of the polypyrimidine tract and the 3′ splice site by U2AF is also indicated; this interaction, which is important for correct splice site recognition is the target of a proofreading activity mediated by DEK [25].

arginine and serine dipeptides that function to stabilize substrate recognition by splicing factors. In conjunction with the recognition of the 5′SS by U1, the BP sequence is recognized by the splicing factor SF1 [33] and the downstream polypyrimidine tract and 3′SS by the U2AF heterodimer [34, 35], with the U2AF^{65} subunit recognizing the polypyrimidine tract and the U2AF^{35} subunit recognizing the 3′SS (Figure 6.2b). Some introns do not require the recognition of the 3′SS at this stage, but recognition of the polypyrimidine tract is critical and overshadows the importance of the BP sequence recognition [36]. In fact, the strength of the polypyrimidine tract can influence a preexisting conformational equilibrium of the U2AF heterodimer, thus regulating downstream recruitment of the U2 snRNP [35] (see below, this section). Together, these initial recognition events, which are targeted by factors that regulate alternative splicing (e.g., Sxl [37]) result in the formation of the E complex.

The E complex can include a weakly associated U2 snRNP [38]. However, tight association of the U2 snRNP with the pre-mRNA requires ATP hydrolysis by the conserved DExD/H-box ATPase Prp5, which in yeast has been shown to remodel the U2 snRNA, including unwinding of a stem loop that utilizes sequences in the loop to sample the BP sequence, after which unwinding of the stem is required to stabilize the interaction with the BP sequence [39]. Base pairing between U2 and the BP sequence is coordinated with the release of SF1 and may be mediated by a second DExD/H-box ATPase UAP56, originally identified as a factor that interacts with U2AF^{65} [40] (Figure 6.2a). The association of the U2 snRNP with the BP sequence [41, 42] leads to the formation of the A complex and seems to coincide with the commitment to splice site choice [43]. With the minor class of introns, the U12 snRNP recognizes the branch site by an analogous mechanism [44].

These early stages of spliceosome assembly are promoted by cooperative recognition of intronic elements by collaboration between the U1 snRNP and SF1/U2AF and/or the U2 snRNP [45–47] (Figure 6.2b). This collaboration is mediated in part by SR proteins [48] that function in constitutive splicing but also play a common role in regulated splicing. In a classic example, the SR protein SC35 was shown to interact with both the U1-70K and U2AF65 proteins [49] (Figure 6.2b). Moreover, SC35 can rescue splicing in extracts depleted of the U1 snRNP, in a manner dependent on U2AF [50], or in extracts depleted of U2AF, in a manner dependent on U1 snRNP [51], suggesting that the cooperative binding interactions between the U1 and U2 snRNPs are important for the formation of a stable A complex as well as its conversion to catalytic splicing complexes.

Importantly, the cooperative interactions between U1 and U2AF or U2 can occur either across an intron (intron definition) or across an exon (exon definition) [52]; while intron definition is favored for short introns, exon definition is favored for longer introns – the norm in humans, in which the average exon (<200 nucleotides) is roughly 10 times shorter than the average intron (~1500 nucleotides) [52]. When splice sites are first recognized by exon definition, the interactions must nevertheless switch to a mode in which interactions occur across the intron, before splicing can take place. When exactly this transition occurs in the assembly pathway and even whether there is a general stage for the transition is currently unclear. However, it is clear that spliceosome assembly across an exon can proceed beyond A complex formation [53–55].

Following A complex formation, the U4, U5, and U6 snRNPs of the major spliceosome, or the corresponding $U4_{atac}$, U5, and $U6_{atac}$ snRNPs of the minor spliceosome, are recruited to the A complex as a preassembled tri-snRNP in which the U4 and U6 snRNAs, or $U4_{atac}$ and $U6_{atac}$ snRNAs, are partially base paired. Recruitment of the tri-snRNP leads to the formation of the B complex (Figure 6.2a), which is not yet active for catalysis and still requires dramatic remodeling [56]. In the major spliceosome, the U1 snRNP is dissociated from the substrate, allowing U6 to replace U1 at the 5′SS, and in the minor spliceosome, the U11 snRNP is dissociated from the substrate, allowing $U6_{atac}$ to replace U11 at the 5′SS, yielding the BΔU1 complex (Figure 6.2a). Next, the U4 snRNP is displaced from U6, allowing the formation of key catalytic conformations (see Section 6.5), and in the minor spliceosome, the $U4_{atac}$ snRNP is displaced from $U6_{atac}$. This transition is accompanied by association of a major trans-acting complex, the nineteen complex (NTC) [57], yielding the spliceosomal complex B^{act}. Subsequently, protein components of the U2 snRNP are destabilized, potentially to reveal the BP for chemistry [58, 59], yielding the complex B*, which catalyzes the 5′SS cleavage [60] (Figure 6.2a).

This first transesterification reaction results in remodeling of the RNA and protein components of the B* complex, leading to the formation of the C complex. At this stage, the NTC and core U5 snRNP components compose a salt-stable core of the spliceosome [60]. The C complex is further remodeled to allow for repositioning of the BP and juxtapositioning of the 5′SS and the 3′SS at the catalytic center [61]. The spliceosome then catalyzes exon ligation, resulting in the release of the mRNA as an mRNP, containing factors that specifically mark the location of the splice junctions [62], as a consequence of recruitment by the splicing factor hCwc22 [63, 64]. Deposition of this exon junction complex (EJC) allows coupling to other processes such as mRNA export, localization, surveillance, and translation (reviewed in [65]). After the release of the mRNA product, the spliceosome releases the excised intron and disassembles into the free U2, U6, and U5 snRNPs [56], which are then recycled to reform the tri-snRNP for a new round of splicing (Figure 6.2a).

The stepwise nature of the assembly pathway was first described *in vitro*, in both yeast and mammals. Indeed, the majority of the protein components of the spliceosomal intermediates are conserved from humans to yeast [24], justifying the study of yeast spliceosomes as a model system for human spliceosomes. Recent *in vivo* studies support this model for spliceosome assembly. While chromatin immunoprecipitation following *in vivo* cross-linking (chromatin immunoprecipitation (ChIP)) has most commonly been used to investigate binding of proteins to DNA *in vivo*, because splicing is cotranscriptional, ChIP can also be used to determine the timing of splicing factor association with nascent transcript relative to polymerase position on a gene. Analysis of spliceosome assembly *in vivo* by ChIP in yeast [66, 67] or by live cell imaging in mammals [68] also reveals a stepwise process in which U1 binds first, U2 bindings second, and U5 snRNP (presumably in the context of the tri-snRNP) binds after U2. Perhaps not surprisingly, association of the U1 snRNP depends on and follows the transcription of the 5′SS and U2 snRNP association depends on and follows the transcription of the branch site [66, 67]. Furthermore, each associate in cells depleted of the U5 snRNP, demonstrating that U1 and U2 can act independently of this snRNP [67], again supporting the stepwise model for assembly. Studies of single spliceosomes assembled *in vitro* have strengthened this view, suggesting in addition that individual assembly steps are reversible [69]; such reversibility might be important during the initial pairing of splice sites and thus influence alternative splicing.

However, other observations have led to an alternative assembly model, in which the snRNPs preassociate before substrate binding. These observations include the isolation of a U2.U4/U6.U5 tetra-snRNP in mammals [70] and yeast [71] and of a U1.U2.U4/U6.U5 penta-snRNP from yeast [71]. Consistent with the activity of a penta-snRNP, the 5′SS cross-links to the tri-snRNP prior (reviewed in [72]) to stable U2 association and U2 associates with the E complex before stable U2 association [38], as noted in this section. Because spliceosome assembly is primarily cotranscriptional [73, 74], variables such as nuclear repositioning of splicing factors from nuclear speckles or of genes to nuclear speckles [75], where splicing components are concentrated and perhaps preassembled, may add another dimension to spliceosome assembly *in vivo*. However, most studies have provided evidence against the formation or importance of preassembled spliceosomes and argue in favor of spliceosome assembly on pre-mRNA substrates by stepwise addition of snRNPs (see above, this section). In either case, the RNP rearrangements described in this section would be required to generate catalytically active spliceosomes.

6.4 Control of Spliceosome Assembly and Activation

To ensure correct expression of specific mRNA isoforms as well as to ensure efficient coupling to transcription, splice site choice must be controlled at each step in the splicing cycle – a challenging task in view of the flexibility required for widespread, alternative splicing. The first stage of control during spliceosome assembly is mediated by the RNA polymerase itself. *In vitro*, phosphorylation of the carboxy-terminal domain (CTD) of RNA polymerase II enhances pre-mRNA splicing [76] and serves to promote recruitment of splicing factors to sites of transcription *in vivo* [77]. Moreover, phosphorylation of the CTD has been implicated in the recruitment of SR proteins [78], which have been implicated in coupling splicing to transcription [73], and shown to mediate effects of DNA damage on alternative splicing [79].

Not surprisingly, given the common nature of phosphorylation, this modification is also a prominent mechanism for regulating spliceosome assembly and activation. It was noted very early on that inhibitors of the PP1 and PP2A family of phosphatases block splicing following assembly, implying a requirement for phosphorylation early in splicing and dephosphorylation later in splicing. Indeed, reversible phosphorylation of U1-70K, a U1 snRNP SR protein, is necessary for a precatalytic step following spliceosome assembly [80]. During the formation of the A complex, phosphorylation of U1-70K and of the SR protein SRSF1, a constitutive and alternative splicing factor, enhances their interaction and stabilizes U1 binding to the 5′SS [81]. In addition, phosphorylation of SR proteins on their RS domains is necessary for proper tri-snRNP addition during B complex formation [82]. Although the precise role of the numerous phosphorylation events reported for SR proteins is unclear, experiments where artificial RS domains were tethered to the mRNA in *S. cerevisiae* have led to the proposition that RS domains serve to stabilize RNA–RNA interactions between the pre-mRNA and the snRNP components, and phosphorylation modulates this function [83]. Moreover, phosphorylation of SRp38 can convert it from a repressor to a specific activator, suggesting that SR protein phosphorylation can indeed alter their functions [84]. SR proteins are phosphorylated by the SRPK and Clk/Sty kinases [85]. Although it is poorly understood how phosphorylation of specific factors is regulated by the cell, in the case of SRp38, it has been shown that environmental cues, such as heat shock, can elicit complex signaling cascades that directly regulate phosphatases that act on SR proteins [84].

SR proteins are not the only spliceosome assembly factors that contain an RS domain that can be phosphorylated. For example, phosphorylation of the RS domain of the DExD/H-box ATPase hPrp28 by the kinase SRPK2 is required for the stable association of hPrp28 with the tri-snRNP and for tri-snRNP integration into the B complex [86]. In addition, phosphorylation of hPrp6 and hPrp31 by the PRP4 kinase appears important for stable tri-snRNP association in the B complex, suggesting that multiple phosphorylation events regulate this stage of assembly [87].

Phosphorylation has also been implicated in spliceosome activation and catalysis. Early on, it was noted that phosphorylation of SAP155, a U2 snRNP component that contacts the pre-mRNA around the BP, occurs concomitantly with the 5′SS cleavage [88]. Indeed, recent semiquantitative analysis of spliceosomes identified SAP155 as one of just two factors, of numerous phosphorylated factors, whose phosphorylation state is dynamic [23]. Furthermore, this study confirms phosphorylation appears in the B^{act} complex, before 5′SS cleavage. Conspicuously, at a similar stage, Prp2 has been implicated in destabilizing SAP155 and other components of the SF3 complex, suggesting coupling of phosphorylation to these events. In addition, SAP155, along with the U5 116-kDa protein, is dephosphorylated prior to exon ligation at the same stage that PP1/PP2A phosphatases are required [89], suggesting that the interaction of SAP155 with the BP may rearrange again to facilitate substrate repositioning before the second catalytic step. Significantly, SAP155, also known as SF3B1, has been linked to cancer; the gene encoding this factor is the second most commonly mutated gene in chronic lymphocytic leukemia and almost always a mutated gene in myelodysplastic syndrome with ring sideroblasts [90]. Furthermore, compounds with antitumor activity, such as spliceostatin, target SF3B1 and alter branch site recognition [91–93].

Besides phosphorylation, other modifications have been implicated in the regulation of spliceosome assembly and activation. Work in *S. cerevisiae* has provided evidence that Prp8, a U5 snRNP component, is ubiquitinated in tri-snRNPs, and work in humans has provided evidence that hPrp3 is ubiquitinated in tri-snRNPs [94, 95]. Further evidence indicates that ubiquitin inhibits premature U4/U6 unwinding. Interestingly, the C-terminal region of Prp8, which contains a Jab1/MPN ubiquitin-binding domain that indeed binds ubiquitin [96], has been shown to upregulate U4/U6 unwinding [97], suggesting ubiquitin modification controls unwinding [94]. Because mutations in this region of Prp8 have been implicated in the degenerative disease retinitis pigmentosa, a deficiency in regulating U4/U6 unwinding may contribute to this disease [98, 99].

6.5 Spliceosome Structure and Dynamics

The highly dynamic nature of the spliceosome makes it difficult to obtain preparative amounts of homogenous spliceosomes, both because the individual intermediates are fleeting and because the intermediates are inherently unstable. Furthermore, the large size of the spliceosome renders reconstitution from individual parts essentially impossible, particularly in mammals. Consequently, the spliceosome remains the last major cellular machinery for which no high-resolution structure is currently available. However, to date, several groups have successfully utilized increasingly refined purification methods and mass spectrometry to provide a detailed view of the spliceosome's protein composition at different stages during the splicing cycle. In addition, the use of cryo-EM has begun to shed some light on the structural organization of the spliceosome. Concomitantly, the use of NMR and X-ray crystallography have yielded higher-resolution images of some individual components of the spliceosome [18].

The first view of the spliceosome came early on from experiments in which native spliceosomes assembled on a reporter pre-mRNA were purified from HeLa extracts by gel filtration and visualized by electron microscopy, which revealed massive complexes of 40–60 nm [100]. However, more than 10 years passed before higher-resolution information of the structure of the spliceosome began to emerge, facilitated in part by new purification strategies that also sufficed for mass spectrometry, allowing the definition of a comprehensive parts list for several splicing complexes in mammals.

The A complex was purified by a double-affinity method using tagged protein components, revealing a main body with foot-like and head-like protrusions by cryo-EM [101]. Mammalian B complexes lacking U1 (BΔU1) were purified using antibodies to a U4/U6 component and analyzed by cryo-EM and mass spectrometry. The complex, reconstructed to a resolution of about 40 Å, reveals the presence of a triangular body and a head domain [102] (Figure 6.3a). Subsequent work, in which antibody-conjugated gold particles were used for labeling of complete B complexes, provided some insight into the location of the exons, the intron, and SAP155 in the head domain of the complex [103] (Figure 6.3a). The cryo-EM structure of purified tri-snRNPs and data regarding the location of a U5 snRNA loop [105, 106], which is implicated in aligning the two exons for exon ligation

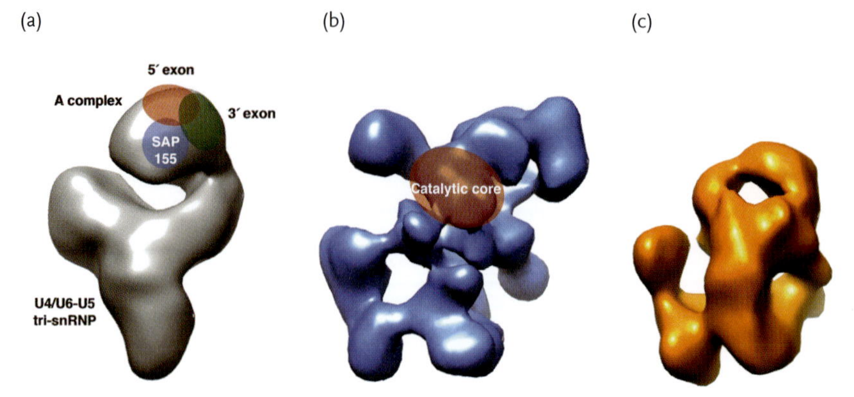

Figure 6.3 Cryo-EM structures of human splicing complexes. (a) The cryo-EM reconstruction of a human BΔU1 complex is shown. The complex was purified by an affinity tag on a specific U4/U6 factor (EMDB 1066, [102]). The likely locations of the exons and the SAP155 protein are also cartooned and are inferred from immunolabeling studies performed on the B complex [103], which exhibits very similar morphology to the BΔU1 complex. (b) The cryo-EM reconstruction of a catalytically active C complex is depicted (EMDB 1846, [104]). The likely locations of its catalytic core were inferred from immunolabeling of the 5′ exon. Notice the significant remodeling compared with the BΔU1 complex. (c) The salt-stable core of the C complex was also reconstructed from cryo-EM data (EMDB 1848, [104]). This complex can be fit into the native C complex, although it is only catalytically active in the presence of added splicing extract, suggesting that it has lost important factors bound less strongly.

[107, 108], suggested that the U2 snRNP components are also located in the head domain of the B complex. Modeling and biochemical work based on the structure of p14 in complex with a portion of SAP155, both U2 snRNP components, suggest that these two proteins may cooperate to form a recognition domain for the BP adenosine [109]. SAP155 has been proposed to undergo conformational rearrangements that could mediate this interaction with the BP, but downstream rearrangements, mediated by the DExD/H-box ATPase Prp2, may serve to undo the SAP155–BP interactions to reveal the branch site for the 5′SS cleavage [58, 110, 111]. Interestingly, SAP155 appears destabilized in purified C complexes [23].

In addition to the B complex, the C complex has been targeted for cryo-EM studies after purification by similar strategies but using a 3′SS mutation to trap spliceosomes before exon ligation [112, 113]. The cryo-EM structure of the C complex confirms the massive rearrangements during catalytic activation and 5′SS cleavage inferred from the mass spectrometry data [113] (Figure 6.3b and see below, this section). Furthermore, various lines of research imply for the first time the possible location of the spliceosomal active site in the C complex (Figure 6.3b). Interestingly, when the 3′ exon is very short, the release of the mRNA following exon ligation is blocked, and this has allowed further isolation of a new postsplicing complex, termed P complex, which contains the mRNA and the excised intron and is similar in composition and morphology to the C complex, although it lacks a putative ATPase factor that may be specifically involved in exon ligation in mammals [114].

While cryo-EM studies have, to date, provided limited insight into the fine architecture of key spliceosomal components, mass spectrometry analyses of purified complexes have provided significant insight into protein dynamics during splicing. A key theme of the transitions from one spliceosomal intermediate to the next is the exchange of one set of protein factors for another. The transition from the A to the B complex is accompanied by the dissociation of 10 proteins, some of which function in alternative splice site choice, and the addition of about 40 new proteins, reflecting in part tri-snRNP integration [60]. A further significant reorganization of the spliceosome occurs from the B to the C complex–most notably, the loss of most U2 and U4/U6 snRNP-specific proteins, the stabilization of the Prp19 complex, and then the binding of trans-acting factors implicated in branch site interactions. Thus, Prp19 complex factors are found in higher abundance, by mass spectrometry, in the C complex compared with the B complex [23]. In addition, components of the SF3 complex, such as SAP155, also appear destabilized during the conversion from the B^{act} to the C complex. The conversion of the B^{act} to the C complex and the SF3B destabilization are mediated by Prp2 [111], whose activity appears important for binding of Cwc25, a factor required for 5′SS cleavage and implicated in interactions with the branch site [111, 115, 116]. However, several factors involved in exon ligation, including hPrp18, hSlu7, or hPrp22, join the spliceosome specifically in the C complex [23] at the same general time that factors required specifically for 5′SS cleavage dissociate [116]. Some of these second step factors are more loosely associated and dissociate at least partially by high salt treatment. Nonetheless, the C complex is generally more resistant to salt treatment than the B complex, suggesting that catalytic activation leads to a stronger network of RNA–RNA, RNA–protein, and protein–protein interactions at the spliceosomal core. Importantly, this salt-resistant core (Figure 6.3c) includes U5 snRNP components and the NTC, suggesting that these factors may play important roles at the catalytic stage [113].

Besides complex protein dynamics, assembled spliceosomes also undergo numerous snRNA and substrate rearrangements. Most of these rearrangements are mediated by a specific class of spliceosomal proteins called DExD/H-box ATPases, which can hydrolyze ATP to remodel RNA and RNA–protein interactions. In the following paragraphs, we highlight some of the best characterized RNA dynamics during the splicing cycle. Although much of the evidence for these events comes from molecular genetics studies performed in *S. cerevisiae*, the relevance of these rearrangements to the mammalian system is supported both by cross-linking studies, the high degree of sequence conservation between yeast and mammalian spliceosomes, and the conservation of large-scale RNA rearrangements.

During spliceosome assembly, RNA dynamics promote substrate recognition. First, RNA dynamics promote branch site recognition. A conserved stem loop of U2 snRNA presents the BP recognition sequencing and permits sampling of a branch site. However, stable base pairing between U2 and the BP requires unwinding of the stem, a rearrangement mediated by Prp5 [39] (see Section 6.3), a factor also implicated in bridging an interaction between the U2 snRNP and the U1 snRNP [47].

Subsequently, the interaction between the U1 snRNP and the 5′SS is disrupted to allow for the establishment of base-pairing interactions between the 5′SS and the ACAGA region of the U6 snRNA [117–119]. In yeast, it has been shown that this switch requires ATP and Prp28 [120]. Interestingly, hyperstabilization of the U1–5′SS interaction or impairment of Prp28 activity in yeast causes accumulation of a complex lacking the U4/U6-U5 tri-snRNP, suggesting that U1 release might be required for stable tri-snRNP addition [120]. In support of this idea, a B complex lacking U1 as well as hPrp28 (BΔU1) [23] could be purified in mammals, indicating that U1 release can precede U4/U6 unwinding (Figure 6.2a; see below, this section).

Prior to catalytic activation of the spliceosome, the base-pairing interaction between the U4 and U6 snRNAs must also be disrupted, allowing U2/U6 annealing. In yeast, it has been shown that the Brr2 DExD/H-box ATPase is necessary for unwinding U4/U6 both *in vitro* and *in vivo* [121, 122], while its mammalian homolog, the U5 200-kDa protein, was shown to unwind U4/U6 *in vitro* [123]. Intriguingly, a recent study has found evidence that another conserved DExD/H-box protein, hUAP56, can contact and promote unwinding of the U4/U6 duplex, raising the possibility that more than one ATPase may mediate this rearrangement [124]. The activity of Brr2 is also necessary for spliceosome disassembly following mRNA release in yeast [125], suggesting that it may play a similar role in unwinding U2/U6 interactions following catalysis. The crucial role of Brr2 in the control of spliceosome assembly and disassembly is underscored by the multiple layers of regulation of its activity. The ATPase and helicase activities of Brr2 are modulated by, in addition to Prp8 (see Section 6.4), the guanine nucleotide state of the only spliceosomal GTPase – Snu114 (the U5 116-kDa protein in mammals), which may also play a role during exon ligation [89, 126]. The activity of Snu114 and the ubiquitination of Prp8, if not also Prp3, appear to converge mechanistically in regulating spliceosome activation and likely disassembly following catalysis.

Defects in U4/U6 unwinding due to mutations in the human 200-kDa protein (the homolog of Brr2) have been recently linked to autosomal dominant retinitis pigmentosa (adRP), an inherited type of blindness [99]. hBrr2 thus joins several other components of the tri-snRNP as hot spots for adRP mutations. However, while the mutations in the other genes have been thought to impact U5 and/or tri-snRNP assembly, the mutations in hBrr2 seem to instead impact U4/U6 unwinding, raising questions as to the fundamental defect in these cases of adRP. Additionally, it remains unclear why mutations in these core splicing factors manifest defects in only one specific tissue.

The DExD/H-box ATPase Prp2 serves to finalize the catalytic activation of the spliceosome by promoting destabilization of the SF3 splicing complex. This activity may serve to destabilize interactions that were required early for branch site recognition but at this stage would prevent chemistry; more specifically, destabilization of these interactions may reveal the branch site nucleophile, as noted in this section. Supporting the potential significance of BP rearrangements at this stage, the action of Prp2 permits the association of Cwc25, which interacts with the branch site and functions specifically in the 5′SS cleavage [115, 116]. It is currently unclear whether this DExD/H-box ATPase, like the others, functions also in remodeling RNA–RNA interactions.

After the 5′SS cleavage, when the substrate and catalytic core must rearrange to accommodate the 3′SS, intramolecular interactions in U2 and intermolecular interactions between U2 and U6 snRNAs appears to be remodeled, consistent with the prominent role of these snRNAs in interacting with key substrate sequences. Within U2, mutually exclusive structures toggle from one conformation (stem IIc) to another (stem IIa) and back again for exon ligation [127]. Because this toggle regulates the interaction between U2 snRNP and the BP during early spliceosome assembly, toggling at the catalytic stage may promote rearrangements of the U2/BP interaction. The interactions between U2 and U6 also appear to rearrange. In this interaction, known as helix I, U6 sequences, situated downstream of the ACAGA sequence that pairs with the 5′SS, base pair with U2 sequences, situated upstream of the BP recognition sequence. This interaction serves to juxtapose the reactants for the 5′SS cleavage and is necessary for splicing in mammals [128] and yeast [129, 130]. In yeast, helix I forms prior to the 5′SS cleavage and is disrupted following the 5′SS cleavage, with the help of the DExD/H-box protein Prp16, which is required for exon ligation [131], but reforms and functions during exon ligation [130]. Because helix I adjoins the U2–branch site interaction, remodeling of this structure could be important for substrate repositioning during the catalytic stage.

Subsequent to exon ligation, DExD/H-box ATPases promote the release of the splicing products. The DExD/H-box ATPase Prp22 is required to release the mRNA from the spliceosome [132, 133] and consequently for export of mRNA to the cytoplasm [134]. Subsequently, the DExD/H-box ATPase Prp43, in conjunction with Brr2 (see above, this section), is required to release the excised intron product and to disassemble the spliceosome for recycling [125, 135]. A number of these DExD/H-box ATPases have not only been shown to function in promoting mRNA production of optimal substrates but also antagonizing production of suboptimal substrates (see Section 6.7).

6.6 The Structure of the Spliceosomal Active Site and the Mechanism of Catalysis

The highly dynamic nature of the spliceosomal snRNAs and proteins and the extensive remodeling of their interactions with the substrate during the splicing cycle have rendered it difficult to characterize the exact components of the active site of the spliceosome. However, even in the absence of a high-resolution crystal structure of an activated spliceosome, biochemical and genetic analyses have thus far provided significant insights into the possible architecture of the active site.

Because the mechanism of pre-mRNA splicing shares the same chemistry with that observed in group II intron self-splicing, it was proposed early on that pre-mRNA splicing may be catalyzed exclusively by RNA [14, 15], as we now know to be the case for another macromolecular RNP machine – the ribosome [136–138]. In the following paragraphs, we will review key evidence to support this hypothesis, and we will end with recent developments that may add an exciting twist to the story.

The discovery of self-splicing group I introns [139] and initial studies implicating metal ion coordination by the RNA phosphate backbone [140] provided evidence that lead to the postulation of a two-metal-ion catalytic mechanism for phosphoryl transfer reactions catalyzed by ribozymes (Figure 6.4a), in analogy to catalytic mechanisms for phosphoryl transfer reactions catalyzed by proteins [141]. In this mechanism, the RNA phosphate backbone oxygens position two metal ions in an appropriate conformation to stabilize the buildup of negative charge that occurs during each transesterification reaction.

The idea that the spliceosome may be a ribozyme was initially supported by the identification, in certain yeast species, of spliceosomal introns in two conserved regions of the U6 snRNA [142, 143] – the ACAGA element, later shown important for 5′SS recognition, and U2/U6 helix I, which is required at the catalytic stage for exon ligation. These introns could have arisen by reverse splicing at the active site, providing an evolutionary record of the presence of U6 at the active site. Indeed, the ACAGA region of U6 was shown to cross-link to the 5′SS in both the pre-mRNA and the lariat intermediate [108], providing physical evidence for the idea that U6 is one of the active site components. Intriguingly, this structure, along with an adjacent intramolecular stem loop (ISL), is similar to the catalytic domain V of group II introns [144] (Figure 6.4c). In addition, a loop in the U5 snRNA was shown to cross-link to the ends of the exons at the exon/intron boundaries [108] and genetic studies showed the functional relevance of such interactions [107], suggesting that the U5 loop serves to align the exons for the second chemical step, in analogy to domain I of the group II intron. Finally, the U2 snRNA was shown first in yeast [41] and then in mammals [145] to base pair to the BP region in a manner that bulges the BP for nucleophilic attack, similar to domain VI of the group II intron. Together, these results provided strong experimental support for the idea that the spliceosomal active site is composed of RNA sequences and that the snRNAs play the critical role in catalysis.

Further support for the ribozyme hypothesis came from initial studies involving phosphorothioate interference experiments in the U6 snRNA [144, 146]. Substitution of backbone oxygens with sulfur at specific positions, including a residue in helix I, resulted in blocks to splicing either before or after the 5′SS cleavage. Because sulfur is expected to coordinate poorly to magnesium, which is required for *in vitro* splicing, this result provided the first evidence supporting the involvement of metal ions in splicing. Subsequent metal rescue analysis in yeast has provided strong evidence that a specific nonbridging oxygen in U6, at the U80 pro-Sp position (Figure 6.4c), does indeed coordinate a metal ion at both steps of splicing [147, 148]. Furthermore, work using HeLa extracts has shown that the leaving groups during both steps of splicing coordinate metal ions [149, 150], as is the case for group II introns [151], strongly supporting the idea that the spliceosome is a metalloenzyme. However, direct functional evidence that the snRNAs coordinate the metals interacting with the splice sites is lacking. Nonetheless, the

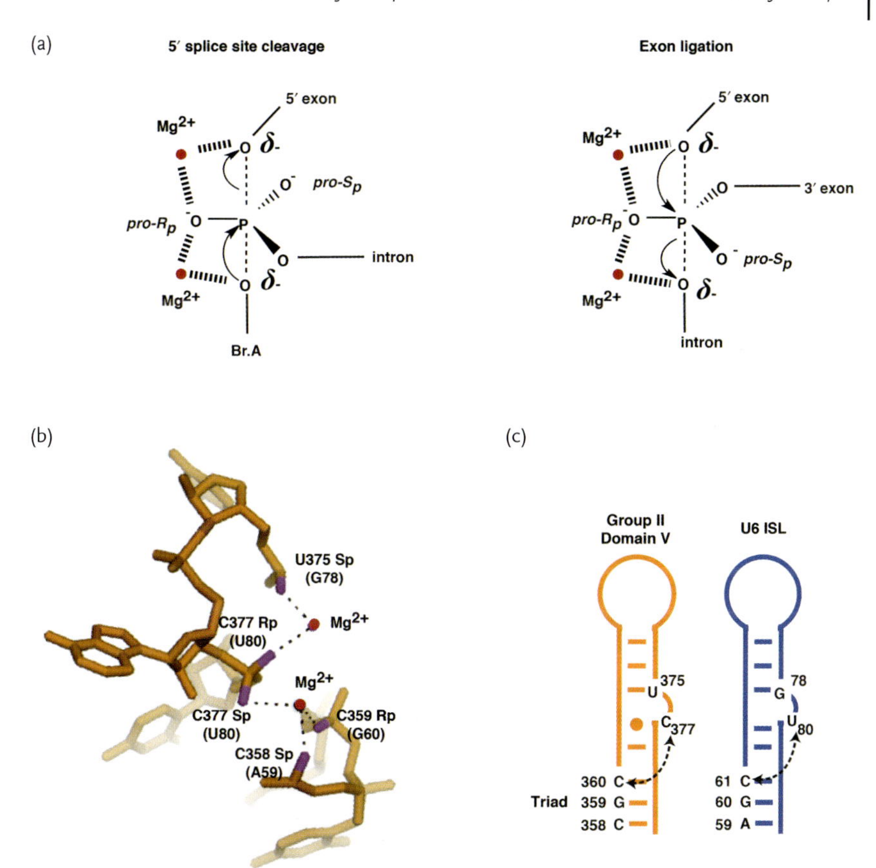

Figure 6.4 Proposed mechanism of pre-mRNA splicing catalysis and structural similarity between the spliceosome and group II introns. (a) The diagrams depict trigonal bipyramidal transition states for the 5′ splice site cleavage and exon ligation. Arrows indicate the direction of electron flow. The partial negative charges that build up during the reaction are stabilized by two Mg^{2+} ions (red circles) [141]. (b) The crystal structure of a self-splicing group II intron in a postcatalytic state shows two Mg^{2+} ions coordinated by ligands in domain V of the intron (PDB 3EOH). Nonbridging oxygens that position the metal ions are colored in magenta. Group II intron residue numbers are shown, with the anticipated corresponding U6 residues indicated in parentheses. (c) The catalytic domain V of group II introns and the spliceosomal U6 snRNA share structural similarity. Notice that both RNAs form similar secondary structures. In U6, this structure is important for bulging out of U80, which has been shown to coordinate a metal ion important for both steps of splicing. The triple between C377 and C360 is shown as a dotted, double-headed arrow; a similar triple interaction has been postulated to occur in U6.

spliceosomal snRNAs have intrinsic catalytic potential, as suggested by the observation that a structure formed by the U2 and U6 snRNAs in the absence of protein can perform a reaction similar to the first step of splicing, albeit at very low efficiency [152, 153]. Nonetheless, whether the protein-free reaction is indeed relevant to splicing in the spliceosome has been called into question [154] (cf. [155]).

Although we lack a structure of the active site, recent crystal structures of a self-splicing group II intron [156] provide a possible glimpse of how the critical residues of the U2, U5, and U6 snRNAs may be arranged. The structures provide support for the two-metal-ion mechanism by showing that the catalytic domain V adopts a stem-loop structure that bends over itself to allow coordination of two magnesium ions by the RNA phosphate backbone (Figure 6.4b). The metals are situated 3.9 Å apart, consistent with the proposed two-metal-ion mechanism. The structures also include the splicing substrate or the ligated exons and shows the reactive phosphates near to the two metal ions, thus complementing previous functional, biochemical studies demonstrating metal stabilization of the splice site leaving groups [151]. However, the precise roles of the domain V ligands that position these metals in the crystal have not been elucidated [157].

Interestingly, both the domain V ISL and AGC triad that coordinate the metals in the group II structure show a high degree of structural conservation in the U6 ISL and AGC triad of helix Ib (Figure 6.4c), the regions previously implicated in metal ion coordination [144, 146–148]. This supports the notion that the group II structure may represent a good approximation of the structure adopted by the U6 snRNA in the active spliceosome. Nevertheless, with the exception of the U80Sp position, the corresponding U6 residues have not been shown experimentally to bind metal ions. Therefore, whether the U6 snRNA mediates a two-metal-ion mechanism during spliceosomal catalysis remains an open question [158].

The group II intron structures have also shown that the catalytic triad in domain V participates in three base triple interactions that converge on the conserved AGC triad and bring together crucial elements of the catalytic core. One triple interaction serves to juxtapose the two metal binding sites (Figure 6.4c). The other two triples buttress the first triple and likely function after the first step to recruit the 3′SS to the catalytic center [156, 159, 160]. All three triples could form in U6 snRNA [158], juxtaposing the conserved bulge in the ISL with the AGC triad in U6, as in domain V, and juxtaposing the ACAGA region with the AGC triad, potentially recruiting the substrate, at least the 5′SS, to the catalytic core. However, though some data are consistent with these interactions forming in the spliceosome [161, 162], data supporting these specific interactions are currently lacking.

Early studies involving phosphorothioate substitution at the splice sites showed that the two steps of splicing were inhibited by the same diastereomers at each splice site [163]. These findings are compatible with a single active site [141] and subsequent studies have shown, as discussed in this section, that both reaction transition states are stabilized by metal ions. Nevertheless, genetic evidence indicates that the spliceosome conformations that catalyze each splicing reaction do differ in some respects [164]. Indeed, it is clear from genetic and biochemical experiments that the spliceosome rearranges between the two steps [130, 131, 165], but whether these rearrangements involve simply repositioning of the lariat intermediate or a more dramatic restructuring of core elements remains to be resolved. The latter possibility, however, is less likely given that, at least in yeast, key structures, such as U2/U6 helix I, function during both steps even though they seem to be disrupted transiently in between the catalytic steps [130, 166].

Finally, the hypothesis that the spliceosome is a ribozyme is complicated by the intimate relationship that proteins have with the spliceosomal snRNAs and the pre-mRNA. Unlike the ribosome, the spliceosome contains proteins in close proximity to reactive regions. In particular, Prp8, the most highly conserved spliceosomal protein, physically interacts with the substrate reactive sites during both steps of splicing in yeast [167]. In addition, its C-terminal domain can be cross-linked to the conserved GU dinucleotide at the 5′SS in the catalytically competent mammalian spliceosome [168] and to residues adjacent to the ACAGAGA sequence of U6 snRNA [169]. Moreover, using *in vivo* reporter assays in yeast, mutations in this domain of Prp8 were shown to suppress mutations at the 5′SS, the BP [164], and the 3′SS [170], implicating Prp8 in the formation of the catalytic core. A recent crystal structure of a large fragment of Prp8 from yeast indicates that the suppressor alleles as well as the X-links to U6 and the BP map to the inside of a large cavity of the protein, suggesting that the catalytic core of the spliceosome may be situated within [171]. Interestingly, Prp8 also contains RNase H-like and endonuclease domains that form part of this cavity. The RNAse H-like domain contains two aspartate residues situated in a semiconserved arrangement reminiscent of RNase H enzymes site, and which are located in the general region that cross-links to the 5′SS [172–174]. Thus, these structures have raised the interesting possibility that Prp8 may function to help position metal ions in the spliceosome, which would render the spliceosome an RNP-zyme, rather than a true ribozyme [175]. Given the intimate relationship between the snRNAs and Prp8, untangling the roles that RNA and protein play in catalysis will prove a complex mechanistic challenge that will require advanced biochemistry, if not also structural analysis.

6.7 Fidelity in Splicing

The spliceosome faces the daunting task of recognizing the precise locations of splice sites having only minimal sequence elements while maintaining the flexibility to skip the same splice sites when enabled to regulate alternative splicing. Such a situation would seem prone for errors. Moreover, 15–35% of human disease are expected to arise from mutations in splicing elements that do or are expected to alter splicing [12], highlighting the danger of splicing errors. A determination of the intrinsic error rate of splicing, however, is confounded by the coupling of splicing to other pre-mRNA processing events such as nonsense-mediated decay, which can accelerate turnover of mis-spliced or alternative messages [176, 177]. Nevertheless, recent studies have begun to provide us with a view of error rates in mammals. Errors in splice site choice have been estimated to be as high as 0.7% [178] while errors resulting in skipping of exons have been estimated to occur at a frequency of roughly 1×10^{-5}, on average [179]. In the latter case, it has been proposed that this limit simply reflects the errors committed by RNA polymerase II, which at similar frequency will transcribe a splice site incorrectly [179]. Recognition and pairing of correct splice sites occurs through a combinatorial set of factors, whose relative contributions have been discussed elsewhere [180]. However, strong discrimination requires selection against suboptimal sites after substrate binding. Here, we will focus briefly on the role of late-acting, trans-acting factors, especially DExD/H-box ATPases, in fidelity.

Initial evidence for a proofreading mechanism that enhances splicing fidelity came from genetic studies in yeast. A mutation in the DExD/H-box ATPase Prp16 suppressed splicing defects of introns carrying BP mutations, allowing utilization of the mutated BP [181]. Suppression was accompanied by an increase in both lariat intermediate and pre-mRNA levels. This observation led to the proposal that ATP hydrolysis by Prp16 causes discard of aberrant splicing species, thus enhancing the fidelity of splicing [181]. Prp16 had also previously been implicated in promoting a rearrangement necessary for exon ligation of optimal substrates [131]. These observations together led to the proposal that kinetic proofreading enhances splicing fidelity of the 5′SS cleavage. In kinetic proofreading, a ubiquitous framework for enhancing fidelity, the fidelity of a process is promoted by a kinetic competition between productive pathways and discard pathways [182] (Figure 6.5). Recent biochemical evidence has provided direct support for this model by showing that Prp16 can compete with and thereby proofread the 5′SS cleavage, rejecting suboptimal spliceosomes and triggering their entry into a discard pathway mediated by Prp43 [148, 183], a DExD/H-box ATPase necessary for spliceosome disassembly, a pathway without which the spliceosome commits errors. Optimal spliceosomes are immune to this rejection pathway and indeed splice faster than

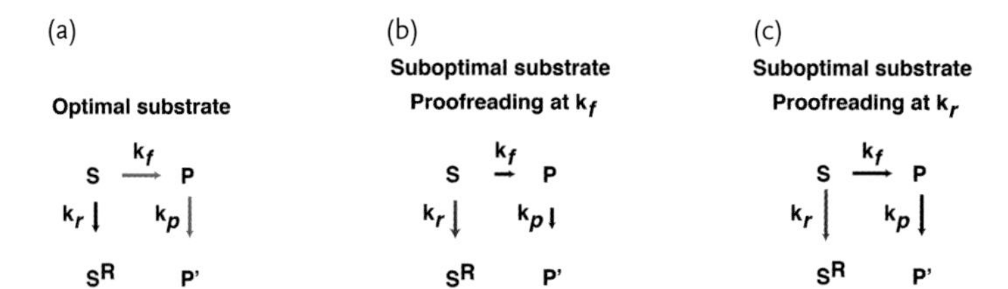

Figure 6.5 Alternative manifestations of kinetic proofreading in the spliceosome. In the case of an optimal substrate (a), the rate of the forward on-pathway conversion of substrate (S) to product (P) (k_f) is faster than the rate of DExD/H-box ATPase-mediated rejection (k_r). Consequently, ATP hydrolysis by the proofreading DExD/H-box ATPase occurs after the conversion and results in transition to a productive conformation (P′) at rate k_p. In the case of a suboptimal substrate (b and c), the rate of rejection effectively competes with the forward rate of conversion to antagonize splicing, resulting in the formation of a rejected conformation (S^R). Two nonmutually exclusive models can explain specificity for the rejection of suboptimal substrates. In proofreading at the conversion step (b), the rate of rejection may be constant for optimal and suboptimal substrates, whereas the conversion rate is sufficiently reduced in the case of a suboptimal substrate, such that the rate of rejection is faster than the rate of conversion. In proofreading at the rejection step (c), the rate of conversion may be constant for optimal and suboptimal substrates, whereas the rate of rejection (k_r) is enhanced for a suboptimal substrate relative to an optimal substrate, similarly leading to a faster rate of rejection than conversion. The sizes of the arrows reflect the magnitude of their rates. Green arrows represent utilization of the productive pathway; red arrows represent utilization of the rejection pathway; black arrows represent unutilized pathways.

suboptimal spliceosomes, suggesting an explicit escape mechanism and satisfying a key prediction of the kinetic proofreading model. However, the optimal substrate would, after completion of the 5′SS cleavage, benefit from the activity of Prp16, which would promote rearrangements that permit exon ligation [131]. In this way, Prp16 not only antagonizes suboptimal substrates but also promotes optimal substrate depending on the stage in splicing at which Prp16 functions, a stage that depends on the identity of the substrate and/or spliceosome.

Two additional DExD/H-box ATPases have been implicated in promoting fidelity at distinct steps in the splicing pathway. In addition to releasing mature mRNA from the spliceosome after exon ligation [133], Prp22 uses an ATP-dependent mechanism to reject suboptimal substrates prior to exon ligation [184]. In a parallel fashion, during spliceosome assembly, Prp5 uses an ATP-dependent mechanism to proofread the U2–branch site interaction when U2 samples a BP, but after consolidation of an interaction between U2 and an optimal BP, Prp5 promotes rearrangements that favors splicing of the substrate [185]. In the case of Prp5, proofreading activity is directly proportional to ATPase activity, also supporting the kinetic proofreading model. Together, these yeast studies point to a general kinetic proofreading mechanism by which DExD/H-box ATPases promote fidelity by inspecting critical substrate interactions at multiple stages along the splicing pathway. Importantly, each of these DExD/H-box ATPases is conserved from yeast to mammals.

The fidelity of splicing is not only supported by kinetic mechanisms but also by thermodynamic mechanisms. Indeed, genetic evidence has shown that Prp16 helps regulate an equilibration between two competing spliceosomal conformations, an equilibration that can help antagonize suboptimal substrates [164, 182]. In this thermodynamic model, fidelity in splicing can be promoted by the relative stability of competing conformations: a mutation that disfavors a productive conformation will in turn favor an unproductive, competing conformation. Furthermore, it has recently been demonstrated that both chemical steps of splicing are reversible under some *in vitro* conditions [186], suggesting that reversal of the chemical steps themselves may allow correction of splicing errors.

Several other factors have been specifically implicated in correct 3′SS choice in mammals. The auxillary U2 factor $U2AF^{35}$ recognizes the AG dinucleotide at the 3′SS very early during spliceosome assembly [187]. This interaction is proofread by DEK [25]. Interestingly, the function of DEK in fidelity requires phosphorylation of the protein, providing yet another opportunity to regulate a step in splicing by posttranslational modification. In addition, correct choice of the AG dinucleotide at the 3′SS is promoted after 5′SS cleavage by human Slu7, whose activity is important for the stabilization of the upstream exon in the C complex [188].

6.8 Concluding Remarks

Although our understanding of the spliceosome lags behind that of the other major machines in gene expression, the RNA polymerase II holoenzyme, and the ribosome, our knowledge of the spliceosome's composition and mechanism has progressed rapidly in the recent past. For example, new affinity purification approaches coupled to advances in mass spectrometry have yielded an essentially complete parts list for the spliceosome [18, 189]. We are beginning to understand the intricate roles of many trans-acting splicing factors such as those mediating mechanisms that enhance the specificity and fidelity of splicing [182]. Our insight into the function of snRNA structures and dynamics has also matured; for example, the structure of the related self-splicing group II intron [156, 158] has strengthened the hypothesis that this ribozyme and the spliceosome share common catalytic strategies, if not also evolutionary origins. Despite these advances, much remains to be revealed in the future. Large-scale structures of spliceosomal intermediates are sorely needed, but such goals will take some time to achieve. The functions of the majority of the splicing factors remain undefined. Whether catalysis is mediated by RNA, protein, or both, also continues to be a pressing question. Additionally, how splicing occurs in the cell will be an important challenge for the future, such as the mechanism for coupling splicing to transcription. Undoubtedly, the splicing field will yield exciting discoveries for some time to come.

Acknowledgments

We thank D. Semlow and K. Nielsen for assistance in figure preparation. We regret that we were unable to describe and/or cite relevant work of colleagues in the field due to space constraints. Work in our laboratory on the mechanism of splicing has been supported by grants from the National Institutes of Health (NIH) (GM062264 and GM088656 to J.P.S.). E.C.S. was partially supported by the NIH grant T32 GM007183. M.M. was partially supported by the NIH grant T32 GM007197.

References

1 Nilsen, T.W. and Graveley, B.R. (2010) Expansion of the eukaryotic proteome by alternative splicing. *Nature*, **463**, 457–463.

2 Wang, E.T., Sandberg, R., Luo, S., Khrebtukova, I., Zhang, L., Mayr, C., Kingsmore, S.F., Schroth, G.P., and Burge, C.B. (2008) Alternative isoform regulation in human tissue transcriptomes. *Nature*, **456**, 470–476.

3 Chen, M. and Manley, J.L. (2009) Mechanisms of alternative splicing regulation: insights from molecular and genomics approaches. *Nat. Rev. Mol. Cell Biol.*, **10**, 741–754.

4 Furger, A., O'Sullivan, J.M., Binnie, A., Lee, B.A., and Proudfoot, N.J. (2002) Promoter proximal splice sites enhance transcription. *Genes Dev.*, **16**, 2792–2799.

5 Valencia, P., Dias, A.P., and Reed, R. (2008) Splicing promotes rapid and efficient mRNA export in mammalian cells. *Proc. Natl. Acad. Sci. U.S.A.*, **105**, 3386–3391.

6 Le Hir, H., Gatfield, D., Braun, I.C., Forler, D., and Izaurralde, E. (2001) The protein Mago provides a link between splicing and mRNA localization. *EMBO Rep.*, **2**, 1119–1124.

7 Moore, M.J. and Proudfoot, N.J. (2009) Pre-mRNA processing reaches back to

transcription and ahead to translation. *Cell*, **136**, 688–700.
8 Pawlicki, J.M. and Steitz, J.A. (2010) Nuclear networking fashions pre-messenger RNA and primary microRNA transcripts for function. *Trends Cell Biol.*, **20**, 52–61.
9 Wahl, M.C., Will, C.L., and Lührmann, R. (2009) The spliceosome: design principles of a dynamic RNP machine. *Cell*, **136**, 701–718.
10 Wang, G.-S. and Cooper, T.A. (2007) Splicing in disease: disruption of the splicing code and the decoding machinery. *Nat. Rev. Genet.*, **8**, 749–761.
11 Keren, H., Lev-Maor, G., and Ast, G. (2010) Alternative splicing and evolution: diversification, exon definition and function. *Nat. Rev. Genet.*, **11**, 345–355.
12 Singh, R.K. and Cooper, T.A. (2012) Pre-mRNA splicing in disease and therapeutics. *Trends Mol. Med.*, **18**, 472–482.
13 Sharp, P.A. (1987) Splicing of messenger RNA precursors. *Science*, **235**, 766–771.
14 Cech, T.R. (1987) The chemistry of self-splicing RNA and RNA enzymes. *Science*, New Series **236**, 1532–1539.
15 Cech, T.R. (1986) The generality of self-splicing RNA: relationship to nuclear mRNA splicing. *Cell*, **44**, 207–210.
16 Sharp, P.A. and Eisenberg, D. (1987) The evolution of catalytic function. *Science*, New Series **238**, 729–730+807.
17 Patel, A.A. and Steitz, J.A. (2003) Splicing double: insights from the second spliceosome. *Nat. Rev. Mol. Cell Biol.*, **4**, 960–970.
18 Will, C.L. and Lührmann, R. (2010) Spliceosome structure and function. *Cold Spring Harb. Perspect. Biol.*, **3**, a003707.
19 Tarn, W.Y. and Steitz, J.A. (1996) A novel spliceosome containing U11, U12, and U5 snRNPs excises a minor class (AT-AC) intron in vitro. *Cell*, **84**, 801–811.
20 Tarn, W.Y. and Steitz, J.A. (1996) Highly diverged U4 and U6 small nuclear RNAs required for splicing rare AT-AC introns. *Science*, **273**, 1824–1832.
21 Frilander, M.J. and Steitz, J.A. (2001) Dynamic exchanges of RNA interactions leading to catalytic core formation in the U12-dependent spliceosome. *Mol. Cell*, **7**, 217–226.
22 Fairman-Williams, M.E., Guenther, U.-P., and Jankowsky, E. (2010) SF1 and SF2 helicases: family matters. *Curr. Opin. Struct. Biol.*, **20**, 313–324.
23 Agafonov, D.E., Deckert, J., Wolf, E., Odenwälder, P., Bessonov, S., Will, C.L., Urlaub, H., and Lührmann, R. (2011) Semiquantitative proteomic analysis of the human spliceosome via a novel two-dimensional gel electrophoresis method. *Mol. Cell. Biol.*, **31**, 2667–2682.
24 Fabrizio, P., Dannenberg, J., Dube, P., Kastner, B., Stark, H., Urlaub, H., and Lührmann, R. (2009) The evolutionarily conserved core design of the catalytic activation step of the yeast spliceosome. *Mol. Cell*, **36**, 593–608.
25 Soares, L.M.M., Zanier, K., Mackereth, C., Sattler, M., and Valcárcel, J. (2006) Intron removal requires proofreading of U2AF/3' splice site recognition by DEK. *Science*, **312**, 1961–1965.
26 Zhuang, Y. and Weiner, A.M. (1986) A compensatory base change in U1 snRNA suppresses a 5' splice site mutation. *Cell*, **46**, 827–835.
27 Bindereif, A. and Green, M.R. (1987) An ordered pathway of snRNP binding during mammalian pre-mRNA splicing complex assembly. *EMBO J.*, **6**, 2415–2424.
28 Siliciano, P.G. and Guthrie, C. (1988) 5' splice site selection in yeast: genetic alterations in base-pairing with U1 reveal additional requirements. *Genes Dev.*, **2**, 1258–1267.
29 Pomeranz Krummel, D.A., Oubridge, C., Leung, A.K.W., Li, J., and Nagai, K. (2009) Crystal structure of human spliceosomal U1 snRNP at 5.5 A resolution. *Nature*, **458**, 475–480.
30 Roca, X., Akerman, M., Gaus, H., Berdeja, A., Bennett, C.F., and Krainer, A.R. (2012) Widespread recognition of 5' splice sites by noncanonical base-pairing to U1 snRNA involving bulged nucleotides. *Genes Dev.*, **26**, 1098–1109.
31 Zhong, X.-Y., Wang, P., Han, J., Rosenfeld, M.G., and Fu, X.-D. (2009)

SR proteins in vertical integration of gene expression from transcription to RNA processing to translation. *Mol. Cell*, **35**, 1–10.

32 Tarn, W.Y. and Steitz, J.A. (1994) SR proteins can compensate for the loss of U1 snRNP functions in vitro. *Genes Dev.*, **8**, 2704–2717.

33 Liu, Z., Luyten, I., Bottomley, M.J., Messias, A.C., Houngninou-Molango, S., Sprangers, R., Zanier, K., Krämer, A., and Sattler, M. (2001) Structural basis for recognition of the intron branch site RNA by splicing factor 1. *Science*, **294**, 1098–1102.

34 Ruskin, B., Zamore, P.D., and Green, M.R. (1988) A factor, U2AF, is required for U2 snRNP binding and splicing complex assembly. *Cell*, **52**, 207–219.

35 Mackereth, C.D., Madl, T., Bonnal, S., Simon, B., Zanier, K., Gasch, A., Rybin, V., Valcárcel, J., and Sattler, M. (2011) Multi-domain conformational selection underlies pre-mRNA splicing regulation by U2AF. *Nature*, **475**, 408–411.

36 Berglund, J.A., Abovich, N., and Rosbash, M. (1998) A cooperative interaction between U2AF65 and mBBP/SF1 facilitates branchpoint region recognition. *Genes Dev.*, **12**, 858–867.

37 Valcárcel, J., Singh, R., Zamore, P.D., and Green, M.R. (1993) The protein sex-lethal antagonizes the splicing factor U2AF to regulate alternative splicing of transformer pre-mRNA. *Nature*, **362**, 171–175.

38 Das, R., Zhou, Z., and Reed, R. (2000) Functional association of U2 snRNP with the ATP-independent spliceosomal complex E. *Mol. Cell*, **5**, 779–787.

39 Perriman, R. and Ares, M. (2010) Invariant U2 snRNA nucleotides form a stem loop to recognize the intron early in splicing. *Mol. Cell*, **38**, 416–427.

40 Fleckner, J., Zhang, M., Valcárcel, J., and Green, M.R. (1997) U2AF65 recruits a novel human DEAD box protein required for the U2 snRNP-branchpoint interaction. *Genes Dev.*, **11**, 1864–1872.

41 Parker, R., Siliciano, P.G., and Guthrie, C. (1987) Recognition of the TACTAAC box during mRNA splicing in yeast involves base pairing to the U2-like snRNA. *Cell*, **49**, 229–239.

42 Zhuang, Y. and Weiner, A.M. (1989) A compensatory base change in human U2 snRNA can suppress a branch site mutation. *Genes Dev.*, **3**, 1545–1552.

43 Lim, S.R. and Hertel, K.J. (2004) Commitment to splice site pairing coincides with A complex formation. *Mol. Cell*, **15**, 477–483.

44 Frilander, M.J. and Steitz, J.A. (1999) Initial recognition of U12-dependent introns requires both U11/5' splice-site and U12/branchpoint interactions. *Genes Dev.*, **13**, 851–863.

45 Michaud, S. and Reed, R. (1993) A functional association between the 5" and 3" splice site is established in the earliest prespliceosome complex (E) in mammals. *Genes Dev.*, **7**, 1008–1020.

46 Dönmez, G., Hartmuth, K., Kastner, B., Will, C.L., and Lührmann, R. (2007) The 5' end of U2 snRNA is in close proximity to U1 and functional sites of the pre-mRNA in early spliceosomal complexes. *Mol. Cell*, **25**, 399–411.

47 Shao, W., Kim, H.-S., Cao, Y., Xu, Y.-Z., and Query, C.C. (2012) A U1-U2 snRNP interaction network during intron definition. *Mol. Cell. Biol.*, **32**, 470–478.

48 Förch, P., Merendino, L., Martínez, C., and Valcárcel, J. (2003) U2 small nuclear ribonucleoprotein particle (snRNP) auxiliary factor of 65 kDa, U2AF65, can promote U1 snRNP recruitment to 5' splice sites. *Biochem. J.*, **372**, 235–240.

49 Wu, J.Y. and Maniatis, T. (1993) Specific interactions between proteins implicated in splice site selection and regulated alternative splicing. *Cell*, **75**, 1061–1070.

50 Crispino, J.D., Blencowe, B.J., and Sharp, P.A. (1994) Complementation by SR proteins of pre-mRNA splicing reactions depleted of U1 snRNP. *Science*, **265**, 1866–1869.

51 MacMillan, A.M., McCaw, P.S., Crispino, J.D., and Sharp, P.A. (1997) SC35-mediated reconstitution of splicing in U2AF-depleted nuclear extract. *Proc. Natl. Acad. Sci. U.S.A.*, **94**, 133–136.

52 Robberson, B.L., Cote, G.J., and Berget, S.M. (1990) Exon definition may facilitate splice site selection in RNAs with multiple exons. *Mol. Cell. Biol.*, **10**, 84–94.

53 House, A.E. and Lynch, K.W. (2006) An exonic splicing silencer represses spliceosome assembly after ATP-dependent exon recognition. *Nat. Struct. Mol. Biol.*, **13**, 937–944.

54 Schneider, M., Will, C., Anokhina, M., Tazi, J., and Urlaub, H. (2010) Exon definition complexes contain the tri-snRNP and can be directly converted into B-like precatalytic splicing complexes. *Mol. Cell*, **38**, 223–235.

55 Sharma, S., Maris, C., Allain, F.H.-T., and Black, D.L. (2011) U1 snRNA directly interacts with polypyrimidine tract-binding protein during splicing repression. *Mol. Cell*, **41**, 579–588.

56 Makarov, E.M., Makarova, O.V., Urlaub, H., Gentzel, M., Will, C.L., Wilm, M., and Lührmann, R. (2002) Small nuclear ribonucleoprotein remodeling during catalytic activation of the spliceosome. *Science*, **298**, 2205–2208.

57 Chan, S.-P., Kao, D.-I., Tsai, W.-Y., and Cheng, S.-C. (2003) The Prp19p-associated complex in spliceosome activation. *Science*, **302**, 279–282.

58 Lardelli, R.M., Thompson, J.X., Yates, J.R., and Stevens, S.W. (2010) Release of SF3 from the intron branchpoint activates the first step of pre-mRNA splicing. *RNA*, **16**, 516–528.

59 Hegele, A., Kamburov, A., Grossmann, A., Sourlis, C., Wowro, S., Weimann, M., Will, C.L., Pena, V., Lührmann, R., and Stelzl, U. (2012) Dynamic protein-protein interaction wiring of the human spliceosome. *Mol. Cell*, **45**, 567–580.

60 Bessonov, S., Anokhina, M., Krasauskas, A., Golas, M.M., Sander, B., Will, C.L., Urlaub, H., Stark, H., and Lührmann, R. (2010) Characterization of purified human Bact spliceosomal complexes reveals compositional and morphological changes during spliceosome activation and first step catalysis. *RNA*, **16**, 2384–2403.

61 Konarska, M.M., Vilardell, J., and Query, C.C. (2006) Repositioning of the reaction intermediate within the catalytic center of the spliceosome. *Mol. Cell*, **21**, 543–553.

62 Le Hir, H., Izaurralde, E., Maquat, L.E., and Moore, M.J. (2000) The spliceosome deposits multiple proteins 20–24 nucleotides upstream of mRNA exon-exon junctions. *EMBO J.*, **19**, 6860–6869.

63 Alexandrov, A., Colognori, D., Shu, M.-D., and Steitz, J.A. (2012) Human spliceosomal protein CWC22 plays a role in coupling splicing to exon junction complex deposition and nonsense-mediated decay. *Proc. Natl. Acad. Sci. U.S.A.*, **109**, 21313–21318.

64 Barbosa, I., Haque, N., Fiorini, F., Barrandon, C., Tomasetto, C., Blanchette, M., and Le Hir, H. (2012) Human CWC22 escorts the helicase eIF4AIII to spliceosomes and promotes exon junction complex assembly. *Nat. Struct. Mol. Biol.*, **19**, 983–990.

65 Dreyfuss, G., Kim, V.N., and Kataoka, N. (2002) Messenger-RNA-binding proteins and the messages they carry. *Nat. Rev. Mol. Cell Biol.*, **3**, 195–205.

66 Görnemann, J., Kotovic, K.M., Hujer, K., and Neugebauer, K.M. (2005) Cotranscriptional spliceosome assembly occurs in a stepwise fashion and requires the cap binding complex. *Mol. Cell*, **19**, 53–63.

67 Tardiff, D.F., Lacadie, S.A., and Rosbash, M. (2006) A genome-wide analysis indicates that yeast pre-mRNA splicing is predominantly posttranscriptional. *Mol. Cell*, **24**, 917–929.

68 Huranová, M., Ivani, I., Benda, A., Poser, I., Brody, Y., Hof, M., Shav-Tal, Y., Neugebauer, K.M., and Stanek, D. (2010) The differential interaction of snRNPs with pre-mRNA reveals splicing kinetics in living cells. *J. Cell Biol.*, **191**, 75–86.

69 Hoskins, A.A., Friedman, L.J., Gallagher, S.S., Crawford, D.J., Anderson, E.G., Wombacher, R., Ramirez, N., Cornish, V.W., Gelles, J., and Moore, M.J. (2011) Ordered and dynamic assembly of single spliceosomes. *Science*, **331**, 1289–1295.

70 Konarska, M.M. and Sharp, P.A. (1988) Association of U2, U4, U5, and U6 small nuclear ribonucleoproteins in a spliceosome-type complex in absence of precursor RNA. *Proc. Natl. Acad. Sci. U.S.A.*, **85**, 5459–5462.

71 Stevens, S.W., Ryan, D.E., Ge, H.Y., Moore, R.E., Young, M.K., Lee, T.D., and Abelson, J. (2002) Composition and functional characterization of the yeast spliceosomal penta-snRNP. *Mol. Cell*, **9**, 31–44.

72 Rino, J. and Carmo-Fonseca, M. (2009) The spliceosome: a self-organized macromolecular machine in the nucleus? *Trends Cell Biol.*, **19**, 375–384.

73 Das, R., Yu, J., Zhang, Z., Gygi, M.P., Krainer, A.R., Gygi, S.P., and Reed, R. (2007) SR proteins function in coupling RNAP II transcription to pre-mRNA splicing. *Mol. Cell*, **26**, 867–881.

74 Girard, C., Will, C.L., Peng, J., Makarov, E.M., Kastner, B., Lemm, I., Urlaub, H., Hartmuth, K., and Lührmann, R. (2012) Post-transcriptional spliceosomes are retained in nuclear speckles until splicing completion. *Nat. Commun.*, **3**, 994.

75 Misteli, T., Cáceres, J.F., and Spector, D.L. (1997) The dynamics of a pre-mRNA splicing factor in living cells. *Nature*, **387**, 523–527.

76 Hirose, Y., Tacke, R., and Manley, J.L. (1999) Phosphorylated RNA polymerase II stimulates pre-mRNA splicing. *Genes Dev.*, **13**, 1234–1239.

77 Misteli, T. and Spector, D.L. (1999) RNA polymerase II targets pre-mRNA splicing factors to transcription sites in vivo. *Mol. Cell*, **3**, 697–705.

78 Lin, S., Coutinho-Mansfield, G., Wang, D., Pandit, S., and Fu, X.-D. (2008) The splicing factor SC35 has an active role in transcriptional elongation. *Nat. Struct. Mol. Biol.*, **15**, 819–826.

79 Muñoz, M.J., Pérez Santangelo, M.S., Paronetto, M.P., de la Mata, M., Pelisch, F., Boireau, S., Glover-Cutter, K., Ben-Dov, C., Blaustein, M., Lozano, J.J., *et al.* (2009) DNA damage regulates alternative splicing through inhibition of RNA polymerase II elongation. *Cell*, **137**, 708–720.

80 Tazi, J., Kornstädt, U., Rossi, F., Jeanteur, P., Cathala, G., Brunel, C., and Lührmann, R. (1993) Thiophosphorylation of U1-70K protein inhibits pre-mRNA splicing. *Nature*, **363**, 283–286.

81 Xiao, S.H. and Manley, J.L. (1997) Phosphorylation of the ASF/SF2 RS domain affects both protein-protein and protein-RNA interactions and is necessary for splicing. *Genes Dev.*, **11**, 334–344.

82 Roscigno, R.F. and Garcia-Blanco, M.A. (1995) SR proteins escort the U4/U6.U5 tri-snRNP to the spliceosome. *RNA*, **1**, 692–706.

83 Shen, H. and Green, M.R. (2006) RS domains contact splicing signals and promote splicing by a common mechanism in yeast through humans. *Genes Dev.*, **20**, 1755–1765.

84 Shin, C., Feng, Y., and Manley, J.L. (2004) Dephosphorylated SRp38 acts as a splicing repressor in response to heat shock. *Nature*, **427**, 553–558.

85 Ghosh, G. and Adams, J.A. (2011) Phosphorylation mechanism and structure of serine-arginine protein kinases. *FEBS J.*, **278**, 587–597.

86 Mathew, R., Hartmuth, K., Möhlmann, S., Urlaub, H., Ficner, R., and Lührmann, R. (2008) Phosphorylation of human PRP28 by SRPK2 is required for integration of the U4/U6-U5 tri-snRNP into the spliceosome. *Nat. Struct. Mol. Biol.*, **15**, 435–443.

87 Schneider, M., Hsiao, H., Will, C., and Giet, R. (2010) Human PRP4 kinase is required for stable tri-snRNP association during spliceosomal B complex formation. *Nat. Struct. Mol. Biol.*, **17**, 216–221.

88 Wang, C., Chua, K., Seghezzi, W., Lees, E., Gozani, O., and Reed, R. (1998) Phosphorylation of spliceosomal protein SAP 155 coupled with splicing catalysis. *Genes Dev.*, **12**, 1409–1414.

89 Shi, Y., Reddy, B., and Manley, J.L. (2006) PP1/PP2A phosphatases are required for the second step of pre-mRNA splicing and target specific snRNP proteins. *Mol. Cell*, **23**, 819–829.

90 Damm, F., Nguyen-Khac, F., Fontenay, M., and Bernard, O.A. (2012) Spliceosome and other novel mutations in chronic lymphocytic leukemia and myeloid malignancies. *Leukemia*, **26**, 2027–2031.

91 Kotake, Y., Sagane, K., Owa, T., Mimori-Kiyosue, Y., Shimizu, H.,

Uesugi, M., Ishihama, Y., Iwata, M., and Mizui, Y. (2007) Splicing factor SF3b as a target of the antitumor natural product pladienolide. *Nat. Chem. Biol.*, **3**, 570–575.

92 Corrionero, A., Miñana, B., and Valcárcel, J. (2011) Reduced fidelity of branch point recognition and alternative splicing induced by the anti-tumor drug spliceostatin A. *Genes Dev.*, **25**, 445–459.

93 Folco, E.G., Coil, K.E., and Reed, R. (2011) The anti-tumor drug E7107 reveals an essential role for SF3b in remodeling U2 snRNP to expose the branch point-binding region. *Genes Dev.*, **25**, 440–444.

94 Bellare, P., Small, E.C., Huang, X., Wohlschlegel, J.A., Staley, J.P., and Sontheimer, E.J. (2008) A role for ubiquitin in the spliceosome assembly pathway. *Nat. Struct. Mol. Biol.*, **15**, 444–451.

95 Song, E.J., Werner, S.L., Neubauer, J., Stegmeier, F., Aspden, J., Rio, D., Harper, J.W., Elledge, S.J., Kirschner, M.W., and Rape, M. (2010) The Prp19 complex and the Usp4Sart3 deubiquitinating enzyme control reversible ubiquitination at the spliceosome. *Genes Dev.*, **24**, 1434–1447.

96 Bellare, P., Kutach, A.K., Rines, A.K., Guthrie, C., and Sontheimer, E.J. (2006) Ubiquitin binding by a variant Jab1/MPN domain in the essential pre-mRNA splicing factor Prp8p. *RNA*, **12**, 292–302.

97 Maeder, C., Kutach, A.K., and Guthrie, C. (2009) ATP-dependent unwinding of U4/U6 snRNAs by the Brr2 helicase requires the C terminus of Prp8. *Nat. Struct. Mol. Biol.*, **16**, 42–48.

98 McKie, A.B., McHale, J.C., Keen, T.J., Tarttelin, E.E., Goliath, R., van Lith-Verhoeven, J.J., Greenberg, J., Ramesar, R.S., Hoyng, C.B., Cremers, F.P., *et al.* (2001) Mutations in the pre-mRNA splicing factor gene PRPC8 in autosomal dominant retinitis pigmentosa (RP13). *Hum. Mol. Genet.*, **10**, 1555–1562.

99 Zhao, C., Bellur, D.L., Lu, S., Zhao, F., Grassi, M.A., Bowne, S.J., Sullivan, L.S., Daiger, S.P., Chen, L.J., Pang, C.P., *et al.* (2009) Autosomal-dominant retinitis pigmentosa caused by a mutation in SNRNP200, a gene required for unwinding of U4/U6 snRNAs. *Am. J. Hum. Genet.*, **85**, 617–627.

100 Reed, R., Griffith, J., and Maniatis, T. (1988) Purification and visualization of native spliceosomes. *Cell*, **53**, 949–961.

101 Behzadnia, N., Golas, M.M., Hartmuth, K., Sander, B., Kastner, B., Deckert, J., Dube, P., Will, C.L., Urlaub, H., Stark, H., *et al.* (2007) Composition and three-dimensional EM structure of double affinity-purified, human prespliceosomal A complexes. *EMBO J.*, **26**, 1737–1748.

102 Boehringer, D., Makarov, E.M., Sander, B., Makarova, O.V., Kastner, B., Lührmann, R., and Stark, H. (2004) Three-dimensional structure of a pre-catalytic human spliceosomal complex B. *Nat. Struct. Mol. Biol.*, **11**, 463–468.

103 Wolf, E., Kastner, B., Deckert, J., Merz, C., Stark, H., and Lührmann, R. (2009) Exon, intron and splice site locations in the spliceosomal B complex. *EMBO J.*, **28**, 2283–2292.

104 Golas, M.M., Sander, B., Bessonov, S., Grote, M., Wolf, E., Kastner, B., Stark, H., and Lührmann, R. (2010) 3D cryo-EM structure of an active step I spliceosome and localization of its catalytic core. *Mol. Cell*, **40**, 927–938.

105 Sander, B., Golas, M.M., Makarov, E.M., Brahms, H., Kastner, B., Lührmann, R., and Stark, H. (2006) Organization of core spliceosomal components U5 snRNA loop I and U4/U6 Di-snRNP within U4/U6.U5 tri-snRNP as revealed by electron cryomicroscopy. *Mol. Cell*, **24**, 267–278.

106 Häcker, I., Sander, B., Golas, M.M., Wolf, E., Karagöz, E., Kastner, B., Stark, H., Fabrizio, P., and Lührmann, R. (2008) Localization of Prp8, Brr2, Snu114 and U4/U6 proteins in the yeast tri-snRNP by electron microscopy. *Nat. Struct. Mol. Biol.*, **15**, 1206–1212.

107 Newman, A. and Norman, C. (1991) Mutations in yeast U5 snRNA alter the specificity of 5' splice-site cleavage. *Cell*, **65**, 115–123.

108 Sontheimer, E.J. and Steitz, J.A. (1993) The U5 and U6 small nuclear RNAs as active site components of the spliceosome. *Science*, **262**, 1989–1996.

109 Schellenberg, M.J., Edwards, R.A., Ritchie, D.B., Kent, O.A., Golas, M.M., Stark, H., Lührmann, R., Glover, J.N.M., and Macmillan, A.M. (2006) Crystal structure of a core spliceosomal protein interface. *Proc. Natl. Acad. Sci. U.S.A.*, **103**, 1266–1271.

110 Spadaccini, R., Reidt, U., Dybkov, O., Will, C., Frank, R., Stier, G., Corsini, L., Wahl, M.C., Lührmann, R., and Sattler, M. (2006) Biochemical and NMR analyses of an SF3b155-p14-U2AF-RNA interaction network involved in branch point definition during pre-mRNA splicing. *RNA*, **12**, 410–425.

111 Warkocki, Z., Odenwälder, P., Schmitzová, J., Platzmann, F., Stark, H., Urlaub, H., Ficner, R., Fabrizio, P., and Lührmann, R. (2009) Reconstitution of both steps of *Saccharomyces cerevisiae* splicing with purified spliceosomal components. *Nat. Struct. Mol. Biol.*, **16**, 1237–1243.

112 Jurica, M.S., Sousa, D., Moore, M.J., and Grigorieff, N. (2004) Three-dimensional structure of C complex spliceosomes by electron microscopy. *Nat. Struct. Mol. Biol.*, **11**, 265–269.

113 Bessonov, S., Anokhina, M., Will, C.L., Urlaub, H., and Lührmann, R. (2008) Isolation of an active step I spliceosome and composition of its RNP core. *Nature*, **452**, 846–850.

114 Ilagan, J.O., Chalkley, R.J., Burlingame, A.L., and Jurica, M.S. (2013) Rearrangements within human spliceosomes captured after exon ligation. *RNA*, **19**, 1–14.

115 Chiu, Y.-F., Liu, Y.-C., Chiang, T.-W., Yeh, T.-C., Tseng, C.-K., Wu, N.-Y., and Cheng, S.-C. (2009) Cwc25 is a novel splicing factor required after Prp2 and Yju2 to facilitate the first catalytic reaction. *Mol. Cell. Biol.*, **29**, 5671–5678.

116 Tseng, C.-K., Liu, H.-L., and Cheng, S.-C. (2010) DEAH-box ATPase Prp16 has dual roles in remodeling of the spliceosome in catalytic steps. *RNA*, **17**, 1–10.

117 Wassarman, D.A. and Steitz, J.A. (1992) Interactions of small nuclear RNA's with precursor messenger RNA during in vitro splicing. *Science*, **257**, 1918–1925.

118 Kandels-Lewis, S. and Séraphin, B. (1993) Role of U6 snRNA in 5′ splice site selection. *Science*, **262**, 2035–2039.

119 Lesser, C. and Guthrie, C. (1993) Mutations in U6 snRNA that alter splice site specificity: implications for the active site. *Science*, **262**, 1982–1988.

120 Staley, J.P. and Guthrie, C. (1999) An RNA switch at the 5' splice site requires ATP and the DEAD box protein Prp28p. *Mol. Cell*, **3**, 55–64.

121 Raghunathan, P.L. and Guthrie, C. (1998) RNA unwinding in U4/U6 snRNPs requires ATP hydrolysis and the DEIH-box splicing factor Brr2. *Curr. Biol.*, **8**, 847–855.

122 Kim, D.H. and Rossi, J.J. (1999) The first ATPase domain of the yeast 246-kDa protein is required for in vivo unwinding of the U4/U6 duplex. *RNA*, **5**, 959–971.

123 Laggerbauer, B., Achsel, T., and Lührmann, R. (1998) The human U5-200kD DEXH-box protein unwinds U4/U6 RNA duplices in vitro. *Proc. Natl. Acad. Sci. U.S.A.*, **95**, 4188–4192.

124 Shen, H., Zheng, X., Shen, J., Zhang, L., Zhao, R., and Green, M.R. (2008) Distinct activities of the DExD/H-box splicing factor hUAP56 facilitate stepwise assembly of the spliceosome. *Genes Dev.*, **22**, 1796–1803.

125 Small, E.C., Leggett, S.R., Winans, A.A., and Staley, J.P. (2006) The EF-G-like GTPase Snu114p regulates spliceosome dynamics mediated by Brr2p, a DExD/H box ATPase. *Mol. Cell*, **23**, 389–399.

126 Liu, Z.R., Laggerbauer, B., Lührmann, R., and Smith, C.W. (1997) Crosslinking of the U5 snRNP-specific 116-kDa protein to RNA hairpins that block step 2 of splicing. *RNA*, **3**, 1207–1219.

127 Hilliker, A.K., Mefford, M.A., and Staley, J.P. (2007) U2 toggles iteratively between the stem IIa and stem IIc conformations to promote pre-mRNA splicing. *Genes Dev.*, **21**, 821–834.

128 Sun, J.S. and Manley, J.L. (1995) A novel U2-U6 snRNA structure is necessary for mammalian mRNA splicing. *Genes Dev.*, **9**, 843–854.

129 Madhani, H.D. and Guthrie, C. (1992) A novel base-pairing interaction between U2 and U6 snRNAs suggests a mechanism for the catalytic activation of the spliceosome. *Cell*, **71**, 803–817.

130 Mefford, M.A. and Staley, J.P. (2009) Evidence that U2/U6 helix I promotes both catalytic steps of pre-mRNA splicing and rearranges in between these steps. *RNA*, **15**, 1386–1397.

131 Schwer, B. and Guthrie, C. (1992) A conformational rearrangement in the spliceosome is dependent on PRP16 and ATP hydrolysis. *EMBO J.*, **11**, 5033–5039.

132 Company, M., Arenas, J., and Abelson, J. (1991) Requirement of the RNA helicase-like protein PRP22 for release of messenger RNA from spliceosomes. *Nature*, **349**, 487–493.

133 Schwer, B. (2008) A conformational rearrangement in the spliceosome sets the stage for Prp22-dependent mRNA release. *Mol. Cell*, **30**, 743–754.

134 Ohno, M. and Shimura, Y. (1996) A human RNA helicase-like protein, HRH1, facilitates nuclear export of spliced mRNA by releasing the RNA from the spliceosome. *Genes Dev.*, **10**, 997–1007.

135 Martin, A., Schneider, S., and Schwer, B. (2002) Prp43 is an essential RNA-dependent ATPase required for release of lariat-intron from the spliceosome. *J. Biol. Chem.*, **277**, 17743–17750.

136 Hiller, D.A., Singh, V., Zhong, M., and Strobel, S.A. (2011) A two-step chemical mechanism for ribosome-catalysed peptide bond formation. *Nature*, **476**, 236–239.

137 Kuhlenkoetter, S., Wintermeyer, W., and Rodnina, M.V. (2011) Different substrate-dependent transition states in the active site of the ribosome. *Nature*, **476**, 351–354.

138 Moore, P.B. and Steitz, T.A. (2011) The roles of RNA in the synthesis of protein. *Cold Spring Harb. Perspect. Biol.*, **3**, a003780.

139 Cech, T.R., Zaug, A.J., and Grabowski, P.J. (1981) In vitro splicing of the ribosomal RNA precursor of *Tetrahymena*: involvement of a guanosine nucleotide in the excision of the intervening sequence. *Cell*, **27**, 487–496.

140 Piccirilli, J.A., Vyle, J.S., Caruthers, M.H., and Cech, T.R. (1993) Metal ion catalysis in the *Tetrahymena* ribozyme reaction. *Nature*, **361**, 85–88.

141 Steitz, T.A. and Steitz, J.A. (1993) A general two-metal-ion mechanism for catalytic RNA. *Proc. Natl. Acad. Sci. U.S.A.*, **90**, 6498–6502.

142 Tani, T. and Ohshima, Y. (1989) The gene for the U6 small nuclear RNA in fission yeast has an intron. *Nature*, **337**, 87–90.

143 Tani, T. and Ohshima, Y. (1991) mRNA-type introns in U6 small nuclear RNA genes: implications for the catalysis in pre-mRNA splicing. *Genes Dev.*, **5**, 1022–1031.

144 Yu, Y.T., Maroney, P.A., Darzynkiwicz, E., and Nilsen, T.W. (1995) U6 snRNA function in nuclear pre-mRNA splicing: a phosphorothioate interference analysis of the U6 phosphate backbone. *RNA*, **1**, 46–54.

145 Query, C.C., Moore, M.J., and Sharp, P.A. (1994) Branch nucleophile selection in pre-mRNA splicing: evidence for the bulged duplex model. *Genes Dev.*, **8**, 587–597.

146 Fabrizio, P. and Abelson, J. (1992) Thiophosphates in yeast U6 snRNA specifically affect pre-mRNA splicing in vitro. *Nucleic Acids Res.*, **20**, 3659–3664.

147 Yean, S.L., Wuenschell, G., Termini, J., and Lin, R.J. (2000) Metal-ion coordination by U6 small nuclear RNA contributes to catalysis in the spliceosome. *Nature*, **408**, 881–884.

148 Koodathingal, P., Novak, T., Piccirilli, J.A., and Staley, J.P. (2010) The DEAH box ATPases Prp16 and Prp43 cooperate to proofread 5' splice site cleavage during pre-mRNA splicing. *Mol. Cell*, **39**, 385–395.

149 Sontheimer, E.J., Sun, S., and Piccirilli, J.A. (1997) Metal ion catalysis during splicing of premessenger RNA. *Nature*, **388**, 801–805.

150 Gordon, P.M., Sontheimer, E.J., and Piccirilli, J.A. (2000) Metal ion catalysis during the exon-ligation step of nuclear pre-mRNA splicing: extending the

parallels between the spliceosome and group II introns. *RNA*, **6**, 199–205.
151 Sontheimer, E.J., Gordon, P., and Piccirilli, J. (1999) Metal ion catalysis during group II intron self-splicing: parallels with the spliceosome. *Genes Dev.*, **13**, 1729–1741.
152 Valadkhan, S. and Manley, J.L. (2001) Splicing-related catalysis by protein-free snRNAs. *Nature*, **413**, 701–707.
153 Jaladat, Y., Zhang, B., Mohammadi, A., and Valadkhan, S. (2011) Splicing of an intervening sequence by protein-free human snRNAs. *RNA Biol.*, **8**, 372–377.
154 Smith, D.J. and Konarska, M.M. (2009) A critical assessment of the utility of protein-free splicing systems. *RNA*, **15**, 1–3.
155 Valadkhan, S. and Manley, J.L. (2009) The use of simple model systems to study spliceosomal catalysis. *RNA*, **15**, 4–7.
156 Marcia, M. and Pyle, A.M. (2012) Visualizing group II intron catalysis through the stages of splicing. *Cell*, **151**, 497–507.
157 Gordon, P.M. and Piccirilli, J.A. (2001) Metal ion coordination by the AGC triad in domain 5 contributes to group II intron catalysis. *Nat. Struct. Biol.*, **8**, 893–898.
158 Toor, N., Keating, K.S., Taylor, S.D., and Pyle, A.M. (2008) Crystal structure of a self-spliced group II intron. *Science*, **320**, 77–82.
159 Mikheeva, S., Murray, H.L., Zhou, H., Turczyk, B.M., and Jarrell, K.A. (2000) Deletion of a conserved dinucleotide inhibits the second step of group II intron splicing. *RNA*, **6**, 1509–1515.
160 Keating, K.S., Toor, N., Perlman, P.S., and Pyle, A.M. (2010) A structural analysis of the group II intron active site and implications for the spliceosome. *RNA*, **16**, 1–9.
161 Madhani, H.D. and Guthrie, C. (1994) Randomization-selection analysis of snRNAs in vivo: evidence for a tertiary interaction in the spliceosome. *Genes Dev.*, **8**, 1071–1086.
162 Ryan, D.E., Kim, C.H., Murray, J.B., Adams, C.J., Stockley, P.G., and Abelson, J. (2004) New tertiary constraints between the RNA components of active yeast spliceosomes: a photo-crosslinking study. *RNA*, **10**, 1251–1265.
163 Moore, M.J. and Sharp, P.A. (1993) Evidence for two active sites in the spliceosome provided by stereochemistry of pre-mRNA splicing. *Nature*, **365**, 364–368.
164 Query, C.C. and Konarska, M.M. (2004) Suppression of multiple substrate mutations by spliceosomal prp8 alleles suggests functional correlations with ribosomal ambiguity mutants. *Mol. Cell*, **14**, 343–354.
165 Liu, L., Query, C.C., and Konarska, M.M. (2007) Opposing classes of prp8 alleles modulate the transition between the catalytic steps of pre-mRNA splicing. *Nat. Struct. Mol. Biol.*, **14**, 519–526.
166 Hilliker, A.K. and Staley, J.P. (2004) Multiple functions for the invariant AGC triad of U6 snRNA. *RNA*, **10**, 921–928.
167 Teigelkamp, S., Newman, A.J., and Beggs, J.D. (1995) Extensive interactions of PRP8 protein with the 5′ and 3′ splice sites during splicing suggest a role in stabilization of exon alignment by U5 snRNA. *EMBO J.*, **14**, 2602–2612.
168 Reyes, J.L., Gustafson, E.H., Luo, H.R., Moore, M.J., and Konarska, M.M. (1999) The C-terminal region of hPrp8 interacts with the conserved GU dinucleotide at the 5' splice site. *RNA*, **5**, 167–179.
169 Sha, M., Levy, T., Kois, P., and Konarska, M.M. (1998) Probing of the spliceosome with site-specifically derivatized 5' splice site RNA oligonucleotides. *RNA*, **4**, 1069–1082.
170 Umen, J.G. and Guthrie, C. (1995) A novel role for a U5 snRNP protein in 3' splice site selection. *Genes Dev.*, **9**, 855–868.
171 Galej, W.P., Oubridge, C., Newman, A.J., and Nagai, K. (2013) Crystal structure of Prp8 reveals active site cavity of the spliceosome. *Nature*, **493**, 638–643.
172 Pena, V., Rozov, A., Fabrizio, P., Lührmann, R., and Wahl, M.C. (2008) Structure and function of an RNase H domain at the heart of the spliceosome. *EMBO J.*, **27**, 2929–2940.

173 Ritchie, D.B., Schellenberg, M.J., Gesner, E.M., Raithatha, S.A., Stuart, D.T., and Macmillan, A.M. (2008) Structural elucidation of a PRP8 core domain from the heart of the spliceosome. *Nat. Struct. Mol. Biol.*, **15**, 1199–1205.

174 Yang, K., Zhang, L., Xu, T., Heroux, A., and Zhao, R. (2008) Crystal structure of the beta-finger domain of Prp8 reveals analogy to ribosomal proteins. *Proc. Natl. Acad. Sci. U.S.A.*, **105**, 13817–13822.

175 Abelson, J. (2008) Is the spliceosome a ribonucleoprotein enzyme? *Nat. Struct. Mol. Biol.*, **15**, 1235–1237.

176 Lejeune, F. and Maquat, L.E. (2005) Mechanistic links between nonsense-mediated mRNA decay and pre-mRNA splicing in mammalian cells. *Curr. Opin. Cell Biol.*, **17**, 309–315.

177 McGlincy, N.J. and Smith, C.W.J. (2008) Alternative splicing resulting in nonsense-mediated mRNA decay: what is the meaning of nonsense? *Trends Biochem. Sci.*, **33**, 385–393.

178 Pickrell, J.K., Pai, A.A., Gilad, Y., and Pritchard, J.K. (2010) Noisy splicing drives mRNA isoform diversity in human cells. *PLoS Genet.*, **6**, e1001236.

179 Fox-Walsh, K.L. and Hertel, K.J. (2009) Splice-site pairing is an intrinsically high fidelity process. *Proc. Natl. Acad. Sci. U.S.A.*, **106**, 1766–1771.

180 Hertel, K.J. (2008) Combinatorial control of exon recognition. *J. Biol. Chem.*, **283**, 1211–1215.

181 Burgess, S.M. and Guthrie, C. (1993) A mechanism to enhance mRNA splicing fidelity: the RNA-dependent ATPase Prp16 governs usage of a discard pathway for aberrant lariat intermediates. *Cell*, **73**, 1377–1391.

182 Semlow, D.R. and Staley, J.P. (2012) Staying on message: ensuring fidelity in pre-mRNA splicing. *Trends Biochem. Sci.*, **37**, 263–273.

183 Mayas, R.M., Maita, H., Semlow, D.R., and Staley, J.P. (2010) Spliceosome discards intermediates via the DEAH box ATPase Prp43p. *Proc. Natl. Acad. Sci. U.S.A.*, **107**, 10020–10025.

184 Mayas, R.M., Maita, H., and Staley, J.P. (2006) Exon ligation is proofread by the DExD/H-box ATPase Prp22p. *Nat. Struct. Mol. Biol.*, **13**, 482–490.

185 Xu, Y.-Z. and Query, C.C. (2007) Competition between the ATPase Prp5 and branch region-U2 snRNA pairing modulates the fidelity of spliceosome assembly. *Mol. Cell*, **28**, 838–849.

186 Tseng, C.-K. and Cheng, S.-C. (2008) Both catalytic steps of nuclear pre-mRNA splicing are reversible. *Science*, **320**, 1782–1784.

187 Wu, S., Romfo, C.M., Nilsen, T.W., and Green, M.R. (1999) Functional recognition of the 3' splice site AG by the splicing factor U2AF35. *Nature*, **402**, 832–835.

188 Chua, K. and Reed, R. (1999) The RNA splicing factor hSlu7 is required for correct 3' splice-site choice. *Nature*, **402**, 207–210.

189 Cvitkovic, I. and Jurica, M.S. (2013) Spliceosome database: a tool for tracking components of the spliceosome. *Nucleic Acids Res.*, **41**, D132–D141.

7
Splicing Decisions Shape Neuronal Protein Function across the Transcriptome

Jill A. Dembowski and Paula J. Grabowski

7.1
Introduction

Alternative precursor mRNA (pre-mRNA) splicing is widely used to diversify and regulate protein functions throughout the transcriptome. These mechanisms specify how various exonic sequences from a single gene are combined to generate multiple mRNA isoforms and their corresponding proteins as a function of cell type, developmental stage, and external stimulus. In recent years, it has become evident that neuronal functions can be fine-tuned in important ways through alternative splicing by the precise regulation of the kinetics of ion channel gating, coupling of synaptic activity to downstream cellular events, interactions of cell adhesion molecules, membrane or synaptic targeting, and regulation of enzyme activity. Yet key questions remain about the mechanisms of splicing regulation and their biological impact at the protein and physiological levels. (i) How are the patterns of tissue specificity and developmental control determined without sacrificing the fidelity of the reading frame of the mRNA? How are similar and divergent mechanisms coordinated across the transcriptome? (ii) To what extent does alternative splicing impact changes at the level of the proteome leading to changes in cellular or physiological function? By contrast, how much is biological noise? (iii) How do inducible splicing events respond to external stimuli? What are the pathways of communication between events detected at the cell membrane and relevant splicing factors in the nucleus?

In this review, we will discuss the current state of progress and experimental approaches used to understand alternative splicing in the nervous system with an emphasis on the central roles of RNA-binding proteins as mediators of these events (Figure 7.1a–d). Not only do these proteins participate directly in the recognition of splicing codes on pre-mRNA targets, but they are involved in regulatory networks in which they are themselves regulated by a variety of post-transcriptional mechanisms, including splicing. Mechanisms of splicing control can have important repercussions on protein modular functions in ways that facilitate synaptic function and plasticity, as well as the development of neuronal circuitry.

Posttranscriptional Gene Regulation: RNA Processing in Eukaryotes, First Edition. Edited by Jane Wu.

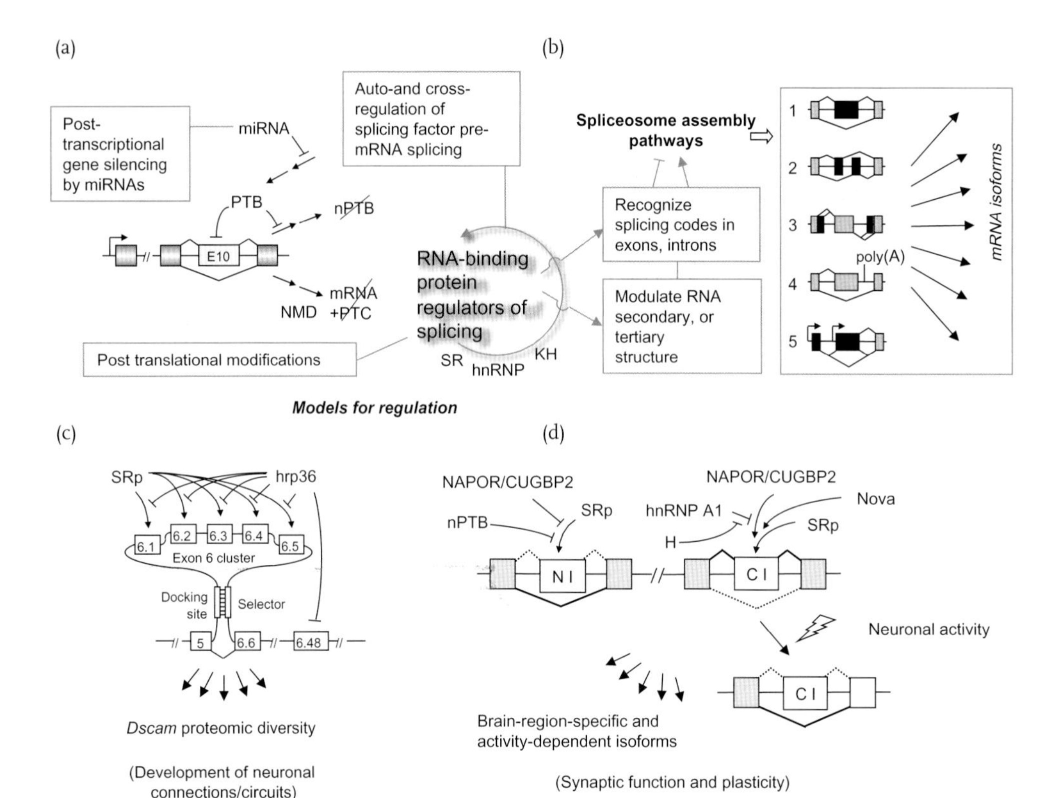

Figure 7.1 RNA-binding proteins from the hnRNP, SR, and KH families play central roles in mechanisms of alternative pre-mRNA splicing in the nervous system. (a) Cell-specific expression of splicing regulators (upstream regulation of splicing factors) can be mediated by negative feedback or cross-regulatory loops at the level of alternative splicing and by other posttranscriptional mechanisms involving the actions of microRNAs. Schematic illustrates the role of PTB in silencing exon 10 inclusion of the paralog, nPTB. Exon 10 skipping leads to a frameshift and exposure of a premature termination codon (PTC) leading to the destruction of the mRNA transcript by non-sense-codon-mediated mRNA decay (NMD). (b) RNA-binding proteins play diverse roles in the regulation of spliceosome assembly pathways (regulation and coordination of splicing decisions across the transcriptome). Inset illustrates the diverse patterns of alternative splicing including exon inclusion/skipping (1), mutually exclusive exon inclusion (2), alternative 5′ or 3′ splice site selection (3), alternative 3′ exon/poly(A) site selection (4), and alternative promoter/5′ exon selection (5). Differential intron retention is yet another mechanism (not shown). (c) The *Dscam* pre-mRNA model of mutually exclusive exon 6 splicing is based on comparative genomics, which identified highly conserved complementary sequences (docking site and selector). Enhancing roles of SR splicing factors (SRp) and silencing roles of hrp36 are indicated. The diversity of *Dscam* protein isoforms is important in establishing neuronal connections and circuits in the developing *Drosophila* nervous system. (d) The NMDA R1 receptor pre-mRNA model for combinatorial control. The schematic illustrates the splicing patterns in the forebrain where NI exon skipping and CI exon inclusion predominate. Enhancing roles of splicing factors are indicated by curved arrows; silencing roles of hnRNP A1, hnRNP H (H), NAPOR/CUGBP2, and nPTB are indicated. As a result of the excitation of primary cortical neurons (jagged arrowhead), the CI cassette exon switches from predominant exon inclusion to predominant skipping. The spliced isoforms are thought to modulate the roles of the receptor during the development and function of the synapse.

7.2
A Diversity of RNA-Binding Protein Regulators

The human genome contains ~600 RNA-binding proteins, but only a small fraction of these have been characterized as splicing regulators in the nervous system [1]. These splicing factors contact their RNA targets directly by recognition of short regulatory sequence motifs termed exonic or intronic splicing enhancers (ESE or ISE) and silencers (ESS or ISS). Splicing codes are combinations of ESE/S or ISE/S motifs that tune splicing patterns through specific interactions with cognate RNA-binding proteins.

RNA-binding proteins, including members of the serine-arginine (SR)-rich, heterogeneous ribonucleoprotein (hnRNP), and hnRNP K-homology (KH) families, have been shown to regulate alternative splicing events at the level of spliceosome function. The spliceosome is the dynamic molecular machinery that accomplishes the recognition of splice sites and the catalysis of intron removal and exon joining [2]. Spliceosome assembly involves the multistep binding of U1, U2, and U4/5/6 small nuclear ribonucleoprotein (snRNP) particles to the pre-mRNA. For an alternatively spliced exon (cassette exon), spliceosome assembly can proceed by multiple pathways leading to silencing (exon skipping) or enhancement (exon inclusion) (Figure 7.1b, 1), or may lead to other types of splicing decisions (Figure 7.1b, 2–5).

SR splicing factors, such as alternative splicing factor/splicing factor 2 (ASF/SF2), generally enhance exon inclusion and contain an RNA-binding domain (RBD) together with a second characteristic domain enriched in arginine-serine dipeptides. Upon binding to ESE motifs, SR proteins recruit factors such as U1 snRNP or U2 snRNP auxiliary factor (U2AF), or they may antagonize the effects of splicing silencers to facilitate exon definition [3, 4]. The exon definition stage of spliceosome assembly can determine whether the splicing pathway will proceed by exon inclusion or skipping (Figure 7.2).

Proteins in the hnRNP family, such as polypyrimidine tract-binding protein (PTB) or hnRNP A1, contain multiple RBDs and often promote exon skipping by antagonizing the roles of SR proteins or snRNPs [5, 6]. Nonetheless, some hnRNP family members enhance exon inclusion, such as Fox 1 and Fox 2. Others have dual roles capable of both enhancement and silencing, such as members of the CUG-binding protein (CUGBP) and ETR3-like family, termed CELF factors [7]. In the KH domain family, the Nova 1 and Nova 2 splicing factors have been characterized as dual functional regulators of alternative splicing events for transcripts encoding synaptic functions [8].

Proteins from the SR and hnRNP families of RNA-binding proteins are phosphorylated and dephosphorylated to adjust their functions. Phosphorylation can modify the ability of RNA-binding proteins to interact with RNA and other proteins, can alter subcellular localization, and can influence intranuclear localization [9]. Consequently, such adjustments can affect the ability of splicing enhancers and silencers to regulate alternative splicing events and therefore contribute to the complex functionalities of splicing codes, or may couple splicing to signal

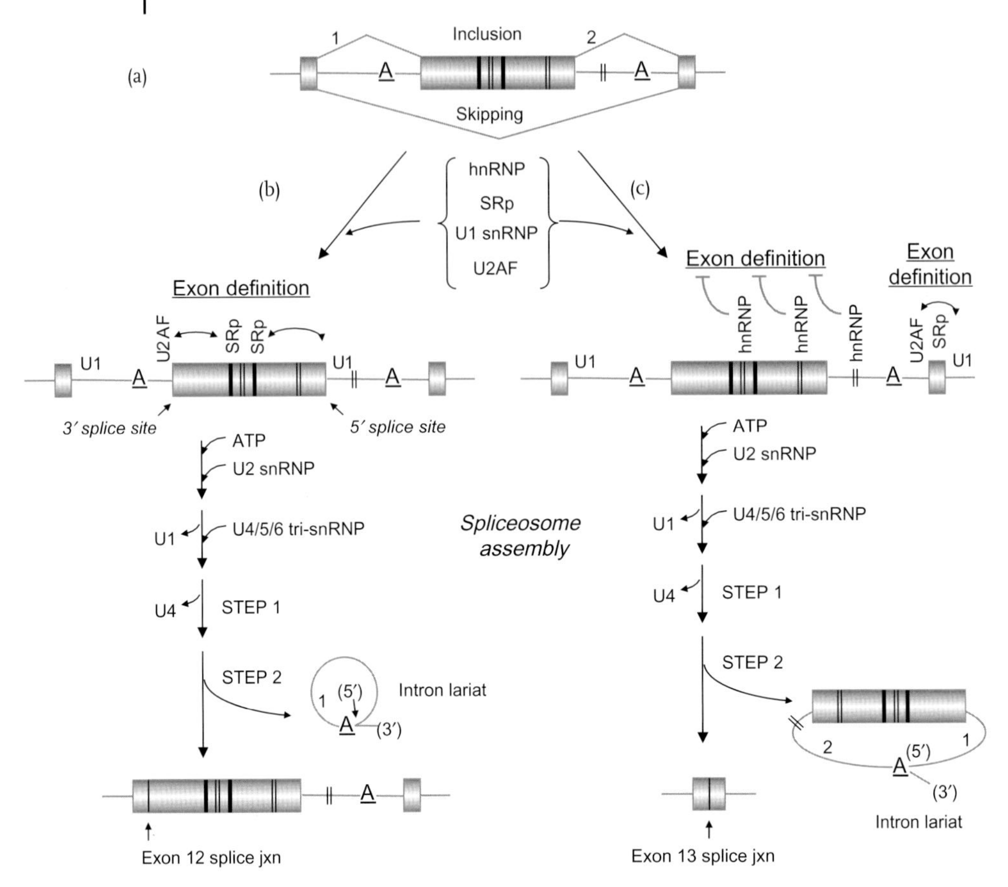

Figure 7.2 Key role for exon definition in modulating splicing decisions. (a) The binary splicing decision (exon inclusion or skipping pathways) for a pre-mRNA substrate is indicated. The branch site adenosines are underscored in introns 1 and 2. ESE (single bars) and ESS motifs (double bars) indicate splicing regulatory motifs in the middle cassette exon. An ISS is indicated by the double bar in intron 2. (b) Inclusion pathway: early spliceosome assembly events mediate exon definition, in which the 5′ splice site of the middle exon is recognized by U1 snRNP, and the polypyrimidine tract and AG dinucleotide of the 3′ splice site are recognized by U2AF. SR splicing factors bound to ESE motifs within the cassette exon facilitate exon definition. U2 snRNP then joins the spliceosome in an ATP-dependent event to recognize the branch site of intron 1. Following this step, U1 snRNP is released, and U4/5/6 tri-snRNP joins the spliceosome. The 5′ splice site of intron 1 is also recognized in these early assembly events and brought into proximity of the branch site of intron 1. Further RNA rearrangements, including the release of U4 snRNP, generate the active spliceosome, which catalyzes chemical steps 1 and 2. The products of the reactions include the intron 1 lariat and the exon1-exon2-intron2-exon3 molecule. (c) Skipping pathway: in the alternative spliceosome assembly pathway, conditions favor the binding of hnRNP splicing silencers to the ESS and ISS motifs. These events antagonize middle exon definition. Exon skipping occurs by the definition of exon 3, and the excision of the large intron lariat with intron 1, the middle exon and intron 2 sequences. The newly formed spliced junctions, and the 5′ and 3′ ends of the intron lariats, are indicated.

transduction pathways. These modifications may serve important roles in mechanisms that integrate splicing responses and extracellular signaling.

7.3 Gene-Specific and Global Experimental Approaches to Splicing Mechanisms

A large variety of experimental tools are in hand to address questions about gene-specific mechanisms and global networks of regulation (Figure 7.3a–d). Large-scale randomization and selection approaches have led to the identification of numerous ESE/S and ISE/S sequences that functionally affect splicing patterns. Several approaches start with a large pool of random sequence variants to select from, and then apply a selection that demands a functional change in the splicing pattern by the use of *in vitro* or *in vivo* splicing assays. The results of these selections have been used to generate ESEFinder, RESCUE-ESE, and FAS-ESS web servers, which offer databases of splicing motifs with search capabilities (ESE-Finder3.0, http://rulai.cshl.edu/cgi-bin/tools/ESE3/esefinder.cgi?process=home; RescueESE, http://genes.mit.edu/burgelab/rescue-ese/; FAS-ESS, http://genes.mit.edu/fas-ess/). For the ESEFinder, the Krainer lab identified regulatory motifs by inserting randomized sequences (6–8 nucleotides) into the middle exons of a large pool of splicing reporters, followed by the selection and sequencing of spliced variant winners [10]. Selection was done iteratively much like the original SELEX (Systematic Evolution of Ligands by Exponential Enrichment) procedure. Winners were confirmed based on their ability to switch the splicing pattern from exon skipping to inclusion.

For the RESCUE-ESE, the Burge lab developed computational approaches to search for hexamers that were preferentially distributed in exons compared with introns together with enrichment in exons with weak versus strong splice sites [11]. In the second step, ESEs were identified by insertion into splicing reporters followed by *in vivo* splicing assays that required the rescue of exon inclusion. For FAS-ESS, splicing silencers were identified in large scale from pools of splicing reporters with random decamers inserted into the middle exon. The key to identifying ESS motifs was to fuse green fluorescent protein (GFP) to the splicing reporter in such a way that fluorescence could be used as a read out for exon skipping. Fluorescence-activated cell sorting was then used to isolate cells expressing the winners, followed by sequence analysis [12].

It is helpful to keep in mind that the roles of these regulatory sequence motifs in natural pre-mRNAs will likely be influenced by the strength of the flanking splice sites, the sequence context, and by the expression levels of the regulatory factors, which may vary in different cells or during development. Another issue is the copy number of the regulatory motifs. A single motif that has a weak effect on splicing can have a strong effect in the presence of additional copies of the same or functionally similar motifs.

Three criteria can be used to experimentally verify a splicing code in a pre-mRNA of interest. (i) Changes in the copy number and/or arrangement of the

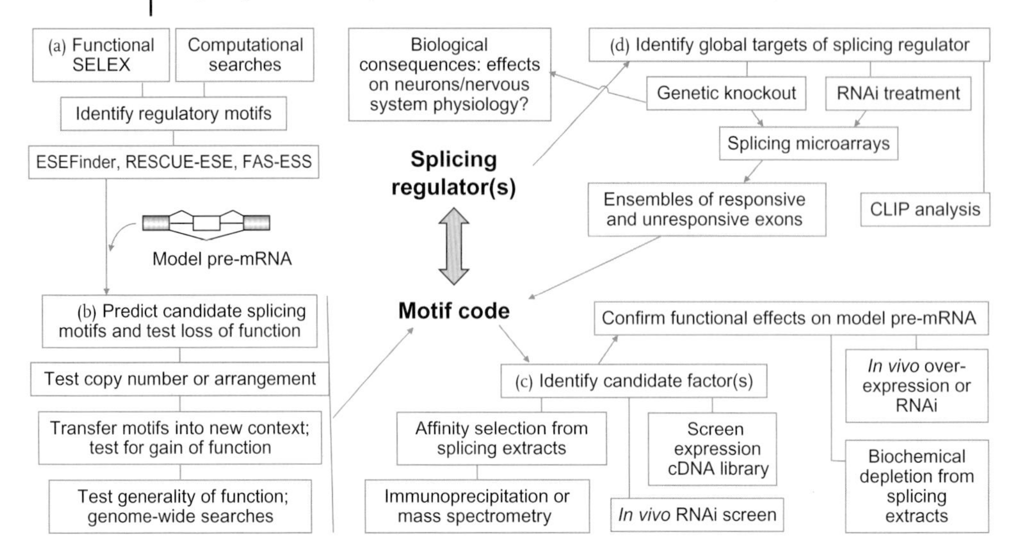

Figure 7.3 Experimental tools for gene-specific and global analysis of splicing codes and factors. (a) The large-scale identification of RNA motifs that regulate splicing patterns by enhancement or silencing have come from functional SELEX and computational approaches. (b) Splicing codes can be identified for a model pre-mRNA (shown schematically) by site-directed mutations that cause a loss of function starting with predictions from web servers (see text). The capacity for the motif pattern to tune the splicing pattern can be assessed by mutations that alter the copy number or spatial arrangement of the motifs. Generality of function can be tested by computational searches for similar motif patterns throughout the transcriptome, followed by functional validation. (c) Splicing factors can be identified from splicing extracts by affinity selection or UV cross-linking to an RNA substrate containing the splicing code, followed by immunoblotting or mass spectrometry. Functional confirmation can be performed by *in vivo* overexpression or RNAi knockdown, or by biochemical depletion from splicing extracts. Expression library screening can be used as an alternative method to identify proteins with regulatory capabilities. (d) Global targets of a splicing regulator can be identified by genetic knockouts or by *in vivo* RNAi treatments together with splicing microarray analysis. This analysis reveals ensembles of responsive and unresponsive exons, which can potentially be analyzed at the sequence level to identify RNA motifs differentially enriched in the responders. CLIP analysis is another large-scale *in vivo* method based on RNA protein binding. Finally, the biological effects of the splicing events under study can potentially be addressed in model organisms, such as mice and flies, in which a targeted gene knockout of the splicing factor has been engineered.

RNA motifs should tune the splicing pattern to a significant degree. (ii) The splicing code should be functionally transferable to a new splicing substrate by gain of function. This will either verify that the complete sequence code has been identified, or serve as an assay to identify the missing components. (iii) Genome-wide searches for the splicing code should identify additional regulatory targets at frequencies greater than random chance. Testing such candidates can further refine the code and may suggest interesting new biological targets.

The motif code offers a handle on the identification of the regulatory factors, which can be manipulated by overexpression or RNA interference (RNAi). Fluorescent splicing reporters have also been used together with cDNA screening to identify splicing factors that affect a particular splicing pattern. Here, the expression of the relevant protein(s) activates a switch to a splicing pattern, which can be detected through its fusion to GFP [13, 14]. Similar fluorescent splicing reporters have been developed for use in mammalian cells and transgenic mice [15–17].

Understanding the roles of splicing factors in gene-specific mechanisms and their biological significance across the transcriptome calls for genetic approaches. Gene knockouts produced in transgenic animals offer a powerful approach to examine functional outcomes in a relevant biological setting. RNAi is another useful way to knock down the expression of a splicing factor in a relevant cell line or primary culture system. A variety of microarray platforms are available for analysis, which can simultaneously profile thousands of splicing patterns [18–21]. Cross-linking and immunoprecipitation (CLIP) analysis is also available as a complementary method for the discovery of candidate targets of a splicing factor based on RNA–protein binding [22]. Massively parallel sequencing, or pyrosequencing, is a novel gene expression profiling approach with potential applications to the splicing field [23]. A number of groups are currently exploring its applications to splicing, since it is an open-ended method with discovery potential.

Proteomics approaches greatly extend classical biochemical approaches for the analysis of splicing mechanisms. These procedures are geared toward discovery based on the use of mass spectrometry of affinity-purified RNA–protein complexes. Proteomic analysis of spliceosomes purified from splicing extracts has identified as many as 311 protein components indicating that this is a vast molecular machine most likely with different functional forms [24–26]. The finding that some of these components are factors involved in transcription, polyadenylation/3′ end formation, and nuclear export of mRNAs indicates that splicing is integrated with other pathways of gene expression. More recently, proteomic analysis has determined that such integration requires SR proteins based on an *in vitro* system that couples RNA polymerase II transcription to splicing [27].

7.4 Alternative Splicing of *Dscam* Pre-mRNA: Mechanism and Significance for the Development of Neuronal Circuits

In the developing *Drosophila* nervous system, the Down syndrome cell adhesion molecule (*Dscam*) transcript has the potential to produce 38 016 unique isoforms by alternative splicing of 95 mutually exclusive exons distributed in four distinct clusters [28]. This demonstrates the greatest potential for molecular diversity generated from a single gene by splicing. That a single exon from each variant cluster is found in any of the mature mRNA transcripts is an unusual feature of this mechanism. Existing models proposed for the splicing of pairs of mutually exclusive exons from other systems have been ruled out based on their inability to explain *Dscam* splicing, which involves as many as 48 variants in the exon 6 cluster. These models involve steric limitations on intron size, specific arrangements of splice sites corresponding to the major (U2-type) and minor (U12-type) snRNPs, or post-splicing mechanisms, such as non-sense-codon-mediated mRNA decay (NMD).

For the exon 6 cluster, the question is not only how a particular exon variant is selected, but also how the 47 remaining variants are excluded. To gain insight into possible models for the exon 6 cluster, a comparative genomics approach was used to identify two types of conserved sequences in *Dscam* genes from four groups of phylogenetically related insects [29]. The most highly conserved sequence element identified from the sequence alignments is called the docking site, which is located in the intron just downstream of the constitutive exon 5. The selector sequence is a second type of conserved sequence motif found in the intron just upstream from each of the exon 6 variants. Notably, each selector sequence bears complementarity to the docking site (Figure 7.1c). The current model holds that the selection of a particular exon variant occurs by Watson–Crick base pairing between the docking site and the selector site of that exon variant. Additional evidence for this model comes from the honeybee, which shows compensatory base pair changes that retain base pairing between the docking and selector sites.

How the remaining exon 6 variants are excluded has been proposed to involve RNA-binding proteins that execute exon silencing in a mechanism that covers the entire exon 6 cluster. Recently, evidence for the role of the RNA-binding protein, hrp36, was shown by a large-scale screen for splicing mistakes using an RNAi approach [30]. RNAi knockdown of hrp36 from *Drosophila* S2 cells resulted in the inclusion of multiple exon 6 variants in each mRNA isoform. This defect was specific for the exon 6 cluster as opposed to the three other exon clusters. Throughout the exon 6 cluster of the *Drosophila melanogaster Dscam* gene, potential hrp36 binding sites were found by sequence searches to be particularly enriched in the exons of that region. Similar trends were observed in 12 related species, suggesting that hrp36 controls mutually exclusive splicing in a cluster-wide manner. The mechanism of action of hrp36 is thought to involve antagonism of SR splicing factors. In contrast to the results in S2 cells, exon 6 splicing occurs normally in *hrp36* null flies. This apparent discrepancy may be explained by the functional redundancy of hnRNP proteins in the fly or by differences in the balance or tissue specificity of protein levels.

Pairing of the docking and selector sites together with antagonism between SR and hnRNP splicing factors provides a plausible model to explain the fidelity of mutually exclusive splicing, yet important questions still remain. To what extent does complementarity between the docking and selector sites affect the relative preference for the splicing of exon 6 variants? If this is a major determinant for selection, mutation of one or several nucleotides in the docking site should shift the splicing patterns based on those selector sequences with the highest base-pairing potential. A related question is whether the interactions between the docking and selector sites are regulated by protein or RNA mediators, and if so how? The question why hrp36 is essential for this mechanism in S2 cells but not the fly is also an important issue for future experiments.

Recent evidence supports a significant biological role for *Dscam* alternative splicing in neural circuit assembly. Hattori *et al.* used homologous recombination to reduce the *Dscam* allele to a single isoform (*Dscam*^{single}). They generated three homozygous mutant flies, each expressing a different *Dscam*^{single} allele [31]. Like *Dscam*^{null} alleles, all three *Dscam*^{single} alleles were found to have a recessive lethal phenotype. While homozygous *Dscam*^{null} animals survived to late larval stages with disorganized axonal pathways, the homozygous *Dscam*^{single} animals died early in larval development with drastically disorganized axonal pathways. Thus, the diversity of *Dscam* isoforms is essential, and reducing expression to a single *Dscam* isoform in all neurons is highly detrimental in the fly. This study also demonstrated that multiple *Dscam* isoforms were essential for a population of neurons. In support of this, several recent studies have shown that interactions between identical *Dscam* isoforms (homophilic interactions) trigger cellular responses that enable self-recognition and self-avoidance of sister dendrites within the same cell [32–34]. Together, these studies are consistent with a role of *Dscam* in providing each neuron with a unique cell surface identity that is essential for neuronal circuit assembly in developing *Drosophila*. Despite these remarkable findings for *Drosophila*, alternative splicing does not diversify the mammalian *Dscam* homologs. Instead, synaptic proteins, such as protocadherins and neurexins, facilitate these functions in higher organisms with the involvement of alternative splicing [28].

7.5 The NMDA R1 Receptor: Brain-Region-Specific and Activity-Dependent Splicing

N-methyl-D-aspartate (NMDA) receptors are calcium permeable glutamate receptors present at the postsynaptic membrane of excitatory synapses in the mammalian central nervous system. These receptors are important for neuronal survival and learning and memory, while abnormalities in receptor activity have been implicated in neurological disorders. These receptors allow calcium influx that regulates signaling pathways involved in synaptic plasticity in response to the binding of glutamate and glycine under depolarizing conditions.

Alternative splicing of three exons of the NR1 subunit of the NMDA receptor (*GRIN1*) pre-mRNA generates eight receptor isoforms with different functional characteristics [35, 36]. The NI cassette exon (exon 4b), which encodes a portion of the extracellular domain, modulates receptor sensitivity to zinc ions, protons, and polyamines. The CI cassette exon (exon 19), which encodes an intracellular region, regulates membrane localization of the receptor, as well as intracellular signaling from the receptor [37–39]. In addition, mutually exclusive splicing of C2 and C2′ exons adjusts receptor trafficking to the synapse [40]. Thus, alternative splicing fine-tunes discrete NMDA receptor properties.

The brain-region-specific splicing factor, NAPOR/CUGBP2, silences the NI cassette exon, and enhances the CI cassette exon of the *GRIN1* transcript (Figure 7.1d) [41]. These dual functions in splicing regulation, together with the forebrain-enriched and hindbrain-deficient pattern of protein expression, are thought to account for the natural distribution of NR1 isoforms in the brain. According to this model, the reciprocal splicing patterns of the NI and CI cassette exons are directed in the forebrain by the presence of NAPOR/CUGBP2, whereas both of these splicing events switch to the opposite patterns in the hindbrain, by loss of function. The molecular mechanisms that explain the dual functions of NAPOR/CUGBP2 are not yet understood, although its binding preference for CUG-related RNA-binding motifs is characteristic of members of the CELF family of splicing factors [42]. CUG-containing RNA motifs have been found in the intron downstream of the CI exon and the intron upstream of the NI exon where they are responsible for the enhancing and silencing activities, respectively, of NAPOR/CUGBP2 [41].

Alternative splicing of the CI cassette exon is under combinatorial control by RNA-binding proteins from the SR, hnRNP, and KH families of splicing factors. The dynamic balance of these splicing factors in different brain regions determines whether CI will be predominantly included or skipped in the population of mature *GRIN1* transcripts (Figure 7.1d). The CI cassette exon is enhanced by the SR family members ASF/SF2 and SC35 [43], hnRNP H [43], and the KH domain protein, Nova [1], in addition to NAPOR/CUGBP2. It is also silenced by hnRNP A1 [43]. In each of these cases, known cis-regulatory motifs within the exon and downstream intron have been shown to contribute to regulation by these factors. A splicing code composed of multiple copies of UAGG motifs in the exon and a GGGG motif adjacent to the 5′ splice site is functionally recognized by hnRNP A1 in the silencing mode and by hnRNP H in the antisilencing mode. The CI cassette exon is a representative example of the complex regulation of an exon by multiple factors, which leads to the fine-tuning of its tissue- and developmental-specific protein activities [44].

7.6 Alternative Splicing Response to Neuronal Excitation

Recent studies have shown that the CI cassette is an inducible exon in living cells, since its splicing pattern shifts to predominant exon skipping after membrane depolarization or as a consequence of overexpression of a constitutively active form of Ca^{2+}/calmodulin-dependent protein kinase (CaMK) IV [45]. The CI cassette exon encodes an intracellular peptide region that is believed to mediate intracellular signaling from the cell membrane to the nucleus. In this way, the decrease in CI cassette exon inclusion may be part of a feedback loop that reduces signaling in neurons to stabilize neuronal excitability during chronic excitation. This scenario may represent the involvement of splicing in a protective cellular response in agreement with the model of synaptic homeostasis. Two recent papers provide insight into the splicing codes and factors involved in the responsiveness of the CI cassette exon.

An and Grabowski developed a neuron-specific splicing reporter system to study how cell excitation adjusts the inclusion of the CI cassette exon in primary rat cortical neurons [46]. In this system, membrane depolarization stimulated CI cassette exon skipping, and this effect was reversed after KCl washout. The UAGG silencing code previously shown to involve regulation by hnRNP A1 was required for the responsiveness of exon skipping in the neurons. This code was used to predict and identify additional neuronal exons that show sensitivity to depolarization treatment in this culture system. Nuclear extracts derived from depolarized cultures showed enhanced binding of hnRNP A1 in UV cross-linking and affinity chromatography assays, suggesting that hnRNP A1, or factors that regulate A1, should be considered as candidate mediator(s) of these events.

Notably, specific antagonists of the NMDA receptor and several cell-permeable inhibitors of signal transduction pathways, including protein kinase A and CaMKs, were found to attenuate splicing responsiveness in the primary neurons. These findings suggest that one or more signaling pathways from NMDA receptors located at the membrane to factors in the nucleus are responsible for inducing changes at the level of splicing.

A similar study by Lee *et al.* used differentiated P19 cells to identify cis-regulatory elements responsible for depolarization-induced silencing of the endogenous CI cassette exon [47]. In this system, the effects on CI splicing were also found to be reversible, and involved signaling through the CaMK IV pathway. These authors used single-nucleotide scanning mutagenesis to identify two degenerate CaMK IV-responsive RNA elements (CaRREs) within the CI cassette exon. One element, CaRRE type 1, is similar to an element defined previously in the 3′ splice site region of the stress-regulated exon (STREX) of the big potassium (BK) channel, and the second, CaRRE type 2, appears to be a novel motif. Genome-wide searches for additional exons that contain CaRREs successfully identified several new CaMK IV-responsive exons in transcripts with functions such as calcium homeostasis, intracellular signaling, and vesicular transport.

Together these studies highlight the importance of conserved sequence elements in the regulation of cellular responsiveness of groups of coregulated exons, but this work is in an early stage. Many questions remain about the nature of the signal transduction pathways involved in conveying information from the cell membrane to splicing factors in the nucleus, and the nature of the landscape of responsive exons that has yet to be fully characterized. The extent to which the induced splicing changes reflect corresponding changes at the protein level, as opposed to biological noise, is also not known. A recent study has described the RNA splicing capacity of neuronal dendrites in hippocampal cultures. Although speculative, this raises the question of whether certain mRNAs with one or two unspliced introns might be targeted to dendrites for local splicing in response to changes in synaptic activity [48].

7.7 Splicing Silencing by PTB: Versatility of Mechanism and Cross-Regulation of nPTB

PTB, also known as PTBP1 and hnRNP I, is a widely expressed RNA-binding protein that has been well studied for its functions to silence neural- and muscle-specific exons in a variety of cell types [49]. PTB and PTB isoforms bind to pyrimidine-rich motifs related to the core sequence, UCUU, through four RNA binding domains (RBDs) to silence alternative exons [50, 51]. PTB is best characterized for its role in silencing the neural-specific NI exon of the tyrosine kinase, *c-src*, in non-neuronal cells [44]. PTB silences the NI exon by binding to motifs in the flanking introns to loop out the exon and block access of the splicing machinery to the splice sites. PTB can also bind near the 3′ splice site of an alternative exon to antagonize $U2AF^{65}$ binding [52] or sequester the branch site [53]. It can also silence an exon by propagating across the RNA to antagonize binding of splicing enhancers and the general splicing machinery [50]. Two recent studies by the Valcarcel and Black labs demonstrate that PTB can silence alternative exons by inhibiting exon and intron definition events, respectively [54, 55]. Thus, PTB is a remarkably versatile splicing regulator.

Recently, Oberstrass *et al.* solved the NMR structures of each PTB RBD bound to RNA [56]. The structural studies confirm that each RBD binds to CU-rich elements, but that each RBD prefers a slightly different variation of this RNA element. This observation explains the variability in previously defined consensus PTB RNA elements [52, 57]. The structure of RBD34 (RBDs 3 and 4 separated by an interdomain linker) bound to RNA demonstrates that these two domains can bind to immediately adjacent stretches of RNA and suggests a mechanism by which two RBDs can bind in an antiparallel manner to bring distant RNA elements close together. This is supportive of a model that has been suggested through biochemical studies by which a single PTB molecule can function to bend RNA to loop out the branch site or alternative exon [58].

In the nervous system, a brain-enriched paralog of PTB (nPTB or PTBP2) has a decreased ability to bind to and regulate some PTB RNA elements. The switch from PTB to nPTB during brain development is associated with an increase in inclusion of a subset of neuron-specific exons that are silenced by PTB in non-neuronal cells. Recent efforts have been aimed at understanding how the tissue-specific balance of PTB and nPTB is regulated, and the effects of these molecules on the global regulation of ensembles of splicing events in the nervous system.

An interesting characteristic of PTB is that it has the ability to regulate splicing of an alternative exon within its own transcript and within the *nPTB* transcript to control protein expression. PTB binds to intronic and exonic elements within exon 11 of its own pre-mRNA to drive skipping of this exon [59]. Exon 11 skipping introduces a frameshift exposing a premature stop codon in exon 12, which targets PTB for destruction by NMD. In this mode of autoregulation, PTB functions in a negative feedback loop to maintain protein levels at a low concentration in non-neuronal cells. Recently, this type of regulation has been demonstrated for several additional proteins from the hnRNP family [20]. Furthermore, members of the SR family of splicing enhancers have been shown to autoregulate their own expression by the reverse mechanism in which enhancement of an internal exon introduces a premature stop codon to target the transcript for NMD.

In the nervous system, cross-regulation between PTB family members can drive mutually exclusive isoform expression in different cell types. The Black lab recently demonstrated that PTB has the ability to regulate nPTB expression at the level of alternative splicing [60]. In the mouse brain, immunostaining with antibodies specific for each isoform showed that nPTB is exclusively expressed in neurons, while PTB expression is restricted to glial cells. One explanation for this mutually exclusive pattern of expression is that PTB silences splicing of exon 10 of the nPTB transcript by a mechanism similar to autoregulation (Figure 7.1a). Exon 10 skipping causes a shift in reading frame exposing a premature stop codon, thereby targeting the nPTB transcript for NMD. Curiously, this mechanism does not completely explain the lack of nPTB protein expression in non-neuronal cells, since nPTB mRNA with exon 10 is found in these cells. The authors speculate that PTB may also be involved in regulating nPTB at the level of protein synthesis. In this study, exon junction microarrays were also used to identify distinct, yet overlapping, sets of exons regulated by PTB and nPTB. At the protein level, such changes may be associated with the differentiation of neuronal and glial cells during brain development.

7.8 Upstream Regulation of Splicing Factors by MicroRNAs

Surprisingly, the expression of PTB is also regulated through posttranscriptional events involving micro RNAs (miRNAs). miRNAs are a well-studied group of ~22 nucleotide RNAs that bear complementarity to sequences in the 3′ untranslated regions (UTRs) of certain mRNAs where their interactions either repress protein synthesis or promote mRNA decay [61]. The question of how the repression of nPTB by PTB might be relieved in neurons has been addressed by two recent studies demonstrating the role of miRNAs in the repression of PTB expression during cell differentiation. Makeyev *et al.* demonstrated that the neuron-specific miRNA, miR-124, inhibits expression of PTB during neuronal differentiation, and showed that expression of an exogenous form of this miRNA into neuroblastoma cells triggers neuron-like differentiation [62]. They also showed that the switch from PTB to nPTB induced by miR-124 is associated with a global switch from non-neuronal to neuronal splicing patterns. These results are in agreement with immunofluorescence experiments showing a mutually exclusive pattern of expression of PTB and nPTB in cultures derived from the hippocampus or cerebellum [60]. In this study, PTB was found exclusively in glial cells or neural progenitors, whereas nPTB was found exclusively in neurons.

Using a different system, the Black lab demonstrated a role for miR-133 in the regulation of nPTB expression in differentiating muscle cells [63]. They found that nPTB was expressed in proliferating myoblasts, but this was essentially eliminated in differentiated myotubes. Levels of PTB were also reduced during differentiation. Curiously, the differentiated myotubes contained nPTB mRNA but not the corresponding protein. This finding suggested a role for miRNAs, since these molecules can bind to sequence elements in the 3′ UTR of an mRNA template to block protein synthesis. Sequence motifs for miR-133 and miR-1/206 were indeed found in the 3′ UTR of nPTB mRNA, and increased expression of these miRNAs was associated with myoblast differentiation. Functional evidence for the roles of these miRNAs in regulating PTB expression was shown by introducing antisense RNAs to block miRNA function and by transfer of the 3′ UTR sequence elements to luciferase reporters to show gain of function.

These studies suggest that global patterns of alternative splicing can be controlled in different cell types and during development by miRNAs operating upstream of the splicing process itself (Figure 7.1a). How many other splicing factors are targets of miRNA regulation, and what mechanisms control the expression of miRNAs? A number of brain-specific miRNAs have been identified, but their roles have not yet been defined.

7.9
Conclusions and Prospects

Insights into the molecular mechanisms of alternative splicing and their biological impact in neuronal cells and model organisms have been facilitated by a powerful array of experimental technologies. Genomic approaches are now providing ways to map global splicing patterns for discovery potential and to greatly expand the scope of analysis. Recent advances have taught us how sophisticated the interplay of splicing codes and regulatory factors can be in determining outcomes for even simple, binary splicing decisions. We know far less about the nature of the regulatory mechanisms as they occur in their natural, dynamic environments in cells and tissues. Understanding splicing mechanisms and their biological impact will continue to be highly important research endeavors. Moreover, the study of unusual mechanisms, such as that of *Drosophila Dscam*, will fuel the development of further advances in technology as they push existing methodologies to their limits.

Determining the effect(s) of a splicing change at the protein level provides another technical challenge. It is one thing to identify a protein's functions, but quite another to discover the functional contributions of a small peptide region corresponding to an alternatively spliced segment. In one successful study, electrophysiology measurements in neurons showed that the amplitude of N-type currents can be modulated in the calcium channel, Cav2.2, by the mutually exclusive splicing of a pair of exons each of which encode 32 amino acids near the carboxyl terminus of the protein [64]. There are other examples like this. But in the great majority of cases, we have no information on whether a particular splicing change makes a difference at the level of protein function or cell physiology. Techniques that engineer changes in one alternatively spliced protein isoform, such as exon-specific RNAi, will serve as valuable tools to study the contributions of individual exons to protein function [65]. Subtlety is another issue. It is an open question whether some splicing changes represent a type of biological noise that has little or no impact on the protein pool.

Understanding how splicing is integrated with other levels of gene expression and signal transduction pathways will also represent important research avenues for future work. The coupling of splicing to transcription, 3′ end formation, and nuclear export is well documented, but how coupling affects splicing mechanisms is less clearly defined. Networks of autoregulation and cross-regulation are becoming more and more evident. Also, the involvement of miRNAs in the regulation of splicing factor expression opens many new questions about how these networks are integrated with cellular differentiation and development.

Acknowledgments

We thank the members of the Grabowski laboratory for critical comments on the manuscript. Our research efforts are supported by a grant from the NIH (GM068584) to P.J.G.

References

1 Ule, J. and Darnell, R.B. (2006) RNA binding proteins and the regulation of neuronal synaptic plasticity. *Curr. Opin. Neurobiol.*, **16**, 102–110.

2 Brow, D.A. (2002) Allosteric cascade of spliceosome activation. *Annu. Rev. Genet.*, **36**, 333–360.

3 Graveley, B.R. (2000) Sorting out the complexity of SR protein functions. *RNA*, **6**, 1197–1211.

4 Maniatis, T. and Tasic, B. (2002) Alternative pre-mRNA splicing and proteome expansion in metazoans. *Nature*, **418**, 236–243.

5 Burd, C.G., Swanson, M.S., Gorlach, M., and Dreyfuss, G. (1989) Primary structures of the heterogeneous nuclear ribonucleoprotein A2, B1, and C2 proteins: a diversity of RNA binding proteins is generated by small peptide inserts. *Proc. Natl. Acad. Sci. U.S.A.*, **86**, 9788–9792.

6 Ladd, A.N. and Cooper, T.A. (2002) Finding signals that regulate alternative splicing in the post-genomic era. *Genome Biol.*, **3**, reviews0008.

7 Ladd, A.N., Charlet, N., and Cooper, T.A. (2001) The CELF family of RNA binding proteins is implicated in cell-specific and developmentally regulated alternative splicing. *Mol. Cell Biol.*, **21**, 1285–1296.

8 Ule, J., Ule, A., Spencer, J., Williams, A., Hu, J.S., Cline, M., Wang, H., Clark, T., Fraser, C., Ruggiu, M., Zeeberg, B.R., Kane, D., Weinstein, J.N., Blume, J., and Darnell, R.B. (2005) Nova regulates brain-specific splicing to shape the synapse. *Nat. Genet.*, **37**, 844–852.

9 Stamm, S. (2007) Regulation of alternative splicing by reversible protein phosphorylation. *J. Biol. Chem.*, **283**, 1223–1227.

10 Cartegni, L., Chew, S.L., and Krainer, A.R. (2002) Listening to silence and understanding nonsense: exonic mutations that affect splicing. *Nat. Rev. Genet.*, **3**, 285–298.

11 Fairbrother, W.G., Yeh, R.F., Sharp, P.A., and Burge, C.B. (2002) Predictive identification of exonic splicing enhancers in human genes. *Science*, **297**, 1007–1013.

12 Wang, Z., Rolish, M.E., Yeo, G., Tung, V., Mawson, M., and Burge, C.B. (2004) Systematic identification and analysis of exonic splicing silencers. *Cell*, **119**, 831–845.

13 Wu, J.Y., Kar, A., Kuo, D., Yu, B., and Havlioglu, N. (2006) SRp54 (SFRS11),

a regulator for tau exon 10 alternative splicing identified by an expression cloning strategy. *Mol. Cell Biol.*, **26**, 6739–6747.

14 Kar, A., Havlioglu, N., Tarn, W.Y., and Wu, J.Y. (2006) RBM4 interacts with an intronic element and stimulates tau exon 10 inclusion. *J. Biol. Chem.*, **281**, 24479–24488.

15 Nasim, M.T. and Eperon, I.C. (2006) A double-reporter splicing assay for determining splicing efficiency in mammalian cells. *Nat. Protoc.*, **1**, 1022–1028.

16 Orengo, J.P., Bundman, D., and Cooper, T.A. (2006) A bichromatic fluorescent reporter for cell-based screens of alternative splicing. *Nucleic Acids Res.*, **34**, e148.

17 Bonano, V.I., Oltean, S., and Garcia-Blanco, M.A. (2007) A protocol for imaging alternative splicing regulation *in vivo* using fluorescence reporters in transgenic mice. *Nat. Protoc.*, **2**, 2166–2181.

18 Blencowe, B.J. (2006) Alternative splicing: new insights from global analyses. *Cell*, **126**, 37–47.

19 Fagnani, M., Barash, Y., Ip, J.Y., Misquitta, C., Pan, Q., Saltzman, A.L., Shai, O., Lee, L., Rozenhek, A., Mohammad, N., Willaime-Morawek, S., Babak, T., Zhang, W., Hughes, T.R., van der Kooy, D., Frey, B.J., and Blencowe, B.J. (2007) Functional coordination of alternative splicing in the mammalian central nervous system. *Genome Biol.*, **8**, R108.

20 Ni, J.Z., Grate, L., Donohue, J.P., Preston, C., Nobida, N., O'Brien, G., Shiue, L., Clark, T.A., Blume, J.E., and Ares, M., Jr. (2007) Ultraconserved elements are associated with homeostatic control of splicing regulators by alternative splicing and nonsense-mediated decay. *Genes Dev.*, **21**, 708–718.

21 Yeakley, J.M., Fan, J.B., Doucet, D., Luo, L., Wickham, E., Ye, Z., Chee, M.S., and Fu, X.D. (2002) Profiling alternative splicing on fiber-optic arrays. *Nat. Biotechnol.*, **20**, 353–358.

22 Ule, J., Jensen, K., Mele, A., and Darnell, R.B. (2005) CLIP: a method for identifying protein-RNA interaction sites in living cells. *Methods*, **37**, 376–386.

23 Torres, T.T., Metta, M., Ottenwalder, B., and Schlotterer, C. (2008) Gene expression profiling by massively parallel sequencing. *Genome Res.*, **18**, 172–177.

24 Makarov, E.M., Makarova, O.V., Urlaub, H., Gentzel, M., Will, C.L., Wilm, M., and Luhrmann, R. (2002) Small nuclear ribonucleoprotein remodeling during catalytic activation of the spliceosome. *Science*, **298**, 2205–2208.

25 Rappsilber, J., Ryder, U., Lamond, A.I., and Mann, M. (2002) Large-scale proteomic analysis of the human spliceosome. *Genome Res.*, **12**, 1231–1245.

26 Zhou, Z., Licklider, L.J., Gygi, S.P., and Reed, R. (2002) Comprehensive proteomic analysis of the human spliceosome. *Nature*, **419**, 182–185.

27 Das, R., Yu, J., Zhang, Z., Gygi, M.P., Krainer, A.R., Gygi, S.P., and Reed, R. (2007) SR proteins function in coupling RNAP II transcription to pre-mRNA splicing. *Mol. Cell*, **26**, 867–881.

28 Schmucker, D., Clemens, J.C., Shu, H., Worby, C.A., Xiao, J., Muda, M., Dixon, J.E., and Zipursky, S.L. (2000) *Drosophila* Dscam is an axon guidance receptor exhibiting extraordinary molecular diversity. *Cell*, **101**, 671–684.

29 Graveley, B.R. (2005) Mutually exclusive splicing of the insect Dscam pre-mRNA directed by competing intronic RNA secondary structures. *Cell*, **123**, 65–73.

30 Olson, S., Blanchette, M., Park, J., Savva, Y., Yeo, G.W., Yeakley, J.M., Rio, D.C., and Graveley, B.R. (2007) A regulator of Dscam mutually exclusive splicing fidelity. *Nat. Struct. Mol. Biol.*, **14**, 1134–1140.

31 Hattori, D., Demir, E., Kim, H.W., Viragh, E., Zipursky, S.L., and Dickson, B.J. (2007) Dscam diversity is essential for neuronal wiring and self-recognition. *Nature*, **449**, 223–227.

32 Hughes, M.E., Bortnick, R., Tsubouchi, A., Baumer, P., Kondo, M., Uemura, T., and Schmucker, D. (2007) Homophilic Dscam interactions control complex dendrite morphogenesis. *Neuron*, **54**, 417–427.

33 Matthews, B.J., Kim, M.E., Flanagan, J.J., Hattori, D., Clemens, J.C., Zipursky, S.L., and Grueber, W.B. (2007) Dendrite self-avoidance is controlled by Dscam. *Cell*, **129**, 593–604.

34 Soba, P., Zhu, S., Emoto, K., Younger, S., Yang, S.J., Yu, H.H., Lee, T., Jan, L.Y., and Jan, Y.N. (2007) *Drosophila* sensory neurons require Dscam for dendritic self-avoidance and proper dendritic field organization. *Neuron*, **54**, 403–416.

35 Zukin, R.S. and Bennett, M.V. (1995) Alternatively spliced isoforms of the NMDARI receptor subunit. *Trends Neurosci.*, **18**, 306–313.

36 Cull-Candy, S., Brickley, S., and Farrant, M. (2001) NMDA receptor subunits: diversity, development and disease. *Curr. Opin. Neurobiol.*, **11**, 327–335.

37 Ehlers, M.D., Tingley, W.G., and Huganir, R.L. (1995) Regulated subcellular distribution of the NR1 subunit of the NMDA receptor. *Science*, **269**, 1734–1737.

38 Okabe, S., Miwa, A., and Okado, H. (1999) Alternative splicing of the C-terminal domain regulates cell surface expression of the NMDA receptor NR1 subunit. *J. Neurosci.*, **19**, 7781–7792.

39 Bradley, J., Carter, S.R., Rao, V.R., Wang, J., and Finkbeiner, S. (2006) Splice variants of the NR1 subunit differentially induce NMDA receptor-dependent gene expression. *J. Neurosci.*, **26**, 1065–1076.

40 Mu, Y., Otsuka, T., Horton, A.C., Scott, D.B., and Ehlers, M.D. (2003) Activity-dependent mRNA splicing controls ER export and synaptic delivery of NMDA receptors. *Neuron*, **40**, 581–594.

41 Zhang, W., Liu, H., Han, K., and Grabowski, P.J. (2002) Region-specific alternative splicing in the nervous system: implications for regulation by the RNA-binding protein NAPOR. *RNA*, **8**, 671–685.

42 Charlet, B.N., Logan, P., Singh, G., and Cooper, T.A. (2002) Dynamic antagonism between ETR-3 and PTB regulates cell type-specific alternative splicing. *Mol. Cell*, **9**, 649–658.

43 Han, K., Yeo, G., An, P., Burge, C.B., and Grabowski, P.J. (2005) A combinatorial code for splicing silencing: UAGG and GGGG motifs. *PLoS Biol.*, **3**, e158.

44 Black, D.L. and Grabowski, P.J. (2003) Alternative pre-mRNA splicing and neuronal function. *Prog. Mol. Subcell. Biol.*, **31**, 187–216.

45 Li, Q., Lee, J.A., and Black, D.L. (2007) Neuronal regulation of alternative pre-mRNA splicing. *Nat. Rev. Neurosci.*, **8**, 819–831.

46 An, P. and Grabowski, P.J. (2007) Exon silencing by UAGG motifs in response to neuronal excitation. *PLoS Biol.*, **5**, e36.

47 Lee, J.A., Xing, Y., Nguyen, D., Xie, J., Lee, C.J., and Black, D.L. (2007) Depolarization and CaM kinase IV modulate NMDA receptor splicing through two essential RNA elements. *PLoS Biol.*, **5**, e40.

48 Glanzer, J., Miyashiro, K.Y., Sul, J.Y., Barrett, L., Belt, B., Haydon, P., and Eberwine, J. (2005) RNA splicing capability of live neuronal dendrites. *Proc. Natl. Acad. Sci. U.S.A.*, **102**, 16859–16864.

49 Zhang, L., Liu, W., and Grabowski, P.J. (1999) Coordinate repression of a trio of neuron-specific splicing events by the splicing regulator PTB. *RNA*, **5**, 117–130.

50 Wagner, E.J. and Garcia-Blanco, M.A. (2001) Polypyrimidine tract binding protein antagonizes exon definition. *Mol. Cell Biol.*, **21**, 3281–3288.

51 Spellman, R. and Smith, C.W. (2006) Novel modes of splicing repression by PTB. *Trends Biochem. Sci.*, **31**, 73–76.

52 Singh, R., Valcarcel, J., and Green, M.R. (1995) Distinct binding specificities and functions of higher eukaryotic polypyrimidine tract-binding proteins. *Science*, **268**, 1173–1176.

53 Ashiya, M. and Grabowski, P.J. (1997) A neuron-specific splicing switch mediated by an array of pre-mRNA repressor sites: evidence of a regulatory role for the polypyrimidine tract binding protein and a brain-specific PTB counterpart. *RNA*, **3**, 996–1015.

54 Izquierdo, J.M., Majos, N., Bonnal, S., Martinez, C., Castelo, R., Guigo, R., Bilbao, D., and Valcarcel, J. (2005) Regulation of Fas alternative splicing by antagonistic effects of TIA-1 and PTB on exon definition. *Mol. Cell*, **19**, 475–484.

55 Sharma, S., Falick, A.M., and Black, D.L. (2005) Polypyrimidine tract binding protein blocks the 5′ splice site-dependent assembly of U2AF and the prespliceosomal E complex. *Mol. Cell*, **19**, 485–496.

56 Oberstrass, F.C., Auweter, S.D., Erat, M., Hargous, Y., Henning, A., Wenter, P., Reymond, L., Amir-Ahmady, B., Pitsch, S., Black, D.L., and Allain, F.H. (2005) Structure of PTB bound to RNA: specific binding and implications for splicing regulation. *Science*, **309**, 2054–2057.

57 Perez, I., Lin, C.H., McAfee, J.G., and Patton, J.G. (1997) Mutation of PTB binding sites causes misregulation of alternative 3′ splice site selection *in vivo*. *RNA*, **3**, 764–778.

58 Liu, H., Zhang, W., Reed, R.B., Liu, W., and Grabowski, P.J. (2002) Mutations in RRM4 uncouple the splicing repression and RNA-binding activities of polypyrimidine tract binding protein. *RNA*, **8**, 137–149.

59 Wollerton, M.C., Gooding, C., Wagner, E.J., Garcia-Blanco, M.A., and Smith, C.W. (2004) Autoregulation of polypyrimidine tract binding protein by alternative splicing leading to nonsense-mediated decay. *Mol. Cell*, **13**, 91–100.

60 Boutz, P.L., Stoilov, P., Li, Q., Lin, C.H., Chawla, G., Ostrow, K., Shiue, L., Ares, M., Jr., and Black, D.L. (2007) A post-transcriptional regulatory switch in polypyrimidine tract-binding proteins reprograms alternative splicing in developing neurons. *Genes Dev.*, **21**, 1636–1652.

61 He, L. and Hannon, G.J. (2004) MicroRNAs: small RNAs with a big role in gene regulation. *Nat. Rev. Genet.*, **5**, 522–531.

62 Makeyev, E.V., Zhang, J., Carrasco, M.A., and Maniatis, T. (2007) The microRNA miR-124 promotes neuronal differentiation by triggering brain-specific alternative pre-mRNA splicing. *Mol. Cell*, **27**, 435–448.

63 Boutz, P.L., Chawla, G., Stoilov, P., and Black, D.L. (2007) MicroRNAs regulate the expression of the alternative splicing factor nPTB during muscle development. *Genes Dev.*, **21**, 71–84.

64 Bell, T.J., Thaler, C., Castiglioni, A.J., Helton, T.D., and Lipscombe, D. (2004) Cell-specific alternative splicing increases calcium channel current density in the pain pathway. *Neuron*, **41**, 127–138.

65 Celotto, A.M. and Graveley, B.R. (2002) Exon-specific RNAi: a tool for dissecting the functional relevance of alternative splicing. *RNA*, **8**, 718–724.

8
Noncoding RNA: The Major Output of Gene Expression

Matthias Harbers and Piero Carninci

8.1
Introduction

Our present understanding of the utilization of genetic information is based on the established concept that genetic information is stored in protein-coding genes, which are expressed into messenger RNAs (mRNAs) containing an open reading frame (ORF, also referred to as CDS or coding sequence) that holds the necessary information for protein synthesis. In contrast to the use of DNA and RNA in information storage and transportation, proteins with their wider complexity are considered the key regulators and catalysators of almost all cellular functions. However, this classical view has been challenged by the identification of large numbers of RNA transcripts lacking any coding potential [1–5]. Recent evidence suggests that even all nonrepeat portions of eukaryotic genomes could be actively expressed [6–9], with the majority of all transcripts being potentially noncoding RNAs (ncRNAs), in part having regulatory and catalytic functions thus far only considered for proteins. Moreover, our studies also showed that repetitive elements are dynamically expressed [10].

Following concepts that origin of life started from an “RNA World” [11], it is tempting to assume that the discovery of ncRNAs as well as new functions for RNA domains demonstrate an evolutionary maintained importance of RNA to biological processes. Starting originally from individual RNA molecules, RNA adopted multiple functions to act in complex organisms often in direct interaction with other RNA molecules, DNA, and proteins. We are still standing at the very beginning of getting an overview on all RNA functions, but the young field of ncRNA exploration has already made fundamental contributions to our understanding of biological processes. In this chapter, we want to provide the readers with an introduction into our present knowledge on ncRNAs, methods for ncRNA discovery, and give examples on how in particular long ncRNAs can function in the cell.

Posttranscriptional Gene Regulation: RNA Processing in Eukaryotes, First Edition. Edited by Jane Wu.

8.2
What is ncRNA?

In principle, ncRNA should include from our viewpoint any RNA that lacks any coding potential, is not translated into a protein or otherwise cannot be translated, or its functions are unrelated to its coding potential. Such a wide definition goes beyond the most obvious concept, that an ncRNA should per se lack any ORF. However, most primary transcripts are processed in the cell into different functional RNAs, out of which only some may act as ncRNAs. Therefore, a too strict definition of ncRNAs may ignore new concepts on how RNAs could function and possibilities for "multifunctional transcripts" derived from one gene. Ongoing efforts in RNA discovery still keep on finding new potentially noncoding transcripts, whereas genome annotations have largely underestimated the number of such genes [12–16], mostly ignoring noncoding transcripts due to their lack of CDSs and lower sequence conservation. Therefore, it is presently difficult to predict how many ncRNAs could be encoded in genomes, and what type of features they will have to group them into well-defined ncRNA families. Beyond the concept of defining ncRNA families, the community will further need to discuss how to integrate those ncRNA families into new concepts on how to define "genes" in the genome [2].

8.3
Discovery of ncRNAs

Other than classical RNAs like ribosomal RNA (rRNA) and transfer RNA (tRNA), ncRNAs have only lately been identified in large numbers after the discovery of new biological mechanisms and driven by innovative high-throughput technologies. Moreover, many ncRNAs are often expressed at very low levels, which makes their detection difficult. It is still an open discussion how to distinguish between "meaningful transcripts" or "transcriptional noise" for those rarely expressed genes [17]. Not entering into the details of this debate, however, we feel that it is important to record all transcriptional events to enable later studies on new transcripts that may turn out to have important biological functions, are potentially expressed only in few cells, or otherwise are enriched in certain cellular locations [14].

8.3.1
Discovery of ncRNAs by Function

Although Jacob and Monod already in 1961 [18] suggested a role for ncRNAs, research continued focusing primarily on proteins [19]. Therefore, it took some 30 years until regulatory ncRNAs were found in biological studies, in particular in homology-dependent gene silencing. It was not until 1993 that studies in *Caenorhabditis elegans* (*C. elegans*) on the lin-4 gene demonstrated for the first time that a small RNA acts by negatively regulating the expression of the complementary lin-14 gene

[20]. This observation led to the discovery of single-stranded RNA molecules of about 21–23 nucleotides with prominent roles in downregulation of gene expression, later named microRNAs (miRNAs) [21]. The discovery of short regulatory RNAs opened up an entirely new field in life science and made small ncRNAs a major focus of research. Long ncRNAs on the contrary gained less attention, although some long ncRNAs could also be identified in biological studies. Most prominent is the early discovery of Xist, a spliced and polyadenylated long transcript (19 kb in human) exclusively expressed from the inactive X chromosome [22, 23] as a housekeeping long ncRNA [24, 25]. Xist is part of the mammalian dosage compensation system and required to maintain gene dose equality for monoallelic expression of X-linked genes. Xist RNA is stably associated with the silenced X chromosome forming the so-called X or Barr body, where the Xist RNA always acts in cis on the chromosome from which it is expressed. In mammals, X-chromosome inactivation starts commonly at the "X inactivation center" or Xic, which encodes seven protein-coding and five noncoding genes [25]. Out of these, the Xist and Tsix genes are expressed antagonistically [26], and they seem to be important in X-chromosome counting. Downregulation of Tsix leads to increased levels of Xist around the Xic, from where it starts coating the entire inactive X chromosome [27]. Downstream processes lead to chromosome condensation, which include histone H3 methylation, histone H2A ubiquitination, and DNA methylation [28].

Thus far, ncRNAs have mostly been found by chance in biological studies that targeted at specific questions, whereas forward genetic screens searching for new genes commonly failed to give phenotypes related to ncRNAs. Most likely, genes encoding ncRNAs are less sensitive to mutations than coding transcripts where even a single-point mutation can disrupt or otherwise alter an ORF. In addition, it is expected to find more ncRNAs in genome-wide association studies made possible by recent developments in genomic research. Such experiments can show the role of ncRNAs in disease, but further experiments will be needed for understanding of their contribution to etiology. Therefore, functional screening methods for the identification and characterization of ncRNA are still in demand, which could for instance make use of RNA interference (RNAi)-mediated knockdowns [29–32] and/or over expression studies in suitable cell culture models [33, 34]. Due to the low throughput of functional studies, new ncRNA discovery is presently dominated by high-throughput expression profiling in combination with computational data analysis, and methods to predicting RNA structures and sequence conservation in genomes.

8.3.2
Large-Scale cDNA Cloning Projects

Initiated first by efforts to characterize transcriptomes by deep sequencing of the so-called expressed sequence tags (Ests) at the beginning of the 1990s [35], large-scale cDNA cloning projects targeted at the cloning and full-length sequencing of representative cDNA clones for each gene in human [36] and other model organisms like for instance mouse [37] or *Arabidopsis* [38]. During the annotation of cDNA sequences from the FANTOM project [37, 39], it became evident that for as many as ~23 000 transcripts, no clear ORF could be identified. Therefore, annotators set new standards on how to separate potentially noncoding transcripts from coding ones. While analyzing sequences from their mouse cDNA collection, any RNA was annotated as "noncoding" for which no ORF encoding more than 100 amino acids could be identified. In addition, expert curators have reviewed all annotations to distinguish noncoding from truncated RNAs. In a similar search for human ncRNAs, the H-Invitation group reduced the threshold from 100 to 80 amino acids to rescue more peptide sequences [40, 41]. Both cutoffs are based on the assumption that short peptides of only 100 amino acids or less seems no longer able to form stable protein structures (J. Gough, personal communication) and therefore are most likely not encoding functional proteins. Such definitions leave a gray zone to which extent short peptides could be real, and manual curation should take predicted or known peptide structures into account to avoid wrong annotations. All long ncRNAs found during the FANTOM project could be cloned from cDNA libraries prepared by oligo(dT) priming and a 5′ end specific cap selection method [42] as commonly used for mRNA cloning [43]. Therefore, it is likely that many long ncRNAs are derived from RNA-polymerase-II-driven transcription and processed similar to coding mRNAs. It remains to be seen whether other long ncRNAs can be found lacking a cap structure and polyadenylation. However, there is evidence for ncRNAs lacking poly(A) tails [44], and the high portion of non-polyadenylated mRNAs found for instance in tiling array experiments may contain additional thus far unknown ncRNAs [13]. Their cloning would require new cDNA library methods [43], which could use 5′ and 3′ end linker ligation steps similar to the cloning of short RNA (sRNA) (see Section 8.3.4). In addition, subtractive hybridizations can be used to remove abundant "housekeeping" RNAs like for instance rRNAs and tRNAs during the preparation of subtracted cDNA libraries, which have been used in ncRNA discovery [45].

8.3.3
Tag-Based Sequencing Methods

Although large-scale cDNA projects were fundamental for our present understanding on the utilization of genomic information and the annotation of genome sequences, the high cost for cDNA sequencing remains a limiting factor in transcriptome analysis. Therefore, new approaches for transcript identification were developed that only rely on short sequence tags of some 14–27 bp [46]. These tags can be obtained in large numbers at a much lower cost as compared with end sequencing of cDNAs. When obtaining 20 or even 27 bp, tags can be directly analyzed by mapping to genome sequences with high reliability [47, 48]. This opened the way to genome-wide searches for expressed genes, which revealed a much wider gene expression than anticipated from genome annotations [12, 49, 50]. The so-called SAGE (Serial Analysis Gene Expression) [51–53] and CAGE (Cap Analysis Gene Expression) [54, 55] methods have been used on a very large scale to obtain an overview on transcriptional activity. In particular, CAGE and related methods [56–59] obtaining sequence tags from the true 5′ ends of mRNAs allow for genome-wide mapping of transcription start sites (TSS). These defined locations in the genome link transcripts to the promoter regions driving their expression. This link is essential to drive gene network studies and to move forward our understanding on how gene expression is regulated (refer to the following references for more information on CAGE and promoter analysis [50, 60–66]). In particular, for ncRNAs, it had been found that the promoter regions tend to be more conserved than the actual transcribed portions [37], which emphasizes the high importance of CAGE in ncRNA discovery. In addition, CAGE tags have been very important for identifying sense–antisense (S/AS) pairs, which seems to be a general feature in genome organization and regulation of transcription [37, 67–70]. Tag-based approaches will gain even more importance since new sequence technologies became available on the market [71–73] (e.g., the Illumina HiSeq2000 can provide up to 150 mil sequencing reads per lane, or the SOLiD 5500xl sequencer from ABI will achieve some 120 mil reads per lane [74]). With a read length of more than 50 bp, these instruments can obtain at least one sequence tag per read including flanking sequences that can be used to barcode sequence tags, for example, for multiplex sequencing [75, 76] or the use of internal standards. For CAGE, the use of internal standards for calibrating expression studies has been initiated [77], providing a higher reproducibility between individual sample preparations. There is an enormous potential in using these sequence capacities for tag-based approaches, as they allow us to get more reliable data on rarely expressed transcripts while at the same time offering means to quantify gene expression. Measuring relative expression levels between transcripts will help in understanding how regulatory RNAs act on their RNA and DNA targets. However, we should also mention the limitations of tag-based approaches in distinguishing splice variants. New approaches to shotgun sequencing of entire RNA pools, so-called RNA-seq experiments, will hopefully allow us in the future to get more information on exon junctions and alternative splicing products. In addition, methods like CAGE depend on the 5′-end-specific cap structure for tag isolation, which limits CAGE to the detection of RNA polymerase II transcripts.

8.3.4 Short RNA Libraries

Classical cDNA library protocols do not allow for cloning of short RNAs (sRNAs), and therefore, sRNAs, like for instance miRNAs, were missed in large-scale cDNA cloning projects. Commonly, cDNAs shorter than 500 bp are removed during library construction to avoid contamination of libraries with cDNAs derived from structural RNAs, adapter sequences, and supposingly truncated cDNAs. In addition, there was originally little evidence that transcripts shorter than 500 bp could be meaningful. Only with the discovery of short regulatory RNAs [20], new cloning protocols were developed that focus specifically at the cloning of size-fractionated RNAs in the range of 25–500 bp (most commonly 20–35 bp) [78–80], sometimes referred to as "RNomics." These protocols benefit from the fact that most known sRNAs are phosphorylated at their 5′ end and have a free hydroxyl group at the 3′ end. These chemical groups allow for an easy manipulation of the 5′ and 3′ ends to introduce a primer site at the 3′ end for first-strand cDNA synthesis and an additional primer site at the 5′ end for PCR amplification. With the availability of new high-throughput sequencing methods (see above), commonly cDNA fragments are directly sequenced after an amplification step at a very high throughput. These enormous sequencing capacities drive brute-force deep-sequencing efforts for expression profiling and discovery of new sRNAs [81]. Due to the high background in sRNA libraries, computational methods in combination with genome mapping are used to identify and catalog the RNA transcripts. In contrast to classical cloning approaches, these high-throughput methods provide only sequence information for computational analysis but no longer cDNA clones. However, because of their short length, related cDNAs can easily be prepared by chemical synthesis as, for example, needed for functional studies on their targets. Moreover, it has to be noticed that modifications at the 5′ and/or 3′ end can make RNA detection by these approaches impossible. For example, 2′-*O*-methylation of tRNA by a 2′-*O*-methyltransferase is common [82], though it seems that a 2′-*O*-methyl group does not affect RNA cloning. We expect additional discoveries for more short RNAs.

8.3.5 New Microarray Approaches

The discovery of many sRNAs led to the development of dedicated hybridization-based expression-profiling methods [83, 84]. However, classical microarray platforms can only detect transcripts for which sequence information is already known. Therefore, alternative microarray formats had to be invented for the discovery of new transcripts. Here in particular, whole-genome tiling arrays provide remarkable genome-wide expression profiles by indicating all regions in the genome expressed at a given time point [85, 86], and have greatly contributed to widening our knowledge on genome-wide expression, for example, in humans [14–16, 87–89], *C. elegans* [9], rice [90], *Arabidopsis* [91], and yeast [6, 92, 93]. Progress in high-density microarray manufacturing allows today the presentation of the entire nonrepetitive sequence of the human and other genomes on a set of few microarrays such as for example offered by Affymetrix for the human, mouse, *Arabidopsis*, and a growing number of other genomes. Commonly, short nonoverlapping oligonucleotides of 25 bp separated by 5- or 35-bp gaps are used to analyze genomic sequences in short increments (e.g., 35 bp) sufficient for the identification of genomic regions or expressed exons, but rather limiting for a reliable detection of sRNAs. Alternatively, printed tiling arrays can be prepared from PCR products or cloned DNA fragments [94]. Although tiling arrays in principle can detect every transcript, it has to be noticed that expression patterns observed in tiling array experiments do not reveal the actual sequences of individual transcripts, because of (i) their lack of resolution at the edges due to the short oligonucleotides on the array, (ii) a lack of connectivity between the expressed exons, and (iii) their inability to distinguish multiple overlapping transcripts from the same locus. Other approaches, like for example, RACE (Rapid Amplification of cDNA Ends) have to be used to build connections between different exons identified in tiling array experiments [15, 87, 88].

Similar to CAGE experiments, whole-genome tiling array experiments also showed that much larger portions of the human genome are actively transcribed than ever anticipated from genome annotations. Tiling array experiments using polyadenylated and non-polyadenylated RNA showed that over 40% of all mRNA may lack a poly(A) tail, and hence would be missing in all cDNA collections derived from oligo(dT) primed cDNA libraries [43]. In addition, looking at polyadenylated RNA in the nucleus and cytoplasm, whole-genome tiling array experiments provided an entirely new view on RNA processing and nuclear RNA [14]. Some 40% of the nuclear RNA might never be exported into the cytoplasm, indicating new functions for those RNAs in gene regulation. Among these, over 450 000 sRNAs of 20–200 nucleotides were identified that map to both strands for about half of the expressed protein-coding genes at their TSS (promoter-associated sRNAs or PASRs) and transcription termination site (termini-associated sRNAs or TASRs). In addition, PASRs also align with the borders of promoter-associated long RNAs (PALRs), while some TASRs might be generated by the cleavage of longer transcripts. The heterogeneity of those sequences identified in the tiling array experiments argues for distinct but thus far unknown functions. Some of the RNAs may act on their parental and other chromosomes, giving them important functions in chromatin formation and gene expression (see Section 8.5). Moreover, it has been found that mRNAs are regulated by nuclear retention. For example, the CAT2 transcribed nuclear RNA (CTN-RNA) is stored in the nucleus and only exported into the cytoplasm for translation upon stress activation [95]. There is evidence that similar mechanisms could apply to some 1700 additional transcripts, suggesting that nuclear retention might be a general regulatory principle.

8.3.6 Other Experimental Approaches

New features discovered in ncRNAs led to the development of new methods, for instance, for the analysis of their locations and interactions of RNA with DNA and proteins. Although beyond the scope of this chapter, we would like to briefly mention some interesting new developments and methods of general importance: method for visualization of RNA in cells are becoming very important, since RNA molecules are processed in the cell prior to use in defined cellular compartments. For example, *in vivo* detection of RNA has been achieved by *in situ* hybridization methods. The cellular localization of RNA transcripts was studied for instance by miRNA *in situ* hybridization (ISH) [96], miRNA fluorescence *in situ* hybridization (FISH) [97], both approaches similar to (FISH) [98], or the use of fluorescent proteins [99].

For the *in vivo* detection of protein–DNA interactions, various methods have been developed [100], most commonly using specific antibodies against protein factors for immunoprecipitation of protein–DNA complexes and analysis of the DNA fragments by different approaches. Similarly, methods for the detection of RNA–protein interactions have been developed, for instance, using an RNA template for protein enrichment [101], UV cross-linking and immunoprecipitation [102] or immunoprecipitation–microarray (RIP-Chip) [103], visualization of RNA using fluorescence complementation triggered by aptamer–protein interactions (RFAP) [104] or streptomycin-binding aptamers [105], RNA affinity in tandem (RAT) for isolation protein–RNA complexes and protein identification by mass spectrometry [106], or the transcriptional repression assay procedure (TRAP) [107]. Binding sites for protein factors can be further identified by the so-called SELEX (Systematic Evolution of Ligands by Exponential Enrichment) method [108]. In the original SELEX method, an oligonucleotide library was screened with high-affinity ligands to identify recognition sequences. Later, the method has been expanded to the so-called "genomic SELEX" using genomic DNA to identify or confirm natural binding sites in genomes [109, 110]. However, although important, all those methods are rather suitable for RNA analysis than ncRNA discovery.

8.3.7
Computational Methods

Computational methods have been instrumental for large-scale data analysis, the identification and annotation of ncRNAs [111, 112], and establishment of large databases providing dedicated information on ncRNAs (refer to Table 8.1). Originally used in sequence annotations, they revealed transcripts lacking any coding potential, opening the discussion on a widespread expression of potentially noncoding transcripts. Many of those long ncRNAs often lack evolutionary conservation (the homology between mouse and human transcripts is overall below 70%), where in some cases only short patches of up to 50-conserved bp can be found, that may represent functional domains required for interactions or proper folding. Interesting possibilities in genome annotation and the identification of new coding and noncoding transcripts derive from the genome sequencing of multiple related species and comparative genomics as recently done for 12 *Drosophila* genomes [113, 114].

Sequence conservation has always been an important consideration in the computational analysis of sequence data, and has been the driving force behind comparative genome annotations. Hence, functions of genomic elements have been traditionally associated with conservation of genetic information across species. Albeit often being transcribed from conserved promoter regions [37], ncRNAs show frequently relative low conservation [115, 116], suggesting that for ncRNA, conservation of the entire sequence is not always important for some regulatory mechanisms, as for instance transcriptional interference or activation through transcription across regulatory regions (see Section 8.5). In addition, for many S/AS pairs, it has been found that ncRNAs acting in cis do not have to be conserved. It is rather that the "transcription of the ncRNA" is conserved, which was, in some cases, demonstrated by the fact that the promoter regions driving the expression are more conserved than the sequences of the actual transcripts [37, 117]. On the other hand, conservation is surely relevant for some RNAs like for instance those acting in trans on other regions in the genome or having structural functions. For

Table 8.1 Resources on ncRNAs.

Database	URL
lncRNAdb	http://lncrnadb.com/
miRBase	http://www.mirbase.org/
miRDB	http://mirdb.org/miRDB/
Wikiomics	http://openwetware.org/wiki/Wikiomics
RNAdb	http://jsm-research.imb.uq.edu.au/rnadb/
RIKEN FANTOM	http://fantom.gsc.riken.jp/
Rfam	http://rfam.sanger.ac.uk/
NONCODE	http://www.noncode.org/NONCODERv3/
Noncoding RNA database	http://biobases.ibch.poznan.pl/ncRNA/
SnoRNA-LBME-db	www-snorna.biotoul.fr/
SnoRNA database	http://lowelab.ucsc.edu/snoRNAdb/
Yeast snoRNA database	http://people.biochem.umass.edu/fournierlab/snornadb/main.php
Ribosomal Database	http://rdp.cme.msu.edu/index.jsp
SILVA rRNA database	http://www.arb-silva.de/
Comparative RNA Web (CRW) Site	www.rna.ccbb.utexas.edu/
SRPDB (Signal Recognition Particle Database)	http://rnp.uthscsa.edu/rnp/SRPDB/SRPDB.html
TmRNA Website	http://www.indiana.edu/~tmrna/
tmRDB	http://www.ag.auburn.edu/mirror/tmRDB/
Viral RNA Structure Database	http://rna.tbi.univie.ac.at/cgi-bin/virusdb.cgi
Nucleic Acid Database	http://ndbserver.rutgers.edu/
Structural Classification of RNA (SCOR)	http://scor.berkeley.edu/scor.html
GenomeTraFaC	http://genometrafac.cchmc.org/genome-trafac/index.jsp

example, studies on ultraconserved regions in human, which acquired new functions only in the recent human lineage, identified HAR1F, an ncRNA, which is coexpressed with *reelin*, an essential protein to guide neuron migration in Cajal–Retzius neurons [118].

Today, computational methods are used in sequence analysis and structure prediction. An increasing knowledge about ncRNA families will hopefully provide in the future, better characteristics to define ncRNAs and their RNA families by computational analysis. Common applications include strategies for *in silico* mapping sequences to genomes and to use genomic information as the basis for data analysis and annotation, searches for sequence homologies and conservation to identify evolutionary conserved regions in genomes and transcripts, base composition of codons, motive recognition to group RNA to RNA families, or prediction of secondary RNA structures to classify ncRNA [119]. Examples for ncRNA predictions programs include QRNA (detecting novel structural RNA genes) [120], CONC (coding or noncoding: considers peptide length, amino acid composition, predicted secondary structures, and the percentage of exposed residues) [121], or VirGel (considers GC contend at the second position compared with the third position, because in ncRNAs, the lack of codons does not allow to predict nucleotide frequency in those positions) [122]. Since many ncRNAs act by sequence-specific binding to DNA and RNA targets, computational methods are further used for the prediction of potential target sites, which is in particular a hot topic for the analysis of miRNAs [123]. Although essential for data analysis, results from computational studies should be considered as "suggestive," and an experimental validation is necessary. For example, about half of the regions identified during the ENCODE project [8] as transcription factor binding sites are not conserved between genomes (also defined as "neutral") and would have been missed by comparative genome studies.

The fact that initial transcripts can have multiple functional domains and are often processed during maturation further complicates computational analysis, in particular, when ncRNAs are identified only based on sequence information in relationship with genome annotations. Most prominent examples for multifunctional primary transcripts are miRNAs [124, 125] and small nucleolar RNAs (snoRNAs) encoded by introns [126], but other functional elements/regions and RNA structures could act in a similar way [127]. In another example, it had been shown that expression of sense and antisense transcripts not necessarily has to lead to RNA breakdown, but may have other or additional functions in inducing alternative splicing and polyadenylation [128]. Moreover, the ability to act by transcriptional interference or to form double-stranded RNA (dsRNA) to enter into an RNAi pathway does not depend on whether or not a transcript is coding or noncoding. Multifunctionality of transcripts could be a general regulatory concept for many biological processes, and regulatory RNA domains and their functions may provide guidance for new mechanisms on how ncRNAs could work. Such considerations will support new concepts of "multifunctional RNA transcripts" extending on polycystronic coding mRNAs.

8.4
ncRNA Families

The range of ncRNAs is very wide and ranks from classical RNA genes like rRNAs and tRNAs; small RNAs like for instance miRNAs, small interfering RNAs (siRNAs), P-element-induced wimpy testis (PIWI)-interacting RNAs (piRNAs), and snoRNAs; to large numbers of heterogeneous long ncRNAs. Here, we will only briefly list the most prominent ncRNA families and then focus on the examples for functional ncRNAs; for further information on these ncRNA families, refer to the cited reviews and links to the Internet.

8.4.1 tRNA

Transfer RNAs (tRNAs) [129, 130] are an essential part of protein biosynthesis at the ribosomes, where tRNAs provide the matching amino acid for elongation of the polypeptide chain. They consist of sRNA chains of some 75–95 nucleotides that are folded into specific structures exposing the 3′ terminus for amino acid attachment and having an anticodon that can base pair to the corresponding codon in mRNA [131]. Although each tRNA is specific for one amino acid only, and tRNA genes can be grouped into just 49 families, most genomes contain hundreds of tRNA genes (refer to http://lowelab.ucsc.edu/GtRNAdb/). Interestingly, deep-sequencing approaches identified shorter tRNA fragments of reproducible length. Although their function is unknown, they are differentially expressed in cells and tissues, and can be confirmed by Northern blotting (H. Kawaji, personal communication).

8.4.2 rRNA

Ribosomal RNAs (rRNAs) are central components of ribosomes [132], where they have structural and catalytic functions in direct interaction with tRNAs for protein biosynthesis [133, 134]. In a typical cell, rRNA contributes the vast majority of cellular RNA and is transcripted from clusters with multiple repeats. In mammalians, two mitochondrial rRNAs (23S and 16S) and four cytoplasmic rRNAs (28S, 5.8S, 5S for the large subunit, and 18S for the small subunit) are found (refer to http://www.his.se/upload/21749/ribosomal%20rna.pdf for an overview on rRNAs).

8.4.3 snRNA

Small nuclear RNAs (snRNAs) are small RNA molecules of 100–300 nucleotides found in the nucleus of eukaryotic cells. Transcribed by RNA polymerase II or III, they play important roles in RNA splicing, regulation of transcription, and the maintenance of telomeres operating in association with specific proteins (snRNP or small nuclear ribonucleoproteins) [135, 136].

8.4.4 snoRNA

Small nucleolar RNAs (snoRNA) [137, 138] are small RNA molecules that guide the chemical modification of rRNA and other RNA molecules by methylation and pseudouridylation. Similar to snoRNA, small Cajal body-specific RNAs (scaRNAs) have been found that localize in nuclear Cajal bodies [139, 140]. Other than snoRNAs, the U85 scaRNA can guide pseudouridylation and methylation. snoRNAs can be divided into two families based on conserved motifs. The C/D box members commonly guide methylation, whereas H/ACA box members are specific for pseudouridylation reactions. snoRNA genes are frequently encoded by introns of ribosomal genes and expressed by RNA polymerase II, though they can also be expressed independently from their parental gene.

8.4.5 Guide RNA

Guide RNA is a general term for RNAs in RNA/protein complexes that "guide" the complexes by sequence-specific hybridization to matching target sequences [141], where insertion of uridines into mRNA in trypanosomes is known as RNA editing [142]. A special form of guide RNA, the gRNAs, was found in the mitochondria of kinetoplastids, where, at the editosome, stretches of uridylates are inserted or deleted from mRNAs [143].

8.4.6 miRNA

MicroRNA (miRNA) [144, 145] are small RNA molecules that have in part reverse complementary sequences to mRNAs, commonly expressed from a different gene. After maturation, miRNAs form short single-stranded RNA molecules of some 21–23 nucleotides that can block the translation of their target RNAs, or facilitate mRNA cleavage [146]. RNA cleavage commonly requires a perfect homology between the miRNA and the target sequence. In addition, secretion of miRNA (denoted as "exosomal shuttle RNA" or esRNA) [147], transfer of miRNA into the nucleus [148], and involvement in chromosome methylation have been described [149]. miRNAs have been found in single-cell algae [150, 151], lower metazoans, plant, and higher organisms including humans, indicating an evolutionary conserved role of miRNAs in gene regulation. In mammals, miRNAs may interfere with the expression of at least 30% of all protein-coding genes giving them a wide impact on almost every cellular process [146] including animal development [152], malignant transformation [153], metastasis [154], and heart disease [155].

miRNAs are commonly expressed as much longer primary miRNAs (pri-miRNAs), where miRNAs located in introns are expressed by RNA polymerase II [125], and noncoding transcripts containing miRNAs are often expressed by RNA polymerase III [156]. In the nucleus, a protein complex consisting of the ribonuclease Drosha and the dsRNA-binding protein Pasha cleaves the primary transcripts to form 70-nucleotide-long pre-miRNAs with a characteristic stem-loop structure [157, 158]. After export into the cytoplasm, pre-miRNAs are further processed by cleavage with the riboendonuclease Dicer to generate mature miRNA molecules. Dicer also contributes to the formation of the RNA-induced silencing complex (RISC), which is responsible for gene silencing by RNAi. Although best understood for the miRNA pathway, there is evidence that RNA processing could be a general mechanism to generate independent RNA molecules from one parental RNA transcript. Another open question is still the identification of miRNA targets in the cell. Besides using computer alignments, this problem may be addressed experimentally by inhibition of miRNAs and genome-wide protein analysis; for example using locked nucleic acid (LNA)-modified oligonucleotides, refer to [159], or analyzing RNAs associated with RISCs, refer to [160].

Unexpectedly, it was recently found that miRNA can also increase translation during cell quiescence [161]. Upon cell cycle arrest, an AU-rich element in the tumor necrosis factor-alpha (TNFalpha) mRNA is transformed into a translation activation signal after recruiting factors associated with the miRNA miR369-3. Moreover, the miRNA Let-7 and a synthetic miRNA miRcxcr4, known repressors of translation in proliferating cells, could also induce the translation of their target mRNAs on cell cycle arrest. These observations argue that activation of translation could be a common function of miRNAs on cell cycle arrest. It is exciting to speculate whether or not other or even all miRNAs could switch between repression and activation of translation depending on the biological context, greatly widening their potential in gene regulation.

8.4.7 siRNA

Similar to miRNAs, small interfering RNAs (siRNA) [162–164] are dsRNA molecules of 20–25 bp that are derived from cleaved dsRNA viruses or small hairpin RNAs (shRNAs). They are of high importance for the expression of specific genes, RNAi-related pathways, antiviral defense, and formation of chromatin structures. Importantly, natural siRNAs can originate from transposons [165–168] and bidirectionally transcribed repetitive sequences, also referred to as repeat-associated small interfering RNA (rasiRNA) [169, 170]. Alternatively to dsRNA transcripted from inverted DNA repeats, in some organisms, siRNA can be produced in a process involving RNA-dependent RNA polymerases to convert single-stranded RNA into dsRNA, digestion of the resulting long dsRNA molecules by the ribonuclease Dicer to produce short double-stranded siRNA molecules, and integration of the siRNAs into the RISC. Argonaute proteins are required for binding to the RISC, similar to miRNA function. siRNAs are in particular widely found in plants, where, for example, genome-wide searches for siRNAs have be made in *Arabidopsis* [171].

8.4.8 piRNA

PIWI-interacting RNA (piRNA) can be found in the ovaries and testes of flies, fish, and mice, where they form protein/RNA complexes with PIWI proteins [172, 173]. They are structurally distinct from miRNAs and siRNAs containing 26–31 nucleotides and carrying a 2′-*O*-methylation at their 3′ ends. In mice, sRNAs bound to MIWI, the mouse homolog of the PIWI [174], are expressed in the testis of a 14-day-old male, while a slightly shorter variant (~26–28 nt) binds to the MILI protein during early development in primordial germs cells to the pachytene spermatocytes [175]. These RNAs are expressed from hundreds of conserved genomic regions, and most likely derive from single-stranded RNA transcribed across large regions [176]. piRNAs play important roles in gene silencing, in particular of transposons and highly repetitive sequences. In *Drosophila*, PIWI and rasiRNAs promote euchromatic histone modifications, and piRNA transcription in subtelomeric heterochromatin [177]. Silencing can be achieved by corecruitment of accessory factors or through the activity of Argonaute [178, 179], which often has endonucleolytic activity. The PIWI-piRNA pathway provides an adaptive defense against replication of transposons [176] as needed to maintaining germ line DNA integrity. However, the control of repeat elements may not be the only function of piRNAs as, for example, most piRNAs found pachytene spermatocytes do not map to repeat elements.

8.4.9 tasiRNA

Trans-acting small interfering RNAs (tasiRNAs) have been found in plants and are derived from noncoding TAS genes [180]. Their polyadenylated precursor RNAs are cleaved by an miRNA-programmed RISC forming dsRNA molecules, which are further processed by DICER-LIKE 4 to generate mature tasiRNAs of 21 nucleotides. tasiRNAs guide the cleavage of endogenous mRNAs by AGO1 or AGO7. They may be one aspect of essential roles of RNA metabolism pathways in *Arabidopsis* [181].

8.4.10 tmRNA

Transfer-messenger RNAs (tmRNA or 10Sa RNA) combine features of tRNAs and mRNAs and are an important part of the protein translation machinery [182–184]. They are ubiquitously found in all bacteria, in certain phage, and mitochondrial and plastidial genomes, but not in archaeal or eukaryotic nuclear genomes. They bind to ribosomes blocked by incomplete proteins. Entering ribosomes similar to tRNAs, their mRNA portion is translated to add a peptide tag to the incomplete protein. The attached peptide tag at the C-terminus marks the fusion protein for destruction by proteolysis. Further information on tmRNA can be found on the Internet under http://www.indiana.edu/~tmrna/.

8.4.11 Ribozymes

Ribozymes (ribonucleic acid enzyme) are RNA enzymes that can catalyze chemical reactions [185–187]. Natural ribozymes often catalyze either the hydrolysis of phosphodiester bonds or play a role in the aminotransferase activity of the ribosome. Although only few natural ribozymes could be found until today, specific ribozymes can be produced *in vitro*, and artificial ribozymes gained some importance as molecular tools like for instance hammerhead ribozymes [188]. The discovery of ribozymes added an important mechanism on how RNA can function, and initiated theories on early forms of life with cells using RNA as genetic and structural material as well as catalytic molecules before those functions have been divided between DNA and proteins later during evolution. Most prominent, the universally conserved endoribonuclease P (RNase P) is involved in the processing of tRNA precursors besides other targets. Even for the eukaryotic RNase P, RNA cleavage does not require the protein subunit, suggesting that the RNA-based catalytic activity has been preserved during evolution [189]. Other early ribozymes may have rather changed their features during evolution when becoming part of RNA–protein complexes, where their functions could have shifted from "catalysis" to "guidance" of proteins.

8.4.12
Untranslated Regions in mRNA

Although not considered as "ncRNA," mRNAs have untranslated regions (UTR) at their 5′ and 3′ ends containing various functional elements important for RNA stability and regulation/initiation of RNA translation (refer to http://www.ba.itb.cnr.it/UTR/ and http://en.wikipedia.org/wiki/Category:Cis-regulatory_RNA_elements for detailed information). It should also be stressed that UTRs contain binding sites for ncRNAs, in particular for miRNA binding to 3′ UTRs to express their functions on targeted mRNAs. Surprisingly, mRNA modifications needed for RNA translation are also commonly found in long ncRNAs that are transcripted by RNA polymerase II. Although we do not want to discuss in general the functional elements in mRNA, some of them are of general importance to elucidate mechanisms on how ncRNAs could work [190, 191]. Moreover, it is important to note that some of them can be identified as conserved sequence motifs [192], like for example, riboswitches, commonly targeted by computational searches for potential ncRNA candidates.

In particular, riboswitches are important to mention as they contain high-affinity binding sites for small molecules, commonly found in the 5′ UTR of many bacterial mRNAs [193, 194]. The ability of RNA to specifically bind to target molecules similar to protein antibodies adds an important feature on how RNA molecules can function. So-called aptamers, specific RNA-binding molecules, have by now found many technical applications in molecular biology, diagnostics, and therapy [195–197]. Riboswitches can be divided into an aptamer region, which specifically binds a ligand, and an expression platform. In response to structural changes of the aptamer, expression platforms can turn off or on gene expression by different mechanisms. Most known riboswitches have been found in eubacteria, though three thiamine pyrophosphate (TPP) riboswitches have also been found in *Neurospora crassa* [198]. Although not yet found in higher organisms, riboswitches are important for metabolic pathways in bacteria and are considered valuable targets for the development of antibiotics [199]. Moreover, it is interesting to note that RNA interacts with enzymes involved in metabolic processes such as glycolysis,

Krebs cycle, fatty acid metabolism, thymidylate synthesis, and others, arguing that RNA via RNA-binding enzymes still links metabolism to regulatory mechanisms even in higher organisms [200].

8.4.13 Long ncRNAs

As already outlined above, large-scale cDNA sequencing has revealed many long transcripts lacking any coding potential. Some of them seem to derive from very long primary transcripts of up to 100 kbp (macro-ncRNA or macroRNA [44, 116]). Only little is known about their functions, but there is a lot of support arguing for the functionality of these transcripts including tissue/stage-specific regulation of their expression [44, 116, 201]. Many of them derive from complex genomic regions and interlace or overlap with other transcripts in sense/antisense (S/AS) orientation. Although distinct from sRNA, thus far long ncRNAs cannot be grouped into specific RNA families with common motives and/or functions.

8.5 Examples for ncRNA Functions

Overlapping transcription is a frequent feature in gene expression. In particular, long antisense (AS) transcripts overlapping with protein-coding mRNAs were identified in cDNA and CAGE libraries [69]. AS transcription is a common feature found in humans, mice, rats, chickens, fruit flies, nematodes, and plants [202, 203]. S/AS transcripts can be grouped into three different patterns: S/AS transcripts can overlap over their entire length, be divergent with overlapping promoters, or convergent with overlapping 3′ ends of the transcripts. Although previous reports suggested a dominance of convergent S/AS transcription, CAGE [12] and Est data revealed more divergent AS RNAs with spots of AS transcripts derived from first introns and transcribed across promoter regions [204]. Some of those transcripts in part overlap with PALRs on the other strand found in whole-genome tiling array experiments [14].

Examples for differential AS transcription from the opposite strand of many known genes have been found in *Saccharomyces cerevisiae* using tiling arrays, that further identified operon-like transcripts, transcripts from neighboring genes not separated by intergenic regions, and genes with complex transcriptional regulation [6]. Many novel transcripts found in yeast have lower sequence conservation than common protein-coding genes and lower rates of deletion phenotypes, similar to other observations in higher organisms. Often, they could only be found after inactivation of RRP6, a catalytic subunit of the RNA degrading exosome complex [92]. Other RNAs are expressed from intergenic regions such as the SER3 gene. The expression of SER3 is repressed during growth in rich medium by a strong expression of the noncoding SRG1 RNA. Expression of SRG1 is essential for repression SER3 by transcription interference [205–207], where SRG1 transcription across the SER3 promoter interferes with the transcription factor binding to the promoter [208]. Another example for AS-mediated gene regulation in yeast is the expression of IME4 during initiation of meiosis [68]. *S. cerevisiae* has no RNAi machinery, and therefore, both ncRNAs act via a cis-acting mechanism that is distinct from RNAi-mediated gene silencing in eukaryotes. Recently, repression mediated by histone deacetylation has been identified in *S. cerevisiae* for two PHO84 AS transcripts [209], adding another mechanisms on how ncRNAs can mediate their regulatory functions. In contrast to *S. cerevisiae, Schizosaccharomyces pombe* has an RNAi machinery, and pericentromeric repeats are transcribed from antiparallel promoters by RNA polymerase II and subsequently processed by an RNA-dependent RNA polymerase and Dicer to form small dsRNAs of 22–24 bp involved in gene silencing [210].

Expression of complementary RNAs can lead to the formation of dsRNA hybrids as, for example, needed for siRNA synthesis [14]. However, it was shown that long ncRNAs do not necessarily have to be converted into sRNAs but can function by other mechanisms. For instance, for thymidylate synthase and hypoxia inducible factor-1alpha, primary examples of endogenous genes with coding and noncoding transcripts forming S/AS transcripts overlapping at their other 3′ ends, no evidence for duplex formation in the cytoplasm and subsequent activation of RNAi pathways could be found [211]. For divergent S/AS transcription from the opposite strand of promoters [69, 204], there is evidence for transcriptional regulation in cis, where the act of transcription itself, rather than the produced ncRNAs, could either positively or negatively regulate the transcription of the gene on the opposite strand. Since more than 70% of all genes in mouse and human are subjected to bidirectional transcription [69], it is most likely that transcriptional interference could be a common principle in gene regulation as already described for yeast. Also, in *Drosophila*, large regions in the genome are transcribed into long ncRNAs, and the ncRNA bithoraxoid (bxd) represses Ultrabithorax (Ubx) in cis by transcriptional interference [212]. Other recent examples include the detailed characterization of AS transcription in the beta-globin cluster [213]. Long transcripts from the Kcnq1ot1 AS promoter can silence flanking genes more efficiently than shorter AS transcripts by chromatin-specific histone modifications, which can further spread into neighboring nonoverlapping genes [214]. Using the tagging and recovery of associated proteins (TRAP) technology, it was shown that the paternally expressed ncRNA RNA LIT1/KCNQ1OT1 coats the inactivated region [215]. Such observations argue that the degree of silencing is dependent on the length of the overlapping AS transcripts, similar to the effects on the spreading of epigenetic silencing to neighboring genes outside the overlapping region as, for example, required for X-chromosome inactivation by Xist.

Genomic imprinting describes a genetic phenomenon demonstrated in mammals, insects, and plants, where certain genes ("imprinted genes") are either expressed only from the allele inherited from the mother or the father [216]. The silencing of the paternal or maternal alleles in imprinted genes is frequently mediated by S/AS transcription [217], where about 85% of some 2100 potentially imprinted mouse genes show S/AS transcription in CAGE studies [69]. Analyzing about 4000 human genes in cell lines by genome-wide approaches identified allele-specific transcription in clonal cell lines for more than 300 genes [218]. It would be interesting to see whether all of those monoallelic genes are silenced by S/AS transcription. Many long polyadenylated ncRNAs have been identified in *Drosophila* [219]. In *Drosophila*, the imprinting control region upstream of H19 is the key regulatory element conferring monoallelic expression on H19 and Igf2 (insulin-like growth factor 2), and the ncRNAs in the H19 ICR control region mediate gene expression [220]. Overlapping expression of coding and noncoding transcripts has also been found in *Drosophila* for the pseudouridine synthase gene, a complex gene with an interlaced organization [221].

AS transcription through regulatory regions does not only cause repression but can also lead to activation of gene expression as shown for the mouse T-cell receptor-alpha (Tcra) [222]. In this locus, an ncRNA from the T early alpha promoter (TEA) crosses the J-alpha regions located immediately downstream. Transcription of TEA opens the chromatin measured by monitoring histone H3K4 methylation in the immediately downstream regions and favors recombination in these regions, while at the same time repressing downstream transcription from the J-alpha 49 promoter and recombination of the downstream regions. Experimentally truncated TEA transcripts, however, close the chromatin and inhibit recombination of the immediately downstream regions. At the same time, they favor the expression of J-alpha 49 downstream transcripts and recombination of further downstream vJ segments [223]. Activation through transcription of intergenic regions associated with chromatin modification has further been well characterized in the HOXA cluster [224]. The elongation of AS ncRNAs over the silent promoter of the endogenous human retrovirus (HERV)-K18 results in its activation [225]. The activation of the endogenous retrovirus can be abolished by insertion of a poly(A) signal into the AS-transcribed region that causes immature termination of the AS RNA transcription. The regulatory mechanisms of AS transcripts expressed from the opposite strand of promoter regions may vary among the promoters. The fact that interference and activation can be caused by the same transcript suggests that either inhibitory or activating factors are displaced or altered in their binding by the ncRNA transcription on the same genomic strand [223].

It is noteworthy that not all known ncRNAs that are transcribed head to head or from bidirectional promoters shared with coding mRNAs act in cis on the expression of the corresponding coding mRNA. In *Arabidopsis*, a genome-wide screen identified about 1320 putative natural AS transcripts that could act in trans [226]. A notable example is a natural ncRNA named HASNT, which is AS to the hyaluronan synthase 2 (HAS2) and inhibits HAS2 expression upon induced expression of HASNT [227]. Although we do not know if there is sRNA production involved, the mechanism cannot be related to direct chromatin modification through transcription. Another very intriguing example is the regulation of the dihydrofolate reductase (DHFR) gene through formation of a triple helix. In quiescent cells, the transcriptional repression of the major promoter of the gene depends on a noncoding transcript initiated from the upstream minor promoter. The repression is mediated by the formation of a stable complex between ncRNA and the major promoter and direct interaction of the ncRNA with the general transcription factor IIB (TFIIB) leading to dissociation of the preinitiation complex from the major promoter [228]. The regulation of the DHFR gene could be an example on how PASRs may affect gene expression – an important consideration since for about half of all expressed genes, PASRs have been identified [14] arguing for widespread gene regulation by ncRNAs in trans. In this context, it is noteworthy that "antigen RNAs" targeting core promoter regions can cause strong inhibition of transcription [229]. Antigene peptide nucleic acid (agPNA) oligomers or antigene RNA (agRNA) duplexes can be used to target specific TSS for inhibition of gene expression and can be useful tools for silencing gene transcription [230, 231]. Whereas duplex RNAs complementary to promoters within chromosomal DNA are potent gene silencing agents, duplex RNAs complementary to the progesterone receptor (PR) promoter were also shown to increase PR expression after ectopic expression in cultured T47D or MCF7 human breast cancer cells [232].

Another example for regulation in trans is Evf-2, a nuclear ncRNA expressed from an ultraconserved intergenic region between the Dlx-5 and Dlx-6 genes, which specifically cooperates with Dlx-2 to increase the transcriptional activity of the Dlx-5/6 enhancer [233]. In the human HOX, loci expression profiling of ncRNA and mRNA transcripts revealed ncRNA and mRNA transcription in fibroblasts derived from 11 anatomic sites resembling embryonic development events. One of the 231 ncRNAs identified in the HOX clusters, termed HOTAIR, is transcribed from the HOXC cluster and represses in trans the transcription at the HOXD complex, where the ncRNA binds to chromatin for mediating repression [234]. In trans, regulation is not limited to developmental genes. Even the epigenetic state of rRNA genes is controlled by short (150–300 nt) ncRNAs that overlap with promoters on the same strand of transcribed rRNAs and derive from rRNA intergenic transcripts [235]. These ncRNAs inhibit rRNA transcription by interacting with TIP5, the large subunit of the chromatin remodeling complex NoRC. Epigenetics is an important aspect for the development of organisms where chromatin and DNA modifications transfer inherited information without involving changes in the underlying DNA sequence of the organism [236, 237].

More examples for the multiplicity of ncRNAs include steroid receptor RNA activator (SRA), which coregulates MyoD and skeletal muscle differentiation together with RNA helicases p68/p72 [238], or the newly discovered ncRNAs urothelial cancer associated 1 (UCA1 or CUDR), which confer drug resistance to tumor cells by inhibiting apoptosis [239], and eosinophil granule ontogeny (EGO), which regulates the expression of eosinophil granule protein expression [240]. Both ncRNAs, CUDR and EGO, are unlikely to act in cis, although details on their mechanisms of actions are still unclear. An example for the regulation of a tumor suppressor gene by an AS transcript is the silencing p15, a cyclin-dependent kinase inhibitor involved in leukemia. The p15 AS transcript p15AS induces silencing of p15 in cis and in trans through heterochromatin formation but not DNA methylation [241]. All these examples emphasize the importance of S/AS transcription for gene regulation, where certain various distinct mechanisms are functioning in the cell to fine-tune transcription and the utilization of genetic information.

In plants, distinct RNAi-related mechanisms are found. In particular, RNA-directed DNA methylation via an RNAi-mediated pathway in the nucleus is common in plants. Such RNA-directed DNA methylation requires besides DNA methyltransferases, histone-modifying enzymes and RNAi-related proteins, several plant-specific proteins that are essential for this process [242]. Among these are DRD1 (defective in RNA-directed DNA methylation), a putative SWI2/SNF2-like chromatin remodeling protein, as well as DRD2 and DRD3 (or NRPD2a and NRPD1b, respectively) subunits of the putative RNA polymerase IVb. In *Arabidopsis*, RNA-directed DNA methylation targets retrotransposon long terminal repeats (LTRs) with bidirectional promoters, and other types of intergenic transposons and repeat elements. Active DNA demethylation of methylcytosines in *Arabidopsis* is catalyzed by DNA glycosylases of the DEMETER (DME) family [243]. Loss of DNA methylation in *Arabidopsis* causes developmental defects [244], showing its central role in genome stability and gene regulation. In addition to the contribution of RNA to the regulation of chromatin structure and chromatin-induced gene silencing by the RNAi machinery in *S. pombe*, plants, and *Drosophila*, in *Drosophila* and mammals, dosage compensation requires specific ncRNAs (refer Xist above). Recent results in the studies of *Drosophila* S2 cells and chicken liver indicate a general structural role for RNA in eukaryotic chromatin. About 2–5% of all chromatin-associated nucleic acids could be poly(A–) RNA [245], which is interesting to note since some 40% of all mRNA may lack a poly(A) tail [13]. Moreover, chromatin association could in part explain the large portion of RNAs that are not exported from the nucleus [16].

8.6 Perspectives

Currently, we do not understand the relationship between the multitudes of RNA species identified by different approaches. Therefore, it was speculated that mRNA is the most visible manifestation of much more complex "transcriptional clouds," with "hidden layers" of regulatory transcripts in the back, which we just started to understand [246]. The current view on such "RNA clouds" is still incomplete, as long as we do not even know whether our present approaches can detect all RNA transcripts. Progress in expression profiling methods still enables an ongoing discovery of more thus far unknown transcripts of unknown function. This process will certainly continue with the increasing sensitivity of deep-sequencing approaches. Most of those newly discovered transcripts would probably lack any coding potential but rather act by other means in their own right. As outlined above, CAGE and full-length cDNAs identify only 5′-capped RNAs, while tiling arrays commonly exclude repeat elements and may be dependent on the labeling method for the hybridization probes. This is further illustrated by the fact that for most of the >450 000 sRNAs observed in tiling array experiments, no matching sRNAs could be found by sequencing sRNA libraries, which in part may be explained by modified ends of some sRNAs impossible to clone by standard approaches. In addition, expression of "multifunctional transcripts" further greatly extends on the transcriptional complexity, as one transcriptional event can drive various downstream processes.

Although becoming only very recently a focus of life sciences, the field of functional ncRNAs is developing very rapidly and has already made major contributions to our understanding of cellular processes. In particular, the spectacular success of RNAi reagents in research and therapy, and the use of miRNA in cancer diagnostic show the potential for uncovering new regulatory mechanisms. In addition, expression of "toxic RNA" can have pathogenic effects in RNA-dominant diseases [247], and ncRNAs are often downregulated in cancer. A single base substitution or small deletions in at least three large ncRNAs have been specifically correlated with a panel of human cancers [248]. Other long conserved ncRNAs, MALAT-1 in humans and hepcarcin in mice, are cancer markers [249, 250]. Another important example for an oncogenetic ncRNAs is H19, which is a maternally expressed, conserved, and imprinted gene. H19 is upregulated in cancer, and silencing its expression by siRNAs decreased tumorigenicity in mice [251]. However, its molecular mechanisms of action are unclear even though an miRNA sequence has been identified in the H19 transcript [252].

Though ncRNAs have been discovered in large numbers, the field still lacks suitable approaches for functional screens on new ncRNAs. This is marked by the inability of genetic screens to uncover ncRNA functions due to their short length, greater tolerance to point mutations, functional redundancy, and lack of knowledge about their functions. Already the analysis of coding genes has taken a long time, and it is not to be expected that ncRNAs can be studied at a much higher rate. Therefore, it will be important to select promising targets to demonstrate new functions for representative ncRNAs. Analysis of a small fraction of ncRNAs could provide the ground for a more general understanding of their regulatory processes. In addition, an experimental verification of their coding potential only may not address all functions of transcripts. Considering "multifunctional transcripts," the characterization of primary transcripts along with the mature RNAs derived thereof is an important requirement. This puts a strong emphasis on RNA processing and means of its analysis. Many of these processes are very fast, and specific inhibitors of RNA processing are in demand, for example, by knockdown of specific enzymes [92, 181, 209].

A better understanding of ncRNA functions will help to develop knowledge-driven approaches to ncRNA discovery and characterization. Here, functional screening systems are in demand to characterize ncRNAs, or to screen for new ncRNAs in a biological context using knockdown and gain-of-function studies. Functional screens may provide alternatives to genetic screens to better link transcript discovery to functional analysis. Understanding the function of ncRNAs will be the entry into an RNA world with entirely new concepts on biological processes. There are exciting times ahead of us with more surprising discoveries to come in the RNA world.

8.7
Note Added in Proof

The field of ncRNAs is developing very rapidly, mostly driven by new sequencing-based methods. Many important observations have been published since we had originally written this book chapter. Therefore, we want to point the reader at some recent reviews on some important aspects for updated views on these amassing developments.

Genome-wide studies are by now dominated by high-speed sequencing methods as summarized in a recent book *Tag-Based Next Generation Sequencing*, which is also published by Wiley-VCH [253]. Great progress has been made in the characterization of transcriptomes by RNA-seq methods that provide unsupervised views on all transcripts. Since RNA-seq methods offer sequence information on entire transcripts, these approaches can address splicing events or expressed genetic alterations [254, 255]. Approaches to miRNA and sRNA profiling are distinct from RNA-seq [256], and are routinely used to study miRNAs as biomarkers [257]. All of those methods largely depend on computational approaches to ncRNA discovery and annotation [258].

Our understanding on the genetic basis of disease greatly benefited from genome-wide association studies (GWAS), where variations in genome are detected by microarrays or high-speed sequencing [259]. Many variations were found at loci related to ncRNAs, which hopefully will contribute to a better understanding of the functions of the related transcripts.

The problem of functional annotation of ncRNAs has recently been discussed [260], where, for example, large-scale RNAi knockdown experiments on long intergenic noncoding RNAs (lincRNAs) have been done in embryonic stem cells [261]. Specific functions of ncRNAs have been reviewed by many authors, where we just want to point at some work on the control of transcription by ncRNAs [262, 263], long ncRNAs and enhancers [264], RNA–chromatin interactions [265], and role of large ncRNAs in cancer [266, 267], neural development and disease [268], and the regulation of embryogenesis [269]. Finally, we want to refer to a series of reviews on “Riboswitches and the RNA world” [270] with many interesting articles related to the content of our book chapter.

Acknowledgments

M.H. wants to thank the DNAFORM cDNA library team and his close collaborators at RIKEN Omics Science Center. P.C. wants to acknowledge all the colleagues and collaborators from RIKEN Omics Science Center and GERG group, the Fantom consortium, and the Genome Network Project. A grant for the Next Generation World-Leading Researchers(NEXT Program).

References

1 Brosius, J. (2005) Waste not, want not–transcript excess in multicellular eukaryotes. *Trends Genet.*, **21**, 287–288.
2 Carninci, P. and Hayashizaki, Y. (2007) Noncoding RNA transcription beyond annotated genes. *Curr. Opin. Genet. Dev.*, **17**, 139–144.
3 Claverie, J.M. (2005) Fewer genes, more noncoding RNA. *Science*, **309**, 1529–1530.
4 Dennis, C. (2002) The brave new world of RNA. *Nature*, **418**, 122–124.
5 Mattick, J.S. and Makunin, I.V. (2006) Non-coding RNA. *Hum. Mol. Genet.*, **15**, Spec. No. 1, R17–R29.
6 David, L., *et al.* (2006) A high-resolution map of transcription in the yeast genome. *Proc. Natl. Acad. Sci. U.S.A.*, **103**, 5320–5325.
7 Willingham, A.T. and Gingeras, T.R. (2006) TUF love for "junk" DNA. *Cell*, **125**, 1215–1220.
8 ENCODE Project Consortium (2007) Identification and analysis of functional elements in 1% of the human genome by the ENCODE pilot project. *Nature*, **447**, 799–816.
9 He, H., *et al.* (2007) Mapping the *C. elegans* noncoding transcriptome with a whole-genome tiling microarray. *Genome Res.*, **17**, 1471–1477.
10 Faulkner, G.J., *et al.* (2009) The regulated retrotransposon transcriptome of mammalian cells. *Nat. Genet.*, **41** (5), 563–571.
11 Copley, S.D., Smith, E., and Morowitz, H.J. (2007) The origin of the RNA world: co-evolution of genes and metabolism. *Bioorg. Chem.*, **35**, 430–443.
12 Carninci, P. (2006) Tagging mammalian transcription complexity. *Trends Genet.*, **22**, 501–510.
13 Cheng, J., *et al.* (2005) Transcriptional maps of 10 human chromosomes at 5-nucleotide resolution. *Science*, **308**, 1149–1154.
14 Kapranov, P., *et al.* (2007) RNA maps reveal new RNA classes and a possible function for pervasive transcription. *Science*, **316**, 1484–1488.
15 Kapranov, P., *et al.* (2005) Examples of the complex architecture of the human transcriptome revealed by RACE and high-density tiling arrays. *Genome Res.*, **15**, 987–997.
16 Kapranov, P., Willingham, A.T., and Gingeras, T.R. (2007) Genome-wide transcription and the implications for genomic organization. *Nat. Rev. Genet.*, **8**, 413–423.
17 Struhl, K. (2007) Transcriptional noise and the fidelity of initiation by RNA polymerase II. *Nat. Struct. Mol. Biol.*, **14**, 103–105.
18 Jacob, F. and Monod, J. (1961) Genetic regulatory mechanisms in the synthesis of proteins. *J. Mol. Biol.*, **3**, 318–356.
19 Ruvkun, G., Wightman, B., and Ha, I. (2004) The 20 years it took to recognize the importance of tiny RNAs. *Cell*, **116**, S93–S96, 2 p following S96.
20 Lee, R.C., Feinbaum, R.L., and Ambros, V. (1993) The *C. elegans* heterochronic gene lin-4 encodes small RNAs with antisense complementarity to lin-14. *Cell*, **75**, 843–854.
21 Ruvkun, G. (2001) Molecular biology. Glimpses of a tiny RNA world. *Science*, **294**, 797–799.
22 Brockdorff, N., *et al.* (1992) The product of the mouse Xist gene is a 15 kb inactive X-specific transcript containing no conserved ORF and located in the nucleus. *Cell*, **71**, 515–526.
23 Brown, C.J., *et al.* (1992) The human XIST gene: analysis of a 17 kb inactive X-specific RNA that contains conserved repeats and is highly localized within the nucleus. *Cell*, **71**, 527–542.
24 Wutz, A. (2007) Xist function: bridging chromatin and stem cells. *Trends Genet.*, **23**, 457–464.
25 Wutz, A. and Gribnau, J. (2007) X inactivation Xplained. *Curr. Opin. Genet. Dev.*, **17**, 387–393.
26 Lee, J.T., Davidow, L.S., and Warshawsky, D. (1999) Tsix, a gene antisense to Xist at the X-inactivation centre. *Nat. Genet.*, **21**, 400–404.

27 Lyon, M.F. (1961) Gene action in the X-chromosome of the mouse (*Mus musculus* L.). *Nature*, **190**, 372–373.
28 Salstrom, J.L. (2007) X-inactivation and the dynamic maintenance of gene silencing. *Mol. Genet. Metab.*, **92**, 56–62.
29 Chang, K., Elledge, S.J., and Hannon, G.J. (2006) Lessons from nature: microRNA-based shRNA libraries. *Nat. Methods*, **3**, 707–714.
30 Clark, J. and Ding, S. (2006) Generation of RNAi libraries for high-throughput screens. *J. Biomed. Biotechnol.*, **2006**, 45716.
31 Du, C., *et al.* (2006) PCR-based generation of shRNA libraries from cDNAs. *BMC Biotechnol.*, **6**, 28.
32 Shirane, D., *et al.* (2004) Enzymatic production of RNAi libraries from cDNAs. *Nat. Genet.*, **36**, 190–196.
33 Carpenter, A.E. and Sabatini, D.M. (2004) Systematic genome-wide screens of gene function. *Nat. Rev. Genet.*, **5**, 11–22.
34 Grimm, S. (2004) The art and design of genetic screens: mammalian culture cells. *Nat. Rev. Genet.*, **5**, 179–189.
35 Adams, M.D., *et al.* (1991) Complementary DNA sequencing: expressed sequence tags and human genome project. *Science*, **252**, 1651–1656.
36 Ota, T., *et al.* (2004) Complete sequencing and characterization of 21 243 full-length human cDNAs. *Nat. Genet.*, **36**, 40–45.
37 Carninci, P., *et al.* (2005) The transcriptional landscape of the mammalian genome. *Science*, **309**, 1559–1563.
38 Seki, M., *et al.* (2002) Functional annotation of a full-length *Arabidopsis* cDNA collection. *Science*, **296**, 141–145.
39 Okazaki, Y., *et al.* (2002) Analysis of the mouse transcriptome based on functional annotation of 60 770 full-length cDNAs. *Nature*, **420**, 563–573.
40 Imanishi, T., *et al.* (2004) Integrative annotation of 21 037 human genes validated by full-length cDNA clones. *PLoS Biol.*, **2**, e162.
41 Yamasaki, C., *et al.* (2008) The H-Invitational Database (H-InvDB), a comprehensive annotation resource for human genes and transcripts. *Nucleic Acids Res.*, **36**, D793–D799.
42 Carninci, P., *et al.* (2003) Targeting a complex transcriptome: the construction of the mouse full-length cDNA encyclopedia. *Genome Res.*, **13**, 1273–1289.
43 Harbers, M. (2008) The current status of cDNA cloning. *Genomics*, **91**, 232–242.
44 Furuno, M., *et al.* (2006) Clusters of internally primed transcripts reveal novel long noncoding RNAs. *PLoS Genet.*, **2**, e37.
45 Mrazek, J., Kreutmayer, S.B., Grasser, F.A., Polacek, N., and Huttenhofer, A. (2007) Subtractive hybridization identifies novel differentially expressed ncRNA species in EBV-infected human B cells. *Nucleic Acids Res.*, **35**, e73.
46 Harbers, M. and Carninci, P. (2005) Tag-based approaches for transcriptome research and genome annotation. *Nat. Methods*, **2**, 495–502.
47 Saha, S., *et al.* (2002) Using the transcriptome to annotate the genome. *Nat. Biotechnol.*, **20**, 508–512.
48 Matsumura, H., *et al.* (2005) SuperSAGE. *Cell. Microbiol.*, **7**, 11–18.
49 Keime, C., Semon, M., Mouchiroud, D., Duret, L., and Gandrillon, O. (2007) Unexpected observations after mapping LongSAGE tags to the human genome. *BMC Bioinformatics*, **8**, 154.
50 Carninci, P., *et al.* (2006) Genome-wide analysis of mammalian promoter architecture and evolution. *Nat. Genet.*, **38**, 626–635.
51 Velculescu, V.E., Zhang, L., Vogelstein, B., and Kinzler, K.W. (1995) Serial analysis of gene expression. *Science*, **270**, 484–487.
52 Powell, J. (2000) SAGE. The serial analysis of gene expression. *Methods Mol. Biol.*, **99**, 297–319.
53 Wang, S.M. (2005) *SAGE: Current Technologies an Applications*, Horizon Bioscience, Norwich.
54 Kodzius, R., *et al.* (2006) CAGE: cap analysis of gene expression. *Nat. Methods*, **3**, 211–222.
55 Shiraki, T., *et al.* (2003) Cap analysis gene expression for high-throughput analysis of transcriptional starting point

and identification of promoter usage. *Proc. Natl. Acad. Sci. U.S.A.*, **100**, 15776–15781.

56 Zhang, Z. and Dietrich, F.S. (2005) Mapping of transcription start sites in *Saccharomyces cerevisiae* using 5′ SAGE. *Nucleic Acids Res.*, **33**, 2838–2851.

57 Hashimoto, S., *et al.* (2004) 5′-End SAGE for the analysis of transcriptional start sites. *Nat. Biotechnol.*, **22**, 1146–1149.

58 Hwang, B.J., Muller, H.M., and Sternberg, P.W. (2004) Genome annotation by high-throughput 5′ RNA end determination. *Proc. Natl. Acad. Sci. U.S.A.*, **101**, 1650–1655.

59 Wei, C.L., *et al.* (2004) 5′ Long serial analysis of gene expression (LongSAGE) and 3′ LongSAGE for transcriptome characterization and genome annotation. *Proc. Natl. Acad. Sci. U.S.A.*, **101**, 11701–11706.

60 Katayama, S., Kanamori-Katayama, M., Yamaguchi, K., Carninci, P., and Hayashizaki, Y. (2007) CAGE-TSSchip: promoter-based expression profiling using the 5′-leading label of capped transcripts. *Genome Biol.*, **8**, R42.

61 Shimokawa, K., *et al.* (2007) Large-scale clustering of CAGE tag expression data. *BMC Bioinformatics*, **8**, 161.

62 Kawaji, H., *et al.* (2006) CAGE basic/analysis databases: the CAGE resource for comprehensive promoter analysis. *Nucleic Acids Res.*, **34**, D632–D636.

63 Bajic, V.B., *et al.* (2006) Mice and men: their promoter properties. *PLoS Genet.*, **2**, e54.

64 Faulkner, G.J., *et al.* (2008) A rescue strategy for multimapping short sequence tags refines surveys of transcriptional activity by CAGE. *Genomics*, **91** (3), 281–288.

65 Frith, M.C., *et al.* (2008) A code for transcription initiation in mammalian genomes. *Genome Res.*, **18**, 1–12.

66 Kawaji, H., *et al.* (2006) Dynamic usage of transcription start sites within core promoters. *Genome Biol.*, **7**, R118.

67 Galante, P.A., Vidal, D.O., de Souza, J.E., Camargo, A.A., and de Souza, S.J. (2007) Sense-antisense pairs in mammals: functional and evolutionary considerations. *Genome Biol.*, **8**, R40.

68 Hongay, C.F., Grisafi, P.L., Galitski, T., and Fink, G.R. (2006) Antisense transcription controls cell fate in *Saccharomyces cerevisiae. Cell*, **127**, 735–745.

69 Katayama, S., *et al.* (2005) Antisense transcription in the mammalian transcriptome. *Science*, **309**, 1564–1566.

70 Wang, X.J., Gaasterland, T., and Chua, N.H. (2005) Genome-wide prediction and identification of cis-natural antisense transcripts in *Arabidopsis thaliana. Genome Biol.*, **6**, R30.

71 Hall, N. (2007) Advanced sequencing technologies and their wider impact in microbiology. *J. Exp. Biol.*, **210**, 1518–1525.

72 Metzker, M.L. (2005) Emerging technologies in DNA sequencing. *Genome Res.*, **15**, 1767–1776.

73 Shendure, J., Mitra, R.D., Varma, C., and Church, G.M. (2004) Advanced sequencing technologies: methods and goals. *Nat. Rev. Genet.*, **5**, 335–344.

74 Chi, K.R. (2008) The year of sequencing. *Nat. Methods*, **5**, 11–14.

75 Binladen, J., *et al.* (2007) The use of coded PCR primers enables high-throughput sequencing of multiple homolog amplification products by 454 parallel sequencing. *PLoS One*, **2**, e197.

76 Nielsen, K.L., Hogh, A.L., and Emmersen, J. (2006) DeepSAGE–digital transcriptomics with high sensitivity, simple experimental protocol and multiplexing of samples. *Nucleic Acids Res.*, **34**, e133.

77 Maeda, N., *et al.* (2008) Development of a DNA barcode tagging method for monitoring dynamic changes in gene expression by using an ultra high-throughput sequencer. *Biotechniques*, **45** (1), 95–97.

78 Huttenhofer, A., Cavaille, J., and Bachellerie, J.P. (2004) Experimental RNomics: a global approach to identifying small nuclear RNAs and their targets in different model organisms. *Methods Mol. Biol.*, **265**, 409–428.

79 Huttenhofer, A. and Vogel, J. (2006) Experimental approaches to identify non-coding RNAs. *Nucleic Acids Res.*, **34**, 635–646.

80 Berezikov, E., Cuppen, E., and Plasterk, R.H. (2006) Approaches to microRNA discovery. *Nat. Genet.*, **38** (Suppl.), S2–S7.

81 Hafner, M., *et al.* (2008) Identification of microRNAs and other small regulatory RNAs using cDNA library sequencing. *Methods*, **44**, 3–12.

82 Wilkinson, M.L., Crary, S.M., Jackman, J.E., Grayhack, E.J., and Phizicky, E.M. (2007) The 2′-O-methyltransferase responsible for modification of yeast tRNA at position 4. *RNA*, **13**, 404–413.

83 Einat, P. (2006) Methodologies for high-throughput expression profiling of microRNAs. *Methods Mol. Biol.*, **342**, 139–157.

84 Yin, J.Q., Zhao, R.C., and Morris, K.V. (2008) Profiling microRNA expression with microarrays. *Trends Biotechnol.*, **26** (2), 70–76.

85 Samanta, M.P., Tongprasit, W., and Stolc, V. (2007) In-depth query of large genomes using tiling arrays. *Methods Mol. Biol.*, **377**, 163–174.

86 Yazaki, J., Gregory, B.D., and Ecker, J.R. (2007) Mapping the genome landscape using tiling array technology. *Curr. Opin. Plant. Biol.*, **10**, 534–542.

87 Kampa, D., *et al.* (2004) Novel RNAs identified from an in-depth analysis of the transcriptome of human chromosomes 21 and 22. *Genome Res.*, **14**, 331–342.

88 Kapranov, P., *et al.* (2002) Large-scale transcriptional activity in chromosomes 21 and 22. *Science*, **296**, 916–919.

89 Bertone, P., *et al.* (2004) Global identification of human transcribed sequences with genome tiling arrays. *Science*, **306**, 2242–2246.

90 Li, L., *et al.* (2006) Genome-wide transcription analyses in rice using tiling microarrays. *Nat. Genet.*, **38**, 124–129.

91 Zhang, X., *et al.* (2006) Genome-wide high-resolution mapping and functional analysis of DNA methylation in *Arabidopsis*. *Cell*, **126**, 1189–1201.

92 Davis, C.A. and Ares, M., Jr. (2006) Accumulation of unstable promoter-associated transcripts upon loss of the nuclear exosome subunit Rrp6p in *Saccharomyces cerevisiae*. *Proc. Natl. Acad. Sci. U.S.A.*, **103**, 3262–3267.

93 Samanta, M.P., Tongprasit, W., Sethi, H., Chin, C.S., and Stolc, V. (2006) Global identification of noncoding RNAs in *Saccharomyces cerevisiae* by modulating an essential RNA processing pathway. *Proc. Natl. Acad. Sci. U.S.A.*, **103**, 4192–4197.

94 Mockler, T.C., *et al.* (2005) Applications of DNA tiling arrays for whole-genome analysis. *Genomics*, **85**, 1–15.

95 Prasanth, K.V., *et al.* (2005) Regulating gene expression through RNA nuclear retention. *Cell*, **123**, 249–263.

96 Wheeler, G., Valoczi, A., Havelda, Z., and Dalmay, T. (2007) *In situ* detection of animal and plant microRNAs. *DNA Cell Biol.*, **26**, 251–255.

97 Silahtaroglu, A.N., *et al.* (2007) Detection of microRNAs in frozen tissue sections by fluorescence *in situ* hybridization using locked nucleic acid probes and tyramide signal amplification. *Nat. Protoc.*, **2**, 2520–2528.

98 Jiang, J. and Gill, B.S. (2006) Current status and the future of fluorescence *in situ* hybridization (FISH) in plant genome research. *Genome*, **49**, 1057–1068.

99 Querido, E. and Chartrand, P. (2008) Using fluorescent proteins to study mRNA trafficking in living cells. *Methods Cell Biol.*, **85**, 273–292.

100 Hawkins, R.D. and Ren, B. (2006) Genome-wide location analysis: insights on transcriptional regulation. *Hum. Mol. Genet.*, **15**, Spec. No. 1, R1–R7.

101 Liu, D.G. and Sun, L. (2005) Direct isolation of specific RNA-interacting proteins using a novel affinity medium. *Nucleic Acids Res.*, **33**, e132.

102 Ule, J., *et al.* (2003) CLIP identifies Nova-regulated RNA networks in the brain. *Science*, **302**, 1212–1215.

103 Keene, J.D., Komisarow, J.M., and Friedersdorf, M.B. (2006) RIP-Chip: the isolation and identification of mRNAs, microRNAs and protein components of ribonucleoprotein complexes from cell extracts. *Nat. Protoc.*, **1**, 302–307.

104 Valencia-Burton, M. and Broude, N.E. (2007) Visualization of RNA using fluorescence complementation triggered by aptamer-protein interactions (RFAP) in live bacterial cells. *Curr. Protoc. Cell Biol.*, Chapter 17, Unit 17.11.

105 Windbichler, N. and Schroeder, R. (2006) Isolation of specific RNA-binding proteins using the streptomycin-binding RNA aptamer. *Nat. Protoc.*, **1**, 637–640.

106 Hogg, J.R. and Collins, K. (2007) RNA-based affinity purification reveals 7SK RNPs with distinct composition and regulation. *RNA*, **13**, 868–880.

107 Paraskeva, E., Atzberger, A., and Hentze, M.W. (1998) A translational repression assay procedure (TRAP) for RNA-protein interactions *in vivo*. *Proc. Natl. Acad. Sci. U.S.A.*, **95**, 951–956.

108 Gold, L., *et al.* (1997) From oligonucleotide shapes to genomic SELEX: novel biological regulatory loops. *Proc. Natl. Acad. Sci. U.S.A.*, **94**, 59–64.

109 Shtatland, T., *et al.* (2000) Interactions of *Escherichia coli* RNA with bacteriophage MS2 coat protein: genomic SELEX. *Nucleic Acids Res.*, **28**, E93.

110 Singer, B.S., Shtatland, T., Brown, D., and Gold, L. (1997) Libraries for genomic SELEX. *Nucleic Acids Res.*, **25**, 781–786.

111 Griffiths-Jones, S. (2007) Annotating noncoding RNA genes. *Annu. Rev. Genomics Hum. Genet.*, **8**, 279–298.

112 Machado-Lima, A., Del Portillo, H.A., and Durham, A.M. (2008) Computational methods in noncoding RNA research. *J. Math. Biol.*, **56**, 15–49.

113 Clark, A.G., *et al.* (2007) Evolution of genes and genomes on the *Drosophila* phylogeny. *Nature*, **450**, 203–218.

114 Stark, A., *et al.* (2007) Discovery of functional elements in 12 *Drosophila* genomes using evolutionary signatures. *Nature*, **450**, 219–232.

115 Pang, K.C., Frith, M.C., and Mattick, J.S. (2006) Rapid evolution of noncoding RNAs: lack of conservation does not mean lack of function. *Trends Genet.*, **22**, 1–5.

116 Ponjavic, J., Ponting, C.P., and Lunter, G. (2007) Functionality or transcriptional noise? Evidence for selection within long noncoding RNAs. *Genome Res.*, **17**, 556–565.

117 Khaitovich, P., *et al.* (2006) Functionality of intergenic transcription: an evolutionary comparison. *PLoS Genet.*, **2**, e171.

118 Pollard, K.S., *et al.* (2006) An RNA gene expressed during cortical development evolved rapidly in humans. *Nature*, **443**, 167–172.

119 Karklin, Y., Meraz, R.F., and Holbrook, S.R. (2005) Classification of non-coding RNA using graph representations of secondary structure. *Pac. Symp. Biocomput.*, 4–15.

120 Rivas, E. and Eddy, S.R. (2001) Noncoding RNA gene detection using comparative sequence analysis. *BMC Bioinformatics*, **2**, 8.

121 Liu, J., Gough, J., and Rost, B. (2006) Distinguishing protein-coding from non-coding RNAs through support vector machines. *PLoS Genet.*, **2**, e29.

122 Cruveiller, S., Clay, O., Jabbari, K., and Bernardi, G. (2007) Simple proteomic checks for detecting noncoding RNA. *Proteomics*, **7**, 361–363.

123 Ghosh, Z., Chakrabarti, J., and Mallick, B. (2007) miRNomics–the bioinformatics of microRNA genes. *Biochem. Biophys. Res. Commun.*, **363**, 6–11.

124 Zeng, Y. (2006) Principles of micro-RNA production and maturation. *Oncogene*, **25**, 6156–6162.

125 Lin, S.L., Miller, J.D., and Ying, S.Y. (2006) Intronic microRNA (miRNA). *J. Biomed. Biotechnol.*, **2006**, 26818.

126 Kiss, T. (2006) SnoRNP biogenesis meets pre-mRNA splicing. *Mol. Cell*, **23**, 775–776.

127 Grosjean, H., Szweykowska-Kulinska, Z., Motorin, Y., Fasiolo, F., and Simos, G. (1997) Intron-dependent enzymatic formation of modified nucleosides in eukaryotic tRNAs: a review. *Biochimie*, **79**, 293–302.

128 Jen, C.H., Michalopoulos, I., Westhead, D.R., and Meyer, P. (2005) Natural antisense transcripts with coding capacity in *Arabidopsis* may have a regulatory role that is not linked to

double-stranded RNA degradation. *Genome Biol.*, **6**, R51.

129 Yoshihisa, T. (2006) tRNA, new aspects in intracellular dynamics. *Cell. Mol. Life Sci.*, **63**, 1813–1818.

130 Clark, B.F. (2006) The crystal structure of tRNA. *J. Biosci.*, **31**, 453–457.

131 Lowe, T.M. and Eddy, S.R. (1997) tRNAscan-SE: a program for improved detection of transfer RNA genes in genomic sequence. *Nucleic Acids Res.*, **25**, 955–964.

132 Yusupov, M.M., *et al.* (2001) Crystal structure of the ribosome at 5.5 A resolution. *Science*, **292**, 883–896.

133 Eickbush, T.H. and Eickbush, D.G. (2007) Finely orchestrated movements: evolution of the ribosomal RNA genes. *Genetics*, **175**, 477–485.

134 Spirin, A.S. (2004) The ribosome as an RNA-based molecular machine. *RNA Biol.*, **1**, 3–9.

135 Barrandon, C., Spiluttini, B., and Bensaude, O. (2008) Non-coding RNAs regulating the transcriptional machinery. *Biol. Cell*, **100**, 83–95.

136 Hopper, A.K. (2006) Cellular dynamics of small RNAs. *Crit. Rev. Biochem. Mol. Biol.*, **41**, 3–19.

137 Bachellerie, J.P., Cavaille, J., and Huttenhofer, A. (2002) The expanding snoRNA world. *Biochimie*, **84**, 775–790.

138 Jady, B.E. and Kiss, T. (2001) A small nucleolar guide RNA functions both in 2′-O-ribose methylation and pseudouridylation of the U5 spliceosomal RNA. *EMBO J.*, **20**, 541–551.

139 Darzacq, X., *et al.* (2002) Cajal body-specific small nuclear RNAs: a novel class of 2′-O-methylation and pseudouridylation guide RNAs. *EMBO J.*, **21**, 2746–2756.

140 Stanek, D. and Neugebauer, K.M. (2006) The Cajal body: a meeting place for spliceosomal snRNPs in the nuclear maze. *Chromosoma*, **115**, 343–354.

141 Huttenhofer, A. and Schattner, P. (2006) The principles of guiding by RNA: chimeric RNA-protein enzymes. *Nat. Rev. Genet.*, **7**, 475–482.

142 Stuart, K.D., Schnaufer, A., Ernst, N.L., and Panigrahi, A.K. (2005) Complex management: RNA editing in trypanosomes. *Trends Biochem. Sci.*, **30**, 97–105.

143 Alfonzo, J.D., Thiemann, O., and Simpson, L. (1997) The mechanism of U insertion/deletion RNA editing in kinetoplastid mitochondria. *Nucleic Acids Res.*, **25**, 3751–3759.

144 Wang, Y., Stricker, H.M., Gou, D., and Liu, L. (2007) MicroRNA: past and present. *Front. Biosci.*, **12**, 2316–2329.

145 Bartel, D.P. (2004) MicroRNAs: genomics, biogenesis, mechanism, and function. *Cell*, **116**, 281–297.

146 Filipowicz, W., Bhattacharyya, S.N., and Sonenberg, N. (2008) Mechanisms of post-transcriptional regulation by microRNAs: are the answers in sight? *Nat. Rev. Genet.*, **9**, 102–114.

147 Valadi, H., *et al.* (2007) Exosome-mediated transfer of mRNAs and microRNAs is a novel mechanism of genetic exchange between cells. *Nat. Cell Biol.*, **9**, 654–659.

148 Hwang, H.W., Wentzel, E.A., and Mendell, J.T. (2007) A hexanucleotide element directs microRNA nuclear import. *Science*, **315**, 97–100.

149 Bao, N., Lye, K.W., and Barton, M.K. (2004) MicroRNA binding sites in *Arabidopsis* class III HD-ZIP mRNAs are required for methylation of the template chromosome. *Dev. Cell*, **7**, 653–662.

150 Molnar, A., Schwach, F., Studholme, D.J., Thuenemann, E.C., and Baulcombe, D.C. (2007) miRNAs control gene expression in the single-cell alga *Chlamydomonas reinhardtii*. *Nature*, **447**, 1126–1129.

151 Zhao, T., *et al.* (2007) A complex system of small RNAs in the unicellular green alga *Chlamydomonas reinhardtii*. *Genes Dev.*, **21**, 1190–1203.

152 Stefani, G. and Slack, F.J. (2008) Small non-coding RNAs in animal development. *Nat. Rev. Mol. Cell Biol.*, **9**, 219–230.

153 Negrini, M., Ferracin, M., Sabbioni, S., and Croce, C.M. (2007) MicroRNAs in human cancer: from research to therapy. *J. Cell Sci.*, **120**, 1833–1840.

154 Huang, Q., *et al.* (2008) The microRNAs miR-373 and miR-520c promote tumour invasion and metastasis. *Nat. Cell Biol.*, **10**, 202–210.

155 Zhao, Y., *et al.* (2007) Dysregulation of cardiogenesis, cardiac conduction, and cell cycle in mice lacking miRNA-1-2. *Cell*, **129**, 303–317.
156 Dieci, G., Fiorino, G., Castelnuovo, M., Teichmann, M., and Pagano, A. (2007) The expanding RNA polymerase III transcriptome. *Trends Genet.*, **23**, 614–622.
157 Denli, A.M., Tops, B.B., Plasterk, R.H., Ketting, R.F., and Hannon, G.J. (2004) Processing of primary microRNAs by the Microprocessor complex. *Nature*, **432**, 231–235.
158 Saini, H.K., Griffiths-Jones, S., and Enright, A.J. (2007) Genomic analysis of human microRNA transcripts. *Proc. Natl. Acad. Sci. U.S.A.*, **104**, 17719–17724.
159 Orom, U.A., Kauppinen, S., and Lund, A.H. (2006) LNA-modified oligonucleotides mediate specific inhibition of microRNA function. *Gene*, **372**, 137–141.
160 Karginov, F.V., *et al.* (2007) A biochemical approach to identifying microRNA targets. *Proc. Natl. Acad. Sci. U.S.A.*, **104**, 19291–19296.
161 Vasudevan, S., Tong, Y., and Steitz, J.A. (2007) Switching from repression to activation: microRNAs can up-regulate translation. *Science*, **318**, 1931–1934.
162 Lu, C., *et al.* (2005) Elucidation of the small RNA component of the transcriptome. *Science*, **309**, 1567–1569.
163 Gunawardane, L.S., *et al.* (2007) A slicer-mediated mechanism for repeat-associated siRNA 5′ end formation in *Drosophila*. *Science*, **315**, 1587–1590.
164 Vagin, V.V., *et al.* (2006) A distinct small RNA pathway silences selfish genetic elements in the germline. *Science*, **313**, 320–324.
165 Aravin, A.A., *et al.* (2003) The small RNA profile during *Drosophila melanogaster* development. *Dev. Cell*, **5**, 337–350.
166 Hamilton, A., Voinnet, O., Chappell, L., and Baulcombe, D. (2002) Two classes of short interfering RNA in RNA silencing. *EMBO J.*, **21**, 4671–4679.
167 Lippman, Z., May, B., Yordan, C., Singer, T., and Martienssen, R. (2003) Distinct mechanisms determine transposon inheritance and methylation via small interfering RNA and histone modification. *PLoS Biol.*, **1**, E67.
168 Vastenhouw, N.L. and Plasterk, R.H. (2004) RNAi protects the *Caenorhabditis elegans* germline against transposition. *Trends Genet.*, **20**, 314–319.
169 Aravin, A.A., *et al.* (2004) Dissection of a natural RNA silencing process in the *Drosophila melanogaster* germ line. *Mol. Cell Biol.*, **24**, 6742–6750.
170 Volpe, T.A., *et al.* (2002) Regulation of heterochromatic silencing and histone H3 lysine-9 methylation by RNAi. *Science*, **297**, 1833–1837.
171 Kasschau, K.D., *et al.* (2007) Genome-wide profiling and analysis of *Arabidopsis* siRNAs. *PLoS Biol.*, **5**, e57.
172 Hartig, J.V., Tomari, Y., and Forstemann, K. (2007) piRNAs–the ancient hunters of genome invaders. *Genes Dev.*, **21**, 1707–1713.
173 Klattenhoff, C. and Theurkauf, W. (2008) Biogenesis and germline functions of piRNAs. *Development*, **135**, 3–9.
174 Girard, A., Sachidanandam, R., Hannon, G.J., and Carmell, M.A. (2006) A germline-specific class of small RNAs binds mammalian Piwi proteins. *Nature*, **442**, 199–202.
175 Aravin, A., *et al.* (2006) A novel class of small RNAs bind to MILI protein in mouse testes. *Nature*, **442**, 203–207.
176 Aravin, A.A., Hannon, G.J., and Brennecke, J. (2007) The Piwi-piRNA pathway provides an adaptive defense in the transposon arms race. *Science*, **318**, 761–764.
177 Yin, H. and Lin, H. (2007) An epigenetic activation role of Piwi and a Piwi-associated piRNA in *Drosophila melanogaster*. *Nature*, **450**, 304–308.
178 Faehnle, C.R. and Joshua-Tor, L. (2007) Argonautes confront new small RNAs. *Curr. Opin. Chem. Biol.*, **11**, 569–577.
179 Hutvagner, G. and Simard, M.J. (2008) Argonaute proteins: key players in RNA silencing. *Nat. Rev. Mol. Cell Biol.*, **9**, 22–32.
180 Vaucheret, H. (2005) MicroRNA-dependent trans-acting siRNA production. *Sci. STKE*, **2005**, pe43.

181 Chekanova, J.A., *et al.* (2007) Genome-wide high-resolution mapping of exosome substrates reveals hidden features in the *Arabidopsis* transcriptome. *Cell*, **131**, 1340–1353.

182 Buchan, J.R. and Stansfield, I. (2007) Halting a cellular production line: responses to ribosomal pausing during translation. *Biol. Cell*, **99**, 475–487.

183 Dulebohn, D., Choy, J., Sundermeier, T., Okan, N., and Karzai, A.W. (2007) Trans-translation: the tmRNA-mediated surveillance mechanism for ribosome rescue, directed protein degradation, and nonstop mRNA decay. *Biochemistry*, **46**, 4681–4693.

184 Moore, S.D. and Sauer, R.T. (2007) The tmRNA system for translational surveillance and ribosome rescue. *Annu. Rev. Biochem.*, **76**, 101–124.

185 Cech, T.R. (2002) Ribozymes, the first 20 years. *Biochem. Soc. Trans.*, **30**, 1162–1166.

186 Scott, W.G. (2007) Ribozymes. *Curr. Opin. Struct. Biol.*, **17**, 280–286.

187 Serganov, A. and Patel, D.J. (2007) Ribozymes, riboswitches and beyond: regulation of gene expression without proteins. *Nat. Rev. Genet.*, **8**, 776–790.

188 Win, M.N. and Smolke, C.D. (2007) RNA as a versatile and powerful platform for engineering genetic regulatory tools. *Biotechnol. Genet. Eng. Rev.*, **24**, 311–346.

189 Kirsebom, L.A. (2007) RNase P RNA mediated cleavage: substrate recognition and catalysis. *Biochimie*, **89**, 1183–1194.

190 Mabeck, C.E. and Andreasen, P.B. (1976) [Gerovital H3, procaine chloride]. *Tidsskr. Nor. Laegeforen.*, **96**, 443–444.

191 Pickering, B.M. and Willis, A.E. (2005) The implications of structured 5′ untranslated regions on translation and disease. *Semin. Cell Dev. Biol.*, **16**, 39–47.

192 Batey, R.T. (2006) Structures of regulatory elements in mRNAs. *Curr. Opin. Struct. Biol.*, **16**, 299–306.

193 Tucker, B.J. and Breaker, R.R. (2005) Riboswitches as versatile gene control elements. *Curr. Opin. Struct. Biol.*, **15**, 342–348.

194 Vitreschak, A.G., Rodionov, D.A., Mironov, A.A., and Gelfand, M.S. (2004) Riboswitches: the oldest mechanism for the regulation of gene expression? *Trends Genet.*, **20**, 44–50.

195 Bunka, D.H. and Stockley, P.G. (2006) Aptamers come of age – at last. *Nat. Rev. Microbiol.*, **4**, 588–596.

196 Lee, J.F., Stovall, G.M., and Ellington, A.D. (2006) Aptamer therapeutics advance. *Curr. Opin. Chem. Biol.*, **10**, 282–289.

197 Mairal, T., *et al.* (2007) Aptamers: molecular tools for analytical applications. *Anal. Bioanal. Chem.*, **390** (4), 989–1007.

198 Cheah, M.T., Wachter, A., Sudarsan, N., and Breaker, R.R. (2007) Control of alternative RNA splicing and gene expression by eukaryotic riboswitches. *Nature*, **447**, 497–500.

199 Blount, K.F. and Breaker, R.R. (2006) Riboswitches as antibacterial drug targets. *Nat. Biotechnol.*, **24**, 1558–1564.

200 Ciesla, J. (2006) Metabolic enzymes that bind RNA: yet another level of cellular regulatory network? *Acta Biochim. Pol.*, **53**, 11–32.

201 Ravasi, T., *et al.* (2006) Experimental validation of the regulated expression of large numbers of non-coding RNAs from the mouse genome. *Genome Res.*, **16**, 11–19.

202 Sun, M., Hurst, L.D., Carmichael, G.G., and Chen, J. (2006) Evidence for variation in abundance of antisense transcripts between multicellular animals but no relationship between antisense transcription and organismic complexity. *Genome Res.*, **16**, 922–933.

203 Numata, K., *et al.* (2007) Comparative analysis of cis-encoded antisense RNAs in eukaryotes. *Gene*, **392**, 134–141.

204 Finocchiaro, G., *et al.* (2007) Localizing hotspots of antisense transcription. *Nucleic Acids Res.*, **35**, 1488–1500.

205 Proudfoot, N.J. (1986) Transcriptional interference and termination between duplicated alpha-globin gene constructs suggests a novel mechanism for gene regulation. *Nature*, **322**, 562–565.

206 Proudfoot, N.J., Gil, A., and Whitelaw, E. (1985) Studies on messenger RNA 3′ end formation in globin genes: a transcriptional interference model for globin gene switching. *Prog. Clin. Biol. Res.*, **191**, 49–65.

207 Shearwin, K.E., Callen, B.P., and Egan, J.B. (2005) Transcriptional interference – a crash course. *Trends Genet.*, **21**, 339–345.

208 Martens, J.A., Laprade, L., and Winston, F. (2004) Intergenic transcription is required to repress the *Saccharomyces cerevisiae* SER3 gene. *Nature*, **429**, 571–574.

209 Camblong, J., Iglesias, N., Fickentscher, C., Dieppois, G., and Stutz, F. (2007) Antisense RNA stabilization induces transcriptional gene silencing via histone deacetylation in *S. cerevisiae*. *Cell*, **131**, 706–717.

210 Zofall, M. and Grewal, S.I. (2006) RNAi-mediated heterochromatin assembly in fission yeast. *Cold Spring Harb. Symp. Quant. Biol.*, **71**, 487–496.

211 Faghihi, M.A. and Wahlestedt, C. (2006) RNA interference is not involved in natural antisense mediated regulation of gene expression in mammals. *Genome Biol.*, **7**, R38.

212 Petruk, S., *et al.* (2006) Transcription of bxd noncoding RNAs promoted by trithorax represses Ubx in cis by transcriptional interference. *Cell*, **127**, 1209–1221.

213 Hu, X., *et al.* (2007) Transcriptional interference among the murine beta-like globin genes. *Blood*, **109**, 2210–2216.

214 Kanduri, C., Thakur, N., and Pandey, R.R. (2006) The length of the transcript encoded from the Kcnq1ot1 antisense promoter determines the degree of silencing. *EMBO J.*, **25**, 2096–2106.

215 Murakami, K., Oshimura, M., and Kugoh, H. (2007) Suggestive evidence for chromosomal localization of non-coding RNA from imprinted LIT1. *J. Hum. Genet.*, **52**, 926–933.

216 Bartolomei, M.S. and Tilghman, S.M. (1997) Genomic imprinting in mammals. *Annu. Rev. Genet.*, **31**, 493–525.

217 Pauler, F.M., Koerner, M.V., and Barlow, D.P. (2007) Silencing by imprinted noncoding RNAs: is transcription the answer? *Trends Genet.*, **23**, 284–292.

218 Gimelbrant, A., Hutchinson, J.N., Thompson, B.R., and Chess, A. (2007) Widespread monoallelic expression on human autosomes. *Science*, **318**, 1136–1140.

219 Tupy, J.L., *et al.* (2005) Identification of putative noncoding polyadenylated transcripts in *Drosophila melanogaster*. *Proc. Natl. Acad. Sci. U.S.A.*, **102**, 5495–5500.

220 Schoenfelder, S., Smits, G., Fraser, P., Reik, W., and Paro, R. (2007) Non-coding transcripts in the H19 imprinting control region mediate gene silencing in transgenic *Drosophila*. *EMBO Rep.*, **8**, 1068–1073.

221 Riccardo, S., Tortoriello, G., Giordano, E., Turano, M., and Furia, M. (2007) The coding/non-coding overlapping architecture of the gene encoding the *Drosophila* pseudouridine synthase. *BMC Mol. Biol.*, **8**, 15.

222 Abarrategui, I. and Krangel, M.S. (2006) Regulation of T cell receptor-alpha gene recombination by transcription. *Nat. Immunol.*, **7**, 1109–1115.

223 Abarrategui, I. and Krangel, M.S. (2007) Noncoding transcription controls downstream promoters to regulate T-cell receptor alpha recombination. *EMBO J.*, **26**, 4380–4390.

224 Sessa, L., *et al.* (2007) Noncoding RNA synthesis and loss of Polycomb group repression accompanies the colinear activation of the human HOXA cluster. *RNA*, **13**, 223–239.

225 Leupin, O., *et al.* (2005) Transcriptional activation by bidirectional RNA polymerase II elongation over a silent promoter. *EMBO Rep.*, **6**, 956–960.

226 Wang, H., Chua, N.H., and Wang, X.J. (2006) Prediction of trans-antisense transcripts in *Arabidopsis thaliana*. *Genome Biol.*, **7**, R92.

227 Chao, H. and Spicer, A.P. (2005) Natural antisense mRNAs to hyaluronan synthase 2 inhibit hyaluronan biosynthesis and cell proliferation. *J. Biol. Chem.*, **280**, 27513–27522.

228 Martianov, I., Ramadass, A., Serra Barros, A., Chow, N., and Akoulitchev, A. (2007) Repression of the human dihydrofolate reductase gene by a non-coding interfering transcript. *Nature*, **445**, 666–670.

229 Janowski, B.A., *et al.* (2005) Inhibiting gene expression at transcription start sites in chromosomal DNA with antigene RNAs. *Nat. Chem. Biol.*, **1**, 216–222.

230 Janowski, B.A. and Corey, D.R. (2005) Inhibiting transcription of chromosomal DNA using antigene RNAs. *Nucleic Acids Symp. Ser. (Oxf)*, (49), 367–368.

231 Janowski, B.A., Hu, J., and Corey, D.R. (2006) Silencing gene expression by targeting chromosomal DNA with antigene peptide nucleic acids and duplex RNAs. *Nat. Protoc.*, **1**, 436–443.

232 Janowski, B.A., *et al.* (2007) Activating gene expression in mammalian cells with promoter-targeted duplex RNAs. *Nat. Chem. Biol.*, **3**, 166–173.

233 Feng, J., *et al.* (2006) The Evf-2 noncoding RNA is transcribed from the Dlx-5/6 ultraconserved region and functions as a Dlx-2 transcriptional coactivator. *Genes Dev.*, **20**, 1470–1484.

234 Rinn, J.L., *et al.* (2007) Functional demarcation of active and silent chromatin domains in human HOX loci by noncoding RNAs. *Cell*, **129**, 1311–1323.

235 Mayer, C., Schmitz, K.M., Li, J., Grummt, I., and Santoro, R. (2006) Intergenic transcripts regulate the epigenetic state of rRNA genes. *Mol. Cell*, **22**, 351–361.

236 Bird, A. (2007) Perceptions of epigenetics. *Nature*, **447**, 396–398.

237 Sasaki, H. and Matsui, Y. (2008) Epigenetic events in mammalian germ-cell development: reprogramming and beyond. *Nat. Rev. Genet.*, **9**, 129–140.

238 Caretti, G., *et al.* (2006) The RNA helicases p68/p72 and the noncoding RNA SRA are coregulators of MyoD and skeletal muscle differentiation. *Dev. Cell*, **11**, 547–560.

239 Tsang, W.P., Wong, T.W., Cheung, A.H., Co, C.N., and Kwok, T.T. (2007) Induction of drug resistance and transformation in human cancer cells by the noncoding RNA CUDR. *RNA*, **13**, 890–898.

240 Wagner, L.A., *et al.* (2007) EGO, a novel, noncoding RNA gene, regulates eosinophil granule protein transcript expression. *Blood*, **109**, 5191–5198.

241 Yu, W., *et al.* (2008) Epigenetic silencing of tumour suppressor gene p15 by its antisense RNA. *Nature*, **451**, 202–206.

242 Huettel, B., *et al.* (2007) RNA-directed DNA methylation mediated by DRD1 and Pol IVb: a versatile pathway for transcriptional gene silencing in plants. *Biochim. Biophys. Acta*, **1769**, 358–374.

243 Penterman, J., *et al.* (2007) DNA demethylation in the *Arabidopsis* genome. *Proc. Natl. Acad. Sci. U.S.A.*, **104**, 6752–6757.

244 Chan, S.W., *et al.* (2006) RNAi, DRD1, and histone methylation actively target developmentally important non-CG DNA methylation in *Arabidopsis*. *PLoS Genet.*, **2**, e83.

245 Rodriguez-Campos, A. and Azorin, F. (2007) RNA is an integral component of chromatin that contributes to its structural organization. *PLoS One*, **2**, e1182.

246 Mattick, J.S. (2003) Challenging the dogma: the hidden layer of non-protein-coding RNAs in complex organisms. *Bioessays*, **25**, 930–939.

247 Osborne, R.J. and Thornton, C.A. (2006) RNA-dominant diseases. *Hum. Mol. Genet.*, **15**, Spec. No. 2, R162–R169.

248 Perez, D.S., *et al.* (2007) Long, abundantly-expressed non-coding transcripts are altered in cancer. *Hum. Mol. Genet.*, **17** (5), 642–655.

249 Ji, P., *et al.* (2003) MALAT-1, a novel noncoding RNA, and thymosin beta4 predict metastasis and survival in early-stage non-small cell lung cancer. *Oncogene*, **22**, 8031–8041.

250 Lin, R., Maeda, S., Liu, C., Karin, M., and Edgington, T.S. (2007) A large noncoding RNA is a marker for murine hepatocellular carcinomas and a spectrum of human carcinomas. *Oncogene*, **26**, 851–858.

251 Matouk, I.J., *et al.* (2007) The H19 non-coding RNA is essential for human tumor growth. *PLoS One*, **2**, e845.

252 Cai, X. and Cullen, B.R. (2007) The imprinted H19 noncoding RNA is a primary microRNA precursor. *RNA*, **13**, 313–316.

253 Harbers, M. and Günter, K. (2011) *Tag-Based Next Generation Sequencing*, Wiley-VCH Verlag GmbH, Weinheim.

254 Garber, M., *et al.* (2011) Computational methods for transcriptome annotation and quantification using RNA-seq. *Nat. Methods*, **8** (6), 469–477.

255 Costa, V., *et al.* (2010) Uncovering the complexity of transcriptomes with RNA-seq. *J. Biomed. Biotechnol.*, **2010**, 853916.

256 Pritchard, C.C., Cheng, H.H., and Tewari, M. (2012) MicroRNA profiling: approaches and considerations. *Nat. Rev. Genet.*, **13** (5), 358–369.

257 Ciesla, M., *et al.* (2011) MicroRNAs as biomarkers of disease onset. *Anal. Bioanal. Chem.*, **401** (7), 2051–2061.

258 Krzyzanowski, P.M., Muro, E.M., and Andrade-Navarro, M.A. (2012) Computational approaches to discovering noncoding RNA. *Wiley Interdiscip. Rev. RNA*, **3** (4), 567–579.

259 Visscher, P.M., *et al.* (2012) Five years of GWAS discovery. *Am. J. Hum. Genet.*, **90** (1), 7–24.

260 Baker, M. (2011) Long noncoding RNAs: the search for function. *Nat. Methods*, **8** (5), 379–383.

261 Guttman, M., *et al.* (2011) lincRNAs act in the circuitry controlling pluripotency and differentiation. *Nature*, **477** (7364), 295–300.

262 Turner, A.M. and Morris, K.V. (2010) Controlling transcription with noncoding RNAs in mammalian cells. *Biotechniques*, **48** (6), ix–xvi.

263 Kim, E.D. and Sung, S. (2012) Long noncoding RNA: unveiling hidden layer of gene regulatory networks. *Trends Plant Sci.*, **17** (1), 16–21.

264 Orom, U.A. and Shiekhattar, R. (2011) Long non-coding RNAs and enhancers. *Curr. Opin. Genet. Dev.*, **21** (2), 194–198.

265 Chu, C., *et al.* (2011) Genomic maps of long noncoding RNA occupancy reveal principles of RNA-chromatin interactions. *Mol. Cell.*, **44** (4), 667–678.

266 Huarte, M. and Rinn, J.L. (2010) Large non-coding RNAs: missing links in cancer? *Hum. Mol. Genet.*, **19** (R2), R152–R161.

267 Prensner, J.R. and Chinnaiyan, A.M. (2011) The emergence of lncRNAs in cancer biology. *Cancer Discov.*, **1** (5), 391–407.

268 Bian, S. and Sun, T. (2011) Functions of noncoding RNAs in neural development and neurological diseases. *Mol. Neurobiol.*, **44** (3), 359–373.

269 Pauli, A., Rinn, J.L., and Schier, A.F. (2011) Non-coding RNAs as regulators of embryogenesis. *Nat. Rev. Genet.*, **12** (2), 136–149.

270 Breaker, R.R. (2012) Riboswitches and the RNA world. *Cold Spring Harb. Perspect. Biol.*, **4** (2), pii: a003566. doi: 10.1101/cshperspect.a003566. Available at: http://cshperspectives.cshlp.org/site/misc/rna_worlds.xhtml.

9
Noncoding RNAs, Neurodevelopment, and Neurodegeneration

Mengmeng Chen, Jianwen Deng, Mengxue Yang, Kun Zhu, Jianghong Liu, Li Zhu, and Jane Y. Wu

9.1
Introduction

Although noncoding regions occupy the vast majority of the human genome, a comprehensive picture of noncoding RNA (ncRNA) genes in human biology and pathogenesis remains to be constructed [1]. ncRNAs are emerging as versatile players in almost every aspect of cellular life and associated with a wide range of human diseases [2].

In this chapter, we review recent molecular and biochemical studies on ncRNA genes involved in neurodevelopment and neurodegeneration. Because of the large number of studies, it is impossible to cover all aspects within this short chapter. We will only focus on selected examples to illustrate the extreme complexity of ncRNA biology and critical importance of ncRNAs in neurodevelopment and neurodegeneration.

9.2
Expression of ncRNAs in the Nervous System

A myriad of ncRNAs are expressed in the nervous system, although their functional roles are not well understood [3–6]. ncRNAs regulate many processes of neurodevelopment, including neuronal differentiation and cell fate determination. ncRNAs play a variety of regulatory roles in gene expression, from chromatin structural modulation, transcription, posttranscriptional RNA processing, translational control, to epigenetic regulation [7].

9.2.1
Long ncRNAs

Over 30 000 putative full-length, long noncoding RNAs (lncRNAs, >200 nucleotides) have been predicted in mice [8], many of which undergo splicing to

Posttranscriptional Gene Regulation: RNA Processing in Eukaryotes, First Edition. Edited by Jane Wu.

generate even larger numbers of gene products. The biological roles of only a small number of them have been investigated. A study utilizing data from the Allen Brain Atlas examined 1328 lncRNAs examined, and of these, 849 are expressed within the adult mouse brain with almost half (623) exhibiting selective profiles in specific regions, cell types, and subcellular compartments [9]. Three-dimensional studies in the adult mouse brain provide clear evidence that the majority of lncRNAs identified so far are expressed in the nervous system. A complementary study examining a more restricted subset of intergenic lncRNAs enriched for evolutionarily constrained sequences showed that over 200 of these lncRNAs are expressed in the developing and adult mouse brain [10]. Intriguingly, these lncRNAs are largely derived from genomic loci located proximal to protein-coding genes with expression profiles similar to the corresponding protein-coding genes in the brain. Another study identified more than 1000 evolutionarily conserved intergenic lncRNAs in mouse by analyzing chromatin signatures from four mouse cell types, including neural precursor cells (NPCs) [11].

9.2.2 Small ncRNAs

Small ncRNAs have a wide spectrum of regulatory functions, including RNA modification and regulation of gene transcription and protein synthesis. They have been isolated and characterized in many organisms and tissues, from Archaeobacteria to mammals. A number of small ncRNAs have been examined in the mammalian brains with some showing regional or cell type-specific expression.

9.2.2.1 Small Nucleolar RNAs (snoRNAs)

Most mammalian snoRNAs are ubiquitously expressed and can be classified into C/D box and H/ACA box families. The major function of snoRNAs is to guide the chemical modifications of RNAs: C/D box members for methylation and H/ACA box members specific for pseudouridylation. Both C/D box and H/ACA box RNAs are expressed in the brain. Examples including MBII-13, MBII-78, MBII-48, MBII-49, MBII52, and MBII85 are in the brain. HBII-13 and HBII-85, in contrast, are also found in other tissues [12]. *In situ* hybridizations revealed that MBI-36, MBII-48, MBII52, and MBII85 are not expressed uniformly in the brain. They are concentrated in the hippocampus and amygdala, two areas important for spatial learning and fear conditioning. Furthermore, inside the hippocampus, a difference exists between the dorsal and ventral regions: MBI-36, MBII-48, and MBII52 have a higher expression in ventral with respect to dorsal hippocampus, suggesting that these snoRNAs play a differential role [13].

9.2.2.2 Small Cytoplasmic RNA

Brain-enriched cytoplasmic RNAs were discovered over 20 years ago [14, 15]. BC1 RNA is found only in neural tissues [16] and is expressed at different levels in different brain regions with a high level detected in the olfactory bulb, hippocampus, and cortical neurons [17]. Furthermore, BC1 is detected in axons and concentrated in dendrites [18]. BC200 shares the same localization: it is expressed in the neurons and transported to the dendrites [19]. The intracellular localization of BC1 and BC200 suggests their involvement in the regulation of local protein translation in dendrites and synapses.

9.2.2.3 MicroRNAs

A class of small, noncoding transcripts called microRNAs (miRNAs) has recently emerged as important players in posttranscriptional gene regulation [20]. Many miRNAs are highly conserved among different organisms. The miRNA database at Sanger Center lists over 3800 unique mature miRNA sequences from different species. More than 2000 human miRNAs have been identified (www.mirbase.org/cgi-bin/mirna_summary.pl?org=hsa [21]).

A large fraction of mammalian miRNAs are detected in the brain at different levels [22–34]. Some miRNAs are highly expressed neuronal progenitor cells (miR-92b), and others are expressed preferentially in differentiated neurons (miR-124). Cell type-specific expression of miRNAs have been reported, with some preferentially detected in neurons (miR-124, miR-128) and astrocytes (miR-23, miR-26, miR-29), and others (miR-9, miR-125) are detected in both cell types [35, 36].

About 40% of miRNAs in the brain appear to be developmentally regulated. The expression of certain families of miRNAs increases dramatically during cortical development [24, 32]. The temporal and regional specificity of miRNA expression in the central nervous system (CNS) has been surveyed by extensive expression studies. In zebrafish, the distribution of conserved miRNAs expressed in the nervous system was analyzed by *in situ* hybridization. In mammals, an miRNA expression atlas has been generated by large-scale small RNA cloning and sequencing [37]. These studies demonstrate the presence of CNS-specific miRNAs and their developmentally regulated expression patterns. For example, the miR-29a family of miRNAs is absent in embryonic tissues but highly expressed in the adult cortex. Such expression profiling studies provide the basis for future functional studies. These studies have revealed certain unique features of CNS-expressed miRNAs regarding evolutionary conservation. Some miRNAs, such as let-7, are highly conserved from nematodes to primates [25], suggesting that these miRNAs might regulate fundamental aspects of nervous system development. Alternatively, other miRNAs are detected only in primate brains and may have more recently evolved regulatory function(s) [38].

Important neurodevelopmental events such as neuronal differentiation and synapse formation are associated with distinct miRNA profiles [23]. Other conditions in which specific miRNAs are expressed include retinoic-acid-induced neuronal differentiation of embryonic carcinoma cells and stem cells undergoing neural differentiation [32]. The diverse patterns of miRNA expression together with their multiple target genes make neuronal miRNA pathways an extremely powerful mechanism to dynamically modulate the protein contents in different brain regions at the posttranscriptional and posttranslational levels.

9.3
ncRNAs in Neurodevelopment

9.3.1
ncRNA Regulation of Neuronal Differentiation

A number of ncRNAs have been implicated in the modulation of transcriptional networks and play crucial roles in neurogenesis [3, 5, 39, 40]. A well-studied example is miR-124, one of the most abundant miRNAs in the adult brain [26].

Overexpression of miR-124 decreases the expression of non-neuronal genes [41]. In neural progenitor cells, miR-124 is detected at low levels, whereas in differentiating and mature neurons, miR-124 expression is upregulated [42]. Consistent with these observations, decreasing the expression of the repressor element-1 silencing transcription factor (REST) that silences miR-124 expression and increasing miR-124 leads to neuronal gene expression in mouse embryonic carcinoma cells [43]. It has been demonstrated that miR-124 targets NeuroD and regulates early neurogenesis in the optic vesicle and forebrain [44]. Furthermore, miR-124 promotes neuronal differentiation and cell cycle exit of neural stem cells in the subventricular zone of the adult brain [45]. These studies support that miR-124 regulates neurodifferentiation and possibly adult neurogenesis.

9.3.2
ncRNAs Regulate Synaptic Development and Function

In the nervous system, neurons interconnect with each other through synapses. Proper formation and function of synapse is critical for the development and function of the nervous system. Recent evidence indicates a crucial role of ncRNAs in synaptic development, including neurite outgrowth, synapse formation, maturation, and maintenance.

9.3.2.1 ncRNAs in Neurite Outgrowth

A chromatin occupancy study has shown that cAMP-response element binding protein (CREB) targets and regulates miR-132 transcription. Increasing miR-132 expression promotes neurite outgrowth, whereas inhibiting miR-132 decreases neurite outgrowth. Further experiments show that miR-132 mediates neuronal morphogenesis by suppressing the expression of p250GTPase-activating protein (p250GAP) [46]. Another miRNA highly expressed in neurons, miR-124, also promotes neurite outgrowth by regulating cytoskeleton proteins [47]. Recent studies showed that Mef-2-induced expression of miR379-410 cluster is required for dendritogenesis of hippocampal neurons. One of the miRNAs in this cluster, miR-134, promotes dendritic outgrowth via inhibiting expression of the translational repressor Pumilio2 [48]. In contrast, miR-34a suppresses dendritic growth and branching by regulating the expression of synaptic targets including synaptotagmin-1 and syntaxin-1A [49].

9.3.2.2 ncRNAs in Synaptic Formation and Maturation

In Dicer knockout mice, dendritic spines are increased in length. Inactivation of Dicer causes a decreased dendritic complexity [50], suggesting that miRNAs are crucial for dendritic spine development. miR-134 localizes to the synaptic compartments in rat hippocampal neurons and regulates dendritic spine sizes. Moreover, miR-134 regulates synapse formation and maturation at least in part by inhibiting the translation of Limk1 [51]. In a functional screen, miR-138 was identified as a negative regulator of dendritic spine size in rat hippocampal neurons. miR-138 decreases dendritic spine size by regulating the translation of a depalmitoylation enzyme, acyl protein thioesterase 1 (APT1). Knocking down APT1 expression and its downstream α_{13} subunits of G proteins (Gα_{13}) inhibits spine enlargement caused by miR-138 downregulation [52].

9.3.2.3 ncRNAs in Synaptic Function and Plasticity

Many ncRNAs are involved in regulating synaptic plasticity and function in mature synapses. It is well established that long-term memory is dependent on protein synthesis [53, 54]. Because ncRNAs, such as miRNAs, regulate messenger RNA (mRNA) translation, it is not surprising that they modulate synaptic plasticity, learning, and memory. For example, an experience-induced miRNA, miR-132, is essential for visual cortex plasticity. Inhibition of miR-132 in mice affected dendritic spine maturation and prevented ocular dominance plasticity [55]. The expression of brain-specific miR-128b is increased in fear-extinction learning in mice, and miR-128 regulates the formation of fear-extinction memory [56].

It has been proposed that miRNAs regulate glutamate receptor (GluR) subunit composition in *Drosophila*. Postsynaptic GluR subunit mRNA and protein are increased in Dicer-1 knocked-down animals. Knocking out miR-284 increases the expression of GluRIIA and GluRIIB but not GluRIIC. Interestingly, only small changes in synaptic strength were observed. These results suggest that miR-284 may regulate GluR composition instead of overall receptor abundance [57].

In summary, ncRNAs and especially miRNAs are abundant in the nervous system. ncRNAs regulate many steps of neurodevelopment including neuronal differentiation, synapse development, and functions. The majority of studies about ncRNAs in neurodevelopment are focused on miRNAs. miRNAs regulate the translation of target genes and play important roles in many processes during the development of the nervous system. The roles of most other ncRNAs in neurodevelopment remain to be investigated.

9.4 ncRNAs in Neurodevelopmental and Neuropsychiatric Diseases

Accumulating evidence supports that ncRNAs are not only important for normal development of the nervous system but also associated with a wide range of neurodevelopmental and neuropsychiatric disorders.

9.4.1
MicroRNAs and Neurodevelopmental Disorders

A large number of neurological and psychiatric disorders are considered to be neurodevelopmental in origin or to have neurodevelopmental consequences. These include autism spectrum disorders (ASDs), Asperger's syndrome, epilepsy, and intellectual disabilities (IDs) of different types such as fragile X syndrome (FXS) and Down syndrome (DS).

9.4.1.1 Autism Spectrum Disorders

Autism spectrum disorder (ASD) is a group of highly prevalent neurodevelopmental disorders characterized by abnormalities in language development, social interaction, and mental flexibilities. It is a family of multigenic and multifactorial complex diseases with high heritability (reviewed in [58]). In addition to a number of protein coding genes identified, ncRNAs may likely contribute to the pathogenesis of ASD.

Several miRNAs have been reported to show changes in expression in postmortem cerebellar cortex specimens from ASD patients, and some of these miRNAs are predicted to target known ASD-associated genes involved in synaptic plasticity and neural connectivity [59]. Some of the overlapping miRNAs were detected in lymphoblastoid cell lines obtained from autistic and normal adults [60]. Dysregulation of miRNAs, including HEY1, SOX9, miR-486, and miR-181b, has also been reported in ASD [61]. Further investigation is needed to define the functional role of these miRNAs in ASD.

A variety of submicroscopic chromosomal changes named as copy number variations (CNVs) have been associated with ASD, including chromosome microdeletions and/or duplications affecting chromosomes 15, 16, and 17 (reviewed in [58]). Although not reported yet, it is likely that ncRNAs in these regions may contribute to the development of ASD.

9.4.1.2 Rett Syndrome

Rett syndrome (RTT) is an X-linked dominant neurodevelopmental disorder characterized by cognitive impairment, anxiety, autistic behavior, motor disabilities, and seizures [62]. RTT is caused by mutations in the methyl-CpG-binding protein 2 (MeCP2) gene [63]. MeCP2 has been proposed as a component of transcriptional repressor complexes and a regulator of chromatin structure [62, 64–66]. MeCP2 may regulate expression of ncRNAs including miRNAs. Parallel sequencing analyses revealed a number of miRNAs whose expression is altered in the cerebella of MeCP2-null mice before and after the onset of neurological symptoms, including some brain-specific miRNAs [67]. MeCP2 suppresses the expression of an imprinted miRNA, miR-184, in an activity-dependent manner [68]. In contrast, MeCP2 also regulates the expression of BDNF (brain-derived neurotrophic factor) gene by interacting with miRNA-212 [69]. Several studies suggested that miRNAs regulate MeCP2 expression and function. miR-132, miR-155, and miR-802 can regulate MeCP2 expression [70, 71]. A feedback regulation between MeCP2 and miR-132 may be a mechanism by which MeCP2 levels are fine-tuned to maintain the range required for normal neuro-development [70].

9.4.1.3 **Fragile X Syndrome**

Fragile X syndrome (FXS) is a common form of inherited ID, with characteristic physical and behavioral features, including delays in speech and language development [72, 73]. FXS is caused by an expansion of CGG trinucleotide repeats within the regulatory region of the fragile X mental retardation (FMR1) gene, resulting in the loss of fragile X mental retardation protein (FMRP) expression ([74–76]; reviewed in [73]). FMRP belongs to a small and highly conserved RNA-binding protein family that has been implicated in translational control [77–80]. FMRP functions as a suppressor of target mRNA translation by binding to the untranslated regions (UTRs) of target mRNAs [81–84].

Moreover, FMRP interacts with the components of the miRNA pathway biochemically and genetically. In *Drosophila,* FMRP biochemically interacts with RNA-induced silencing complex (RISC) proteins, including miRNAs, dAGO1, dAGO2, and Dicer [85, 86]. In the adult mouse brain, Dicer and eIF2c2 (the mouse homolog of AGO1) interact with FMRP at the postsynaptic densities [87]. FMRP is able to act as an acceptor for Dicer-derived miRNAs. Importantly, the endogenous miRNAs are associated with FMRP in both flies and mammals [88].

FMRP may affect the miRNA pathway at two stages: miRNA processing and miRNA-mediated translational regulation. In the fly brain, dFMRP is associated with miR-124a. The proper processing of pre-miR-124a requires dFMRP, whereas a loss of dFMR1 leads to a reduced level of mature miR-124a and an increased level of pre-miR-124a [89]. Whether FMRP is associated with specific miRNAs in mammalian brains remains to be determined.

9.4.1.4 **DiGeorge Syndrome**

DiGeorge syndrome, also known as velocardiofacial syndrome, is a clinically heterogeneous disorder characterized by developmental brain malformation and cognitive and behavioral abnormalities with an increased risk of psychiatric disorders [90]. DiGeorge syndrome is caused by a microdeletion in the long arm of chromosome 22. DiGeorge critical region gene 8, Dgcr8, forms a microprocessor complex along with Drosha to process the primary miRNAs (pri-miRNAs) [91]. Mature miRNAs were reduced in the brains of mice containing either the Dgcr8 disruption or the systemic hemizygous deletion of the Dgcr8 [91]. The human, mouse, and fly Dgcr8 genes are all essential for miRNA biogenesis [91, 92]. Together, these data argue that the heterozygous loss of Dgcr8 causes abnormal miRNA biogenesis and leads to a deficit in cognitive performance; however, whether specific miRNAs are responsible for cognitive impairment associated with these mutants remains to be determined.

9.4.1.5 Down Syndrome

Down syndrome (DS), which affects 1 in 700 newborns, has a variable phenotype that includes congenital heart defects, craniofacial abnormalities, and cognitive impairment. Recently, the potential contribution of miRNAs to the pathogenesis of DS has been investigated. Bioinformatic analyses revealed that chromosome 21 encodes a number of miRNAs, including miR-99a, let-7c, miR-125b-2, miR-155, and miR-802, all of which are overexpressed in the fetal brain and heart tissues from DS individuals. This suggests that these miRNAs may contribute, at least in part, to the cognitive and cardiac defects in DS [93]. A role for miRNAs in DS is also supported by the finding that miR-155 downregulates a human gene associated with hypertension, angiotensin II type 1 receptor (AGTR1) [94]. Indeed, DS individuals do have lower blood pressure and lower AGTR1 protein levels than normal subjects. Simple associations between each miRNA and specific DS phenotype are likely to be rare, as individual miRNAs are capable of regulating a large number of protein-coding genes [95]. As improved computational and experimental methods continue to uncover new miRNAs, the miRNAs residing on chromosome 21 could be excellent candidates to further study the molecular pathogenesis of DS.

9.4.1.6 Neuropsychiatric Diseases: Schizophrenia and Bipolar Disorder

Psychiatric disorders, including schizophrenia and bipolar disorder, are a major health problem. Approximately 25% of adults in the United States suffer from a major psychiatric illness. Of these, schizophrenia and bipolar disorder are the most psychologically and economically debilitating, accounting for 1.1 and 2.5% of cases, respectively [96].

Schizophrenia and bipolar disorder are highly complex genetic disorders. miRNAs have been implicated in the pathophysiology of these neuropsychiatric disorders. Schizophrenia is characterized by behavioral deficits, accompanied by locomotive defects and cognitive impairment. Schizophrenia is associated with changes in dopamine and glutamate neurotransmission in the brain [97]. A recent genome-wide association study uncovered seven schizophrenia-associated loci, including five new loci (1p21.3, 2q32.3, 8p23.2, 8q21.3, and 10q24.32-q24.33) and two previously implicated loci (6p21.32-p22.1 and 18q21.2) [98]. Interestingly, the strongest association was found with rs1625579 located within an intron of a putative primary transcript for miRNA-137, a known regulator of neuronal development [99, 100]. Changes in Dicer and a subset of miRNAs in the prefrontal cortex, superior temporal gyrus (STG), and dorsolateral prefrontal cortex (DL-PFC) have been found in schizophrenia patients as compared with the control subjects [101–103]. Hemizygous deletions of the 22q11.2 locus in humans result in deficits in attention, learning, executive function, and emotional behavior and may account for up to 2% of all cases of schizophrenia [104]. A mouse model hemizygous for a deletion of a 1.3-Mb region syntenic to 22q11.2 has been developed that exhibits a number of schizophrenia-like phenotypes, including hyperactivity, poor prepulse inhibition, and reduced dendritic spine density. These mice show an increase in pri-miRNA levels and a decrease in mature miRNA levels in the brain [91].

In lymphoblastoid cell lines from bipolar disorder patients, alterations have been detected in the expression of miRNAs in patient samples as compared with that of unaffected siblings [105], including several miRNAs that target genes already associated with genetic risk factors of bipolar disorder. Chronic mood stabilizer treatments of rats by lithium and valproate (VPA) induce changes in the levels of miRNAs in the brain, including let-7b, let-7c, miR-128a, miR-24a, miR-30c, miR-34a, miR-221, and miR-144 [106].

9.4.2 Long ncRNAs in Neurodevelopmental and Psychiatric Disorders

Long ncRNAs (lncRNAs) have been implicated in many processes of gene regulation. lncRNAs have a profound effect on genomic imprinting, exemplified in diseases such as Prader–Willi syndrome (PWS) and Angelman syndrome (AS) [107]. Some of these lncRNAs regulated the imprinted cluster by affecting host genes for small nucleolar RNAs (snoRNAs). Other lncRNAs may regulate gene expression within the imprinted cluster directly.

AS is caused by a deletion or mutation in the maternal allele of the ubiquitin protein ligase E3A (Ube3a). In neurons, the paternal allele of Ube3a is intact but epigenetically silenced. In addition to Ube3a transcript, an antisense lncRNA, Ube3a-as, is expressed and has been proposed to repress paternal Ube3a expression [108]. Intriguingly, a recent study suggests that lncRNAs derived from the PWS-AS domain are retained in the nucleus and may mediate the spatial organization of gene expression through dynamic modulation of nuclear architecture [109].

In FXS or FXTAS, FMR4 is a primate-specific lncRNA that shares a two-way promoter with the FMR1 gene. FMR4 is silenced in FXS patients because of a CGG expansion repeat in the 5′ UTR of the FMR1 gene [110]. Knockdown of FMR4 by specific small interfering RNA (siRNA) resulted in alterations in cell cycle regulation and increased apoptotic cell death, whereas overexpression of FMR4 caused an increase in cell proliferation. Similarly, ASFMR1, another lncRNA derived from the FMR1 locus, may also be important in mediating the complex clinical phenotypes associated with mutations at this genomic locus [111]. Similar to FMR1 and FMR4, ASFMR1 is also silenced in FXS patients and upregulated in premutation carriers, suggesting that a common mechanism may regulate the expression of these transcripts [112, 113].

lncRNAs have also been implicated in the development of brain malformations. Genetic defects of Sox2 cause various syndromes of optic nerve hypoplasia associated with a number of CNS developmental abnormalities. The Sox2OT gene, encompassing the entire Sox2 gene, is dynamically regulated in the CNS during development and is implicated in modulating Sox2 expression [114].

Similarly, the identification of DiGeorge syndrome-associated ncRNA (DGCR5), a REST-regulated lncRNA, suggests a potential role for this lncRNA in mediating neural developmental processes and the phenotype of DiGeorge syndrome [115].

lncRNAs may play a role in the pathobiology of Down's syndrome (DS). NRON is an lncRNA that mediates the cytoplasmic to nuclear shuttling of the NFAT transcription factor [116]. In animal models, deregulation of the DSCR1 and DYRK1A genes act synergistically to prevent the nuclear occupancy of NFATc transcription factor, leading to reduced NFAT activity. This may help explain many features of DS, providing a potential link between NRON activity and DS pathogenesis [117].

A number of psychiatric disorders have also been associated with lncRNAs. Disruption of the DISC genomic locus, which encodes both the DISC1 protein-coding gene and the DISC2 lncRNA, has been linked in a number of genetic analyses to the risk of developing schizophrenia, bipolar disorder, and major depression [118–120]. DISC2 overlaps DISC1 and is transcribed in the opposite direction. As an antisense transcript, DISC2 has been proposed to regulate DISC1 expression, which regulates multiple processes in the development and function of CNS [121].

In summary, ncRNAs contribute to the molecular pathogenesis of a wide range of neurodevelopmental and psychiatric disorders. In addition to dysregulation of ncRNA transcription and biogenesis, CNVs in ncRNA genes, for example, those associated with chromosomal microduplication or deletion, could contribute to disease phenotypes. Furthermore, single-nucleotide polymorphisms (SNPs) may create or disrupt putative ncRNA target sites. Therefore, variations in the target mRNA sequences could also modulate the activity of specific ncRNAs and contribute to phenotypical variations [122, 123]. It is likely that many of these variations will affect neuronal ncRNAs and their target gene expression.

9.5 ncRNAs in Neurodegeneration

Neurodegeneration is the progressive loss of neuronal structure or function and can be caused by multiple genetic defects and/or environmental insults. Recently, changes in miRNA expression levels have been reported in patients of Alzheimer's disease (AD), Huntington's disease (HD), Parkinson's disease (PD), and frontotemporal lobar degeneration (FTLD) (Table 9.1).

9.5.1 Alzheimer's Disease

A number of SNP located in the 3′ UTR of amyloid precursor protein (APP) have been found in familial AD cases and are predicted to affect miRNA binding. In a reporter assay, AD-associated variations affected binding of miR-20a or miR-147 and influenced expression of APP [124]. miR-20a, miR-17-5p, and miR-106b may regulate the levels of APP *in vitro*. Consistent with this notion, miR-106b expression was decreased in AD brains [125]. In Dicer-deficient mice, a moderate change in APP exon 7, 8 inclusion was reported. It was proposed that such an effect on APP splicing was mediated by miR-124, which targets polypyrimidine-binding proteins (PTB1 and PTB2). Furthermore, miR-124 is downregulated in AD brains

Table 9.1 Examples of microRNAs associated with neurodegenerative diseases.

microRNAs	Candidate target genes	Diseases	Function	References
miR-20a, -17, -5b, -106b, -147	APP	AD	Regulating APP expression	[124, 145]
miR-29a/b-1, -107, -298/328	BACE1	AD	Regulating APP cleavage by targeting BACE1	[125, 127, 128]
miR-7	α-Synuclein	PD	Regulating α-synuclein expression	[131]
miR-133b	Pitx3	PD	Negative feedback loop in the maturation and function of DA neurons	[130]
miR-433	FGF20	PD	Regulating FGF20	[132]
miR-659	GRN	FTLD	Regulating GRN expression	[139]
miR-29a, -124a, -132, -135b	REST	HD	Regulating REST	[136]
miR-9, -9*	REST	HD	Regulating REST and Co-REST	[137]
miR-19, -101, -130	ATXN1	SCA1	Regulating ataxin 1 level	[143]

[126]. In addition, the miR-29 cluster (miR-29a, miR-29b-1), miR-107, miR-298, and miR-328 have been reported to regulate the expression of beta-secretase (BACE1), an important regulator of APP cleavage. Consistently, miR-29a, miR-29b-1, and miR-107 are decreased in AD patients, with abnormal expression of BACE1 and amyloid β accumulation in sporadic AD cases [125, 127, 128].

9.5.2
Parkinson's Disease

Parkinson's disease (PD) is the second most common neurodegenerative disease characterized by a loss of dopaminergic neurons in the midbrain, resulting in rigidity, tremor, bradykinesia, and postural instability [129]. Downregulation of miR-133b was found in the midbrain of PD patients. miR-133b is expressed highly in the midbrain and modulates the maturation and function of midbrain dopaminergic neurons. The transcription factor Pitx3 may activate miR-133b, which in turn inhibits Pitx3, forming a negative feedback loop between miR-133b and its target in PD [130]. Accumulation of α-synuclein is one of the pathological characteristics in sporadic PD. miR-7, expressed mainly in the neurons, binds to the 3′ UTR of α-synuclein mRNA to repress its expression and may protect the cells against oxidative stress. Decreased expression of miR-7 led to increased α-synuclein expression in a mouse PD model and in cultured cells [131]. In addition, fibroblast growth factor 20 (FGF20) was identified as a risk factor for PD, with increased expression of FGF20 correlated with increased α-synuclein expression. Functional assays implicated that disrupting the binding site (rs12720280) for miR-433 in the 3′ UTR of FGF20 may increase the translation of FGF20 *in vitro* and *in vivo*, possibly associated with pathogenesis of PD [132].

A large number of mutations have been identified in PD patients in the LRRK2 gene, encoding leucine-rich repeat kinase 2, a multidomain protein with guanosine-5′-triphosphate (GTP)-regulated serine/threonine kinase activity [129]. Recent findings of translational regulation by pathogenic human LRRK2 or dLRRK mutants reveal an additional level of complexity in miRNA regulation. Mutations in LRRK2 lead to elevated levels of the cell-cycle control proteins E2F-1 and DP, because of reduced translational repression by let-7 and miR-184, respectively, suggesting that pathogenic LRRK2 mutations regulate miRNA pathway(s) [133].

9.5.3 Huntington's Disease

Huntington's disease (HD) is characterized by uncontrollable motor movements, cognitive impairment, and emotional deficits. HD is caused by CAG repeat expansion (>35 CAGs) in the huntingtin (Htt) gene, leading to a long polyglutamine (polyQ) expansion in Htt protein. The polyQ expansion makes Htt prone to aggregate formation and causes neurotoxicity, possibly because of conformational changes [134, 135]. Htt protein interacts with and sequesters REST in the cytoplasm. Mutant Htt is unable to bind REST, allowing REST to translocate into the nucleus and dysregulate downstream target genes. Abnormal expression of several specific miRNAs has been detected in HD mouse models and in HD patient samples that are likely the targets of REST. In mouse models, miR-29a, miR-124a, miR-132, and miR-135b were downregulated. Consistently, miR-132 was also decreased expression in HD patients. However, miR-29a and miR-330 were upregulated in HD patient brain samples [136]. Further studies show that brain-enriched and bifunctional miR-9/miR-9* were downregulated in HD patients. miR-9 and miR-9* targeted two components of REST complex with miR-9 targeting REST and miR-9* targeting CoREST, suggesting a double negative feedback loop between the REST silencing complex and its regulated miRNAs [137].

9.5.4 Frontotemporal Lobar Degeneration

Frontotemporal lobar degeneration (FTLD) is characterized by atrophy in the frontal and temporal lobes, representing ~5% of all dementia patients. Its clinical features include impairment of cognitive and memory, language dysfunction, and changes in behavior. FTLD can be classified as ubiquitin-negative tauopathy (FTLD-tau) or ubiquitin-positive proteinopathies (FTLD-u), characterized respectively by the presence of tau-positive, ubiquitin-negative neurofibrillary tangles, or ubiquitin-positive protein inclusions in the affected brain areas [138]. The majority of FTLD-u patients present protein inclusions containing either TAR DNA-binding protein 43 (TDP-43) or fused in sarcoma (FUS) protein, and thus are classified as FTLD-TDP or FTLD-FUS. FTLD-TDP accounts for approximately 10% of early onset dementia patients. FTLD-u can be caused by loss of function mutations in the progranulin (GRN) gene. Interestingly, a common genetic variant, rs5848, is located in the 3′ UTR of the GRN gene at the binding site for miR-659, which represses GRN translation. Patients carrying variant rs5848 show reduced GRN protein levels. Enhanced binding of miR-659 to the 3′ UTR of the GRN gene has been proposed to increase the risk for FTLD-u [139]. Furthermore, both TDP-43 and FUS affect ncRNA expression and function [140–142]. FUS interacts with a number of promoter-associated ncRNAs (pncRNAs) at the cyclin D1 locus [142].

In the spinocerebellar ataxia type 1 (SCA1), caused by an expansion of a translated CAG repeat in ataxin 1, inhibition of miR-19, miR-101, and miR-130 coregulates ataxin 1 levels and enhances the cytotoxicity of polyglutamine-expanded ataxin 1 in human cells [143]. In other neurodegenerative diseases such as prion disease, changes in the levels of several miRNAs have been reported, although their target genes remain unclear [144].

In summary, emerging evidence supports an important role of ncRNAs in neurodegeneration. miRNAs regulate the expression and function of genes critical for neurodegenerative disorders. Changes in ncRNAs in various neurodegenerative disorders may also serve as new biomarkers for these diseases. Recent studies provide hope for manipulating critical ncRNAs in the future development of therapeutics for such complex and devastating diseases.

9.6
Concluding Remarks and Perspectives

Our research on noncoding genes in the genome is just beginning. In particular, our understanding remains very limited about how multiple layers of networks regulate not only expression but also functional activities of different genes, modulating the ultimate phenotypes or disease manifestations. A number of challenging tasks await us. First, a large fraction of functional ncRNA genes remain to be identified and characterized. This objective will be aided by the tremendous advancements in genome-transcriptome sequencing/analysis and emerging computational methods for ncRNA identification. Second, the biological function and pathogenic roles of individual ncRNAs need to be elucidated. This is going to be especially difficult and informative, because many ncRNAs may have multiple target genes and the phenotypes will be dependent on the combined outcome of the multiple interactions. Third, whether certain ncRNA expression patterns specific to particular diseases can be developed into efficient diagnostic tools remain unclear. Many human disorders, including neurodevelopmental and neurodegenerative diseases, are highly heterogeneous with multifactorial etiology. Individual variations in ncRNA genes may not have simple causative relationship with specific diseases, making it difficult to use such genetic variations as efficient diagnostic biomarkers. Finally, it remains to be tested whether therapeutic strategies targeting specific ncRNAs can be developed for relevant human diseases caused by mutations or disruption of regulations of ncRNAs. Promising advancements in experimental methods have established siRNA and small hairpin RNA (shRNA) tools for studying diseases in model organisms. Certainly, there is great potential in translating such fundamental knowledge into clinically applicable diagnostic or therapeutic tools.

Acknowledgments

We apologize to our colleagues for using reviews in many places rather than citing original research articles due to the space limit. We would like to acknowledge the support by the Ministry of Science and Technology (MOST) China 973 Project (2010CB529603, 2009CB825402) and the National Natural Science Foundation of China (91132710) and Chinese Academy of Science (CASNN-GWPPS-2008). J.Y.W. is supported by NIH (RO1AG033004, R56NS074763), ALS Therapy Alliance, and the James S. McDonnell Foundation. We would like to thank David Kuo, Haoming Yu, and members of Wu lab for critical reading of the manuscript.

References

1 Alexander, R.P., Fang, G., Rozowsky, J., Snyder, M., and Gerstein, M.B. (2010) Annotating non-coding regions of the genome. *Nat. Rev. Genet.*, **11**, 559–571.

2 Esteller, M. (2011) Non-coding RNAs in human disease. *Nat. Rev. Genet.*, **12** (12), 861–874.

3 Cao, X.W., Yeo, G., Muotri, A.R., Kuwabara, T., and Gage, F.H. (2006) Noncoding RNAs in the mammalian central nervous system. *Annu. Rev. Neurosci.*, **29**, 77–103.

4 Kosik, K.S. (2006) The neuronal microRNA system. *Nat. Rev. Neurosci.*, **7**, 911–920.

5 Mehler, M.F. and Mattick, J.S. (2007) Noncoding RNAs and RNA editing in brain development, functional diversification, and neurological disease. *Physiol. Rev.*, **87**, 799–823.

6 Saxena, A. and Carninci, P. (2011) Long non-coding RNA modifies chromatin: epigenetic silencing by long non-coding RNAs. *Bioessays*, **33** (11), 830–839.

7 Czech, B. and Hannon, G.J. (2011) Small RNA sorting: matchmaking for Argonautes. *Nat. Rev. Genet.*, **12** (1), 19–31.

8 Carninci, P., Kasukawa, T., Katayama, S., Gough, J., Frith, M.C., Maeda, N., Oyama, R., Ravasi, T., Lenhard, B., Wells, C., *et al.* (2005) The transcriptional landscape of the mammalian genome. *Science*, **309**, 1559–1563.

9 Mercer, T.R., Dinger, M.E., Mariani, J., Kosik, K.S., Mehler, M.F., and Mattick, J.S. (2008) Noncoding RNAs in long-term memory formation. *Neuroscientist*, **14** (5), 434–445.

10 Ponjavic, J., Oliver, P.L., Lunter, G., and Ponting, C.P. (2009) Genomic and transcriptional co-localization of protein-coding and long non-coding RNA pairs in the developing brain. *PLoS Genet.*, **5**, e1000617.

11 Guttman, M., Amit, I., Garber, M., French, C., Lin, M.F., Feldser, D., Huarte, M., Zuk, O., Carey, B.W., Cassady, J.P., *et al.* (2009) Chromatin

signature reveals over a thousand highly conserved large non-coding RNAs in mammals. *Nature*, **458**, 223–227.

12 Huettenhofer, A.G., Cavaille, J., Buiting, K., Kiefman, M., Lalande, M., Brannan, C.I., Horsthemke, B., Bachellerie, J.P., and Brosius, J. (2000) Identification of imprinted, tissue-specific C/D box small nucleolar RNA genes in the Prader-Willi syndrome region. *Am. J. Hum. Genet.*, **67**, 28–28.

13 Rogelj, B., Hartmann, C.E.A., Yeo, C.H., Hunt, S.P., and Giese, K.P. (2003) Contextual fear conditioning regulates the expression of brain-specific small nucleolar RNAs in hippocampus. *Eur. J. Neurosci.*, **18**, 3089–3096.

14 Dechiara, T.M. and Brosius, J. (1987) Neural Bc1 RNA–cDNA clones reveal nonrepetitive sequence content. *Proc. Natl. Acad. Sci. U.S.A.*, **84**, 2624–2628.

15 Martignetti, J.A. and Brosius, J. (1993) BC200 RNA–a neural RNA polymerase-III product encoded by a monomeric Alu element. *Proc. Natl. Acad. Sci. U.S.A.*, **90**, 11563–11567.

16 Sutcliffe, J.G., Milner, R.J., Gottesfeld, J.M., and Reynolds, W. (1984) Control of neuronal gene-expression. *Science*, **225**, 1308–1315.

17 Lin, Y., Brosius, J., and Tiedge, H. (2001) Neuronal BC1 RNA: co-expression with growth-associated protein-43 messenger RNA. *Neuroscience*, **103**, 465–479.

18 Tiedge, H., Fremeau, R.T., Weinstock, P.H., Arancio, O., and Brosius, J. (1991) Dendritic location of neural BC1 RNA. *Proc. Natl. Acad. Sci. U.S.A.*, **88**, 2093–2097.

19 Tiedge, H., Chen, W., and Brosius, J. (1993) Primary structure, neural-specific expression, and dendritic location of human BC200 RNA. *J. Neurosci.*, **13**, 2382–2390.

20 Bartel, D.P. (2009) MicroRNAs: target recognition and regulatory functions. *Cell*, **136** (2), 215–233.

21 Kozomara, A. and Griffiths-Jones, S. (2011) miRBase: integrating microRNA annotation and deep-sequencing data. *NAR*, **39** (Database Issue), D152–D157.

22 Babak, T., Zhang, W., Morris, Q., Blencowe, B.J., and Hughes, T.R. (2004) Probing microRNAs with microarrays: tissue specificity and functional inference. *RNA*, **10**, 1813–1819.

23 Grad, Y., Aach, J., Hayes, G.D., Reinhart, B.J., Church, G.M., Ruvkun, G., and Kim, J. (2003) Computational and experimental identification of *C. elegans* microRNAs. *Mol. Cell*, **11**, 1253–1263.

24 Kim, J., Krichevsky, A., Grad, Y., Hayes, G.D., Kosik, K.S., Church, G.M., and Ruvkun, G. (2004) Identification of many microRNAs that copurify with polyribosomes in mammalian neurons. *Proc. Natl. Acad. Sci. U.S.A.*, **101**, 360–365.

25 Krichevsky, A.M., King, K.S., Donahue, C.P., Khrapko, K., and Kosik, K.S. (2003) A microRNA array reveals extensive regulation of microRNAs during brain development. *RNA*, **9**, 1274–1281.

26 Lagos-Quintana, M., Rauhut, R., Lendeckel, W., and Tuschl, T. (2001) Identification of novel genes coding for small expressed RNAs. *Science*, **294**, 853–858.

27 Lagos-Quintana, M., Rauhut, R., Yalcin, A., Meyer, J., Lendeckel, W., and Tuschl, T. (2002) Identification of tissue-specific microRNAs from mouse. *Curr. Biol.*, **12**, 735–739.

28 Lim, L.P., Lau, N.C., Weinstein, E.G., Abdelhakim, A., Yekta, S., Rhoades, M.W., Burge, C.B., and Bartel, D.P. (2003) The microRNAs of *Caenorhabditis elegans*. *Genes Dev.*, **17**, 991–1008.

29 Liu, C.G., Calin, G.A., Meloon, B., Gamliel, N., Sevignani, C., Ferracin, M., Dumitru, C.D., Shimizu, M., Zupo, S., Dono, M., *et al.* (2004) An oligonucleotide microchip for genome-wide microRNA profiling in human and mouse tissues. *Proc. Natl. Acad. Sci. U.S.A.*, **101**, 9740–9744.

30 Miska, E.A., Alvarez-Saavedra, E., Townsend, M., Yoshii, A., Sestan, N., Rakic, P., Constantine-Paton, M., and Horvitz, H.R. (2004) Microarray analysis of microRNA expression in the developing mammalian brain. *Genome Biol.*, **5**, R68.

31 Nelson, P.T., Hatzigeorgiou, A.G., and Mourelatos, Z. (2004) miRNP : mRNA association in polyribosomes in a human neuronal cell line. *RNA*, **10**, 387–394.

32 Rogelj, B. and Giese, K.P. (2004) Expression and function of brain specific small RNAs. *Rev. Neurosci.*, **15**, 185–198.

33 Sempere, L.F., Freemantle, S., Pitha-Rowe, I., Moss, E., Dmitrovsky, E., and Ambros, V. (2004) Expression profiling of mammalian microRNAs uncovers a subset of brain-expressed microRNAs with possible roles in murine and human neuronal differentiation. *Genome Biol.*, **5**, R13.

34 Bak, M., Silahtaroglu, A., Moller, M., Christensen, M., Rath, M.F., Skryabin, B., Tommerup, N., and Kauppinen, S. (2008) MicroRNA expression in the adult mouse central nervous system. *RNA*, **14**, 432–444.

35 Cheng, L.C., Tavazoie, M., and Doetsch, F. (2005) Stem cells: from epigenetics to microRNAs. *Neuron*, **46**, 363–367.

36 Smirnova, L., Grafe, A., Seiler, A., Schumacher, S., Nitsch, R., and Wulczyn, F.G. (2005) Regulation of miRNA expression during neural cell specification. *Eur. J. Neurosci.*, **21**, 1469–1477.

37 Landgraf, P., Rusu, M., Sheridan, R., Sewer, A., Iovino, N., Aravin, A., Pfeffer, S., Rice, A., Kamphorst, A.O., Landthaler, M., *et al.* (2007) A mammalian microRNA expression atlas based on small RNA library sequencing. *Cell*, **129**, 1401–1414.

38 Berezikov, E., Thuemmler, F., van Laake, L.W., Kondova, I., Bontrop, R., Cuppen, E., and Plasterk, R.H.A. (2006) Diversity of microRNAs in human and chimpanzee brain. *Nat. Genet.*, **38**, 1375–1377.

39 Kosik, K.S. and Krichevsky, A.M. (2005) The elegance of the microRNAs: a neuronal perspective. *Neuron*, **47**, 779–782.

40 Shi, Y., Zhao, X., Hsieh, J., Wichterle, H., Impey, S., Banerjee, S., Neveu, P., and Kosik, K.S. (2010) MicroRNA regulation of neural stem cells and neurogenesis. *J. Neurosci.*, **30**, 14931–14936.

41 Lim, L.P., Lau, N.C., Garrett-Engele, P., Grimson, A., Schelter, J.M., Castle, J., Bartel, D.P., Linsley, P.S., and Johnson, J.M. (2005) Microarray analysis shows that some microRNAs downregulate large numbers of target mRNAs. *Nature*, **433**, 769–773.

42 Deo, M., Yu, J.Y., Chung, K.H., Tippens, M., and Turner, D.L. (2006) Detection of mammalian microRNA expression by *in situ* hybridization with RNA oligonucleotides. *Dev. Dyn.*, **235**, 2538–2548.

43 Conaco, C., Otto, S., Han, J.J., and Mandel, G. (2006) Reciprocal actions of REST and a microRNA promote neuronal identity. *Proc. Natl. Acad. Sci. U.S.A.*, **103**, 2422–2427.

44 Liu, K., Liu, Y., Mo, W., Qiu, R., Wang, X., Wu, J.Y., and He, R. (2011) miR-124 regulates early neurogenesis in the optic vesicle and forebrain, targeting NeuroD1. *Nucleic Acids Res.*, **39** (7), 2869–2879.

45 Cheng, L.C., Pastrana, E., Tavazoie, M., and Doetsch, F. (2009) miR-124 regulates adult neurogenesis in the subventricular zone stem cell niche. *Nat. Neurosci.*, **12**, 399–408.

46 Vo, N., Klein, M.E., Varlamova, O., Keller, D.M., Yamamoto, T., Goodman, R.H., and Impey, S. (2005) A cAMP-response element binding protein-induced microRNA regulates neuronal morphogenesis. *Proc. Natl. Acad. Sci. U.S.A.*, **102**, 16426–16431.

47 Yu, J.Y., Chung, K.H., Deo, M., Thompson, R.C., and Turner, D.L. (2008) MicroRNA miR-124 regulates neurite outgrowth during neuronal differentiation. *Exp. Cell Res.*, **314**, 2618–2633.

48 Fiore, R., Khudayberdiev, S., Christensen, M., Siegel, G., Flavell, S.W., Kim, T.K., Greenberg, M.E., and Schratt, G. (2009) Mef2-mediated transcription of the miR379-410 cluster regulates activity-dependent dendritogenesis by fine-tuning Pumilio2 protein levels. *EMBO J.*, **28** (6), 697–710.

49 Agostini, M., Tucci, P., Steinert, J.R., Shalom-Feuerstein, R., Rouleau, M., Aberdam, D., Forsythe, I.D., Young,

K.W., Ventura, A., Concepcion, C.P., Han, Y.C., Candi, E., Knight, R.A., Mak, T.W., and Melino, G. (2011) microRNA-34a regulates neurite outgrowth, spinal morphology, and function. *Proc. Natl. Acad. Sci. U.S.A.*, **108** (52), 21099–21104.

50 Davis, T.H., Cuellar, T.L., Koch, S.M., Barker, A.J., Harfe, B.D., McManus, M.T., and Ullian, E.M. (2008) Conditional loss of Dicer disrupts cellular and tissue morphogenesis in the cortex and hippocampus. *J. Neurosci.*, **28**, 4322–4330.

51 Schratt, G.M., Tuebing, F., Nigh, E.A., Kane, C.G., Sabatini, M.E., Kiebler, M., and Greenberg, M.E. (2006) A brain-specific microRNA regulates dendritic spine development. *Nature*, **439**, 283–289.

52 Siegel, G., Obernosterer, G., Fiore, R., Oehmen, M., Bicker, S., Christensen, M., Khudayberdiev, S., Leuschner, P.F., Busch, C.J., Kane, C., *et al.* (2009) A functional screen implicates microRNA-138-dependent regulation of the depalmitoylation enzyme APT1 in dendritic spine morphogenesis. *Nat. Cell Biol.*, **11**, 705–716.

53 Bailey, C.H., Kandel, E.R., and Si, K. (2004) The persistence of long-term memory: a molecular approach to self-sustaining changes in learning-induced synaptic growth. *Neuron*, **44** (1), 49–57.

54 Kelleher, R.J., 3rd, Govindarajan, A., and Tonegawa, S. (2004) Translational regulatory mechanisms in persistent forms of synaptic plasticity. *Neuron*, **44**, 59–73.

55 Mellios, N., Sugihara, H., Castro, J., Banerjee, A., Le, C., Kumar, A., Crawford, B., Strathmann, J., Tropea, D., Levine, S.S., Edbauer, D., and Sur, M. (2011) miR-132, an experience-dependent microRNA, is essential for visual cortex plasticity. *Nat. Neurosci.*, **14** (10), 1240–1242.

56 Lin, Q., Wei, W., Coelho, C.M., Li, X., Baker-Andresen, D., Dudley, K., Ratnu, V.S., Boskovic, Z., Kobor, M.S., Sun, Y.E., and Bredy, T.W. (2011) The brain-specific microRNA miR-128b regulates the formation of fear-extinction memory. *Nat. Neurosci.*, **14** (9), 1115–1117.

57 Karr, J., Vagin, V., Chen, K., Ganesan, S., Olenkina, O., Gvozdev, V., and Featherstone, D.E. (2009) Regulation of glutamate receptor subunit availability by microRNAs. *J. Cell Biol.*, **185**, 685–697.

58 Geschwind, D.H. (2011) Genetics of autism spectrum disorders. *Trends Cogn. Sci.*, **15** (9), 409–416.

59 Abu-Elneel, K., Liu, T., Gazzaniga, F.S., Nishimura, Y., Wall, D.P., Geschwind, D.H., Lao, K.Q., and Kosik, K.S. (2008) Heterogeneous dysregulation of microRNAs across the autism spectrum. *Neurogenetics*, **9**, 153–161.

60 Talebizadeh, Z., Butler, M.G., and Theodoro, M.F. (2008) Feasibility and relevance of examining lymphoblastoid cell lines to study role of microRNAs in autism. *Autism Res.*, **1**, 240–250.

61 Seno, M.M., Hu, P., Gwadry, F.G., Pinto, D., Marshall, C.R., Casallo, G., and Scherer, S.W. (2011) Gene and miRNA expression profiles in autism spectrum disorders. *Brain Res.*, **1380**, 85–97.

62 Chahrour, M. and Zoghbi, H.Y. (2007) The story of Rett syndrome: from clinic to neurobiology. *Neuron*, **56**, 422–437.

63 Amir, R.E., Van den Veyver, I.B., Wan, M., Tran, C.Q., Francke, U., and Zoghbi, H.Y. (1999) Rett syndrome is caused by mutations in X-linked MECP2, encoding methyl-CpG-binding protein 2. *Nat. Genet.*, **23**, 185–188.

64 Ghosh, R.P., Horowitz-Scherer, R.A., Nikitina, T., Shlyakhtenko, L.S., and Woodcock, C.L. (2010) MeCP2 binds cooperatively to its substrate and competes with histone H1 for chromatin binding sites. *Mol. Cell Biol.*, **30**, 4656–4670.

65 Skene, P.J., Illingworth, R.S., Webb, S., Kerr, A.R., James, K.D., Turner, D.J., Andrews, R., and Bird, A.P. (2010) Neuronal MeCP2 is expressed at near histone-octamer levels and globally alters the chromatin state. *Mol. Cell*, **37**, 457–468.

66 Cohen, S., Gabel, H.W., Hemberg, M., Hutchinson, A.N., Sadacca, L.A., Ebert,

D.H., Harmin, D.A., Greenberg, R.S., Verdine, V.K., Zhou, Z., Wetsel, W.C., West, A.E., and Greenberg, M.E. (2011) Genome-wide activity-dependent MeCP2 phosphorylation regulates nervous system development and function. *Neuron*, **72** (1), 72–85.

67 Wu, H., Tao, J., Chen, P.J., Shahab, A., Ge, W., Hart, R.P., Ruan, X., Ruan, Y., and Sun, Y.E. (2010) Genome-wide analysis reveals methyl-CpG-binding protein 2-dependent regulation of microRNAs in a mouse model of Rett syndrome. *Proc. Natl. Acad. Sci. U.S.A.*, **107** (42), 18161–18166.

68 Nomura, T., Kimura, M., Horii, T., Morita, S., Soejima, H., Kudo, S., and Hatada, I. (2008) MeCP2-dependent repression of an imprinted miR-184 released by depolarization. *Hum. Mol. Genet.*, **17**, 1192–1199.

69 Im, H.I., Hollander, J.A., Bali, P., and Kenny, P.J. (2010) MeCP2 controls BDNF expression and cocaine intake through homeostatic interactions with microRNA-212. *Nat. Neurosci.*, **13** (9), 1120–1127.

70 Klein, M.E., Lioy, D.T., Ma, L., Impey, S., Mandel, G., and Goodman, R.H. (2007) Homeostatic regulation of MeCP2 expression by a CREB-induced microRNA. *Nat. Neurosci.*, **10**, 1513–1514.

71 Kuhn, D.E., Nuovo, G.J., Terry, A.V., Jr., Martin, M.M., Malana, G.E., Sansom, S.E., Pleister, A.P., Beck, W.D., Head, E., Feldman, D.S., and Elton, T.S. (2010) Chromosome 21-derived microRNAs provide an etiological basis for aberrant protein expression in human Down syndrome brains. *J. Biol. Chem.*, **285** (2), 1529–1543.

72 Janicki, M.P., Henderson, C.M., and Rubin, I.L.; Neurodevelopmental Conditions Study Group. (2008) Neurodevelopmental conditions and aging: report on the Atlanta Study Group Charrette on Neurodevelopmental Conditions and Aging. *Disabil. Health J.*, **1** (2), 116–124.

73 Santoro, M.R., Bray, S.M., and Warren, S.T. (2011) Molecular mechanisms of fragile X syndrome: a twenty-year perspective. *Annu. Rev. Pathol.*, **7**, 219–245.

74 Kremer, E.J., Pritchard, M., Lynch, M., Yu, S., Holman, K., Baker, E., Warren, S.T., Schlessinger, D., Sutherland, G.R., and Richards, R.I. (1991) Mapping of DNA instability at the fragile X to a trinucleotide repeat sequence p(CCG)n. *Science*, **252**, 1711–1714.

75 Oberle, I., Rousseau, F., Heitz, D., Kretz, C., Devys, D., Hanauer, A., Boue, J., Bertheas, M.F., and Mandel, J.L. (1991) Instability of a 550-base pair DNA segment and abnormal methylation in fragile X syndrome. *Science*, **252**, 1097–1102.

76 Verkerk, A.J., Pieretti, M., Sutcliffe, J.S., Fu, Y.H., Kuhl, D.P., Pizzuti, A., Reiner, O., Richards, S., Victoria, M.F., Zhang, F.P., *et al.* (1991) Identification of a gene (FMR-1) containing a CGG repeat coincident with a breakpoint cluster region exhibiting length variation in fragile X syndrome. *Cell*, **65**, 905–914.

77 Ashley, C.T., Jr., Wilkinson, K.D., Reines, D., and Warren, S.T. (1993) FMR1 protein: conserved RNP family domains and selective RNA binding. *Science*, **262** (5133), 563–566.

78 Laggerbauer, B., Ostareck, D., Keidel, E.M., Ostareck-Lederer, A., and Fischer, U. (2001) Evidence that fragile X mental retardation protein is a negative regulator of translation. *Hum. Mol. Genet.*, **10** (4), 329–338.

79 Li, Z., Zhang, Y., Ku, L., Wilkinson, K.D., Warren, S.T., and Feng, Y. (2001) The fragile X mental retardation protein inhibits translation via interacting with mRNA. *Nucleic Acids Res.*, **29** (11), 2276–2283.

80 Feng, Y., Absher, D., Eberhart, D.E., Brown, V., Malter, H.E., and Warren, S.T. (1997) FMRP associates with polyribosomes as an mRNP, and the I304N mutation of severe fragile X syndrome abolishes this association. *Mol. Cell*, **1** (1), 109–118.

81 Darnell, J.C., Fraser, C.E., Mostovetsky, O., Stefani, G., Jones, T.A., Eddy, S.R., and Darnell, R.B. (2005) Kissing complex RNAs mediate interaction between the Fragile-X mental retardation protein KH2 domain and brain polyribosomes. *Genes Dev.*, **19**, 903–918.

82 Darnell, J.C., Jensen, K.B., Jin, P., Brown, V., Warren, S.T., and Darnell, R.B. (2001) Fragile X mental retardation protein targets G quartet mRNAs important for neuronal function. *Cell*, **107**, 489–499.

83 Schaeffer, C., Bardoni, B., Mandel, J.L., Ehresmann, B., Ehresmann, C., and Moine, H. (2001) The fragile X mental retardation protein binds specifically to its mRNA via a purine quartet motif. *EMBO J.*, **20**, 4803–4813.

84 Stefani, G. (2004) Fragile X mental retardation protein is associated with translating polyribosomes in neuronal cells. *J. Neurosci.*, **24**, 7272–7276.

85 Jin, P., Zarnescu, D.C., Ceman, S., Nakamoto, M., Mowrey, J., Jongens, T.A., Nelson, D.L., Moses, K., and Warren, S.T. (2004) Biochemical and genetic interaction between the fragile X mental retardation protein and the microRNA pathway. *Nat. Neurosci.*, **7**, 113–117.

86 Xu, K., Bogert, B.A., Li, W., Su, K., Lee, A., and Gao, F.-B. (2004) The fragile X-related gene affects the crawling behavior of *Drosophila* larvae by regulating the mRNA level of the DEG/ENaC protein pickpocket1. *Curr. Biol.*, **14**, 1025–1034.

87 Lugli, G., Larson, J., Martone, M.E., Jones, Y., and Smalheiser, N.R. (2005) Dicer and eIF2c are enriched at postsynaptic densities in adult mouse brain and are modified by neuronal activity in a calpain-dependent manner. *J. Neurochem.*, **94**, 896–905.

88 Jin, P., Alisch, R.S., and Warren, S.T. (2004) RNA and microRNAs in fragile X mental retardation. *Nat. Cell Biol.*, **6**, 1048–1053.

89 Xu, X.L., Li, Y., Wang, F., and Gao, F.B. (2008) The steady-state level of the nervous-system-specific microRNA-124a is regulated by dFMR1 in *Drosophila*. *J. Neurosci.*, **28**, 11883–11889.

90 Gothelf, D., Frisch, A., Michaelovsky, E., Weizman, A., and Shprintzen, R.J. (2009) Velo-cardio-facial syndrome. *J. Ment. Health Res. Intellect. Disabil.*, **2** (2), 149–167.

91 Stark, K.L., Xu, B., Bagchi, A., Lai, W.S., Liu, H., Hsu, R., Wan, X., Pavlidis, P., Mills, A.A., Karayiorgou, M., *et al.* (2008) Altered brain microRNA biogenesis contributes to phenotypic deficits in a 22q11-deletion mouse model. *Nat. Genet.*, **40**, 751–760.

92 Landthaler, M., Yalcin, A., and Tuschl, T. (2004) The human DiGeorge syndrome critical region gene 8 and its *D. melanogaster* homolog are required for miRNA biogenesis. *Curr. Biol.*, **14** (23), 2162–2167.

93 Kuhn, D.E., Nuovo, G.J., Martin, M.M., Malana, G.E., Pleister, A.P., Jiang, J., Schmittgen, T.D., Terry, A.V., Gardiner, K., Head, E., *et al.* (2008) Human chromosome 21-derived miRNAs are overexpressed in Down syndrome brains and hearts. *Biochem. Biophys. Res. Commun.*, **370**, 473–477.

94 Sethupathy, P., Borel, C., Gagnebin, M., Grant, G.R., Deutsch, S., Elton, T.S., Hatzigeorgiou, A.G., and Antonarakis, S.E. (2007) Human microRNA-155 on chromosome 21 differentially interacts with its polymorphic target in the AGTR1 3′ untranslated region: a mechanism for functional single-nucleotide polymorphisms related to phenotypes. *Am. J. Hum. Genet.*, **81**, 405–413.

95 Bushati, N. and Cohen, S.M. (2007) MicroRNA functions. *Annu. Rev. Cell Dev. Biol.*, **23**, 175–205.

96 Breslau, J., Kendler, K.S., Su, M., Gaxiola-Aguilar, S., and Kessler, R.C. (2005) Lifetime risk and persistence of psychiatric disorders across ethnic groups in the United States. *Psychol. Med.*, **35** (3), 317–327.

97 Winterer, G. and Weinberger, D.R. (2004) Genes, dopamine and cortical signal-to-noise ratio in schizophrenia. *Trends Neurosci.*, **27** (11), 683–690.

98 Ripke, S., *et al.* Schizophrenia Psychiatric Genome-Wide Association Study (GWAS) Consortium (2011) Genome-wide association study identifies five new schizophrenia loci. *Nat. Genet.*, **43** (10), 969–976.

99 Smrt, R.D., *et al.* (2010) MicroRNA miR-137 regulates neuronal maturation by targeting ubiquitin ligase mind bomb-1. *Stem Cells*, **28**, 1060–1070.

100 Szulwach, K.E., *et al.* (2010) Cross talk between microRNA and epigenetic regulation in adult neurogenesis. *J. Cell Biol.*, **189**, 127–141.
101 Perkins, D.O., Jeffries, C.D., Jarskog, L.F., Thomson, J.M., Woods, K., Newman, M.A., Parker, J.S., Jin, J., and Hammond, S.M. (2007) microRNA expression in the prefrontal cortex of individuals with schizophrenia and schizoaffective disorder. *Genome Biol.*, **8** (2), R27.
102 Beveridge, N.J., Gardiner, E., Carroll, A.P., Tooney, P.A., and Cairns, M.J. (2010) Schizophrenia is associated with an increase in cortical microRNA biogenesis. *Mol. Psychiatry*, **15** (12), 1176–1189.
103 Santarelli, D.M., Beveridge, N.J., Tooney, P.A., and Cairns, M.J. (2011) Upregulation of dicer and microRNA expression in the dorsolateral prefrontal cortex Brodmann area 46 in schizophrenia. *Biol. Psychiatry*, **69** (2), 180–187.
104 Lindsay, E.A., Greenberg, F., Shaffer, L.G., Shapira, S.K., Scambler, P.J., and Baldini, A. (1995) Submicroscopic deletions at 22q11.2: variability of the clinical picture and delineation of a commonly deleted region. *Am. J. Med. Genet.*, **56** (2), 191–197.
105 Chen, H., Wang, N., Burmeister, M., and McInnis, M.G. (2009) MicroRNA expression changes in lymphoblastoid cell lines in response to lithium treatment. *Int. J. Neuropsychopharmacol.*, **12** (7), 975–981.
106 Zhou, R., Yuan, P., Wang, Y., Hunsberger, J.G., Elkahloun, A., Wei, Y., Damschroder-Williams, P., Du, J., Chen, G., and Manji, H.K. (2009) Evidence for selective microRNAs and their effectors as common long-term targets for the actions of mood stabilizers. *Neuropsychopharmacology*, **34** (6), 1395–1405.
107 Koerner, M.V., Pauler, F.M., Huang, R., and Barlow, D.P. (2009) The function of non-coding RNAs in genomic imprinting. *Development*, **136**, 1771–1783.
108 Chamberlain, S.J. and Brannan, C.I. (2001) The Prader-Willi syndrome imprinting center activates the paternally expressed murine Ube3a antisense transcript but represses paternal Ube3a. *Genomics*, **73**, 316–322.
109 Vitali, P., Royo, H., Marty, V., Bortolin-Cavaillé, M.-L., and Cavaillé, J. (2010) Long nuclear-retained non-coding RNAs and allele-specific higher-order chromatin organization at imprinted snoRNA gene arrays. *J. Cell Sci.*, **123**, 70–83.
110 Khalil, A.M. and Wahlestedt, C. (2008) Epigenetic mechanisms of gene regulation during mammalian spermatogenesis. *Epigenetics*, **3**, 21–27.
111 Ladd, P.D., Smith, L.E., Rabaia, N.A., Moore, J.M., Georges, S.A., Hansen, R.S., Hagerman, R.J., Tassone, F., Tapscott, S.J., and Filippova, G.N. (2007) An antisense transcript spanning the CGG repeat region of FMR1 is upregulated in premutation carriers but silenced in full mutation individuals. *Hum. Mol. Genet.*, **16**, 3174–3187.
112 Cho, Y.S., Iguchi, N., Yang, J.X., Handel, M.A., and Hecht, N.B. (2005) Meiotic messenger RNA and noncoding RNA targets of the RNA-binding protein translin (TSN) in mouse testis. *Biol. Reprod.*, **73**, 840–847.
113 Filippova, G.N., Thienes, C.P., Penn, B.H., Cho, D.H., Hu, Y.J., Moore, J.M., Klesert, T., Lobanenkov, V.V., and Tapscott, S.J. (2001) CTCF-binding sites flank CTG/CAG repeats and form a methylation-sensitive insulator at the DM1 locus. *Nat. Genet.*, **28**, 335–343.
114 Amaral, P.P., Neyt, C., Wilkins, S.J., Askarian-Amiri, M.E., Sunkin, S.M., Perkins, A.C., and Mattick, J.S. (2009) Complex architecture and regulated expression of the Sox2ot locus during vertebrate development. *RNA*, **15** (11), 2013–2027.
115 Johnson, R., Teh, C.H., Jia, H., Vanisri, R.R., Pandey, T., Lu, Z.H., Buckley, N.J., Stanton, L.W., and Lipovich, L. (2009) Regulation of neural macroRNAs by the transcriptional repressor REST. *RNA*, **15** (1), 85–96.
116 Willingham, A.T., Orth, A.P., Batalov, S., Peters, E.C., Wen, B.G., Aza-Blanc,

P., Hogenesch, J.B., and Schultz, P.G. (2005) A strategy for probing the function of noncoding RNAs finds a repressor of NFAT. *Science*, **309**, 1570–1573.

117 Arron, J.R., Winslow, M.M., Polleri, A., Chang, C.P., Wu, H., Gao, X., Neilson, J.R., Chen, L., Heit, J.J., Kim, S.K., *et al.* (2006) NFAT dysregulation by increased dosage of DSCR1 and DYRK1A on chromosome 21. *Nature*, **441**, 595–600.

118 Millar, J.K., Wilson-Annan, J.C., Anderson, S., Christie, S., Taylor, M.S., Semple, C.A., Devon, R.S., St Clair, D.M., Muir, W.J., Blackwood, D.H., and Porteous, D.J. (2000) Disruption of two novel genes by a translocation co-segregating with schizophrenia. *Hum. Mol. Genet.*, **9** (9), 1415–1423.

119 Chubb, J.E., Bradshaw, N.J., Soares, D.C., Porteous, D.J., and Millar, J.K. (2008) The DISC locus in psychiatric illness. *Mol. Psychiatry*, **13**, 36–64.

120 Williams, J.M., Beck, T.F., Pearson, D.M., Proud, M.B., Cheung, S.W., and Scott, D.A. (2009) A 1q42 deletion involving DISC1, DISC2, and TSNAX in an autism spectrum disorder. *Am. J. Med. Genet. A*, **149A**, 1758–1762.

121 Brandon, N.J., Millar, J.K., Korth, C., Sive, H., Singh, K.K., and Sawa, A. (2009) Understanding the role of DISC1 in psychiatric disease and during normal development. *J. Neurosci.*, **29**, 12768–12775.

122 Abelson, J.F., Kwan, K.Y., O'Roak, B.J., Baek, D.Y., Stillman, A.A., Morgan, T.M., Mathews, C.A., Pauls, D.A., Rasin, M.R., Gunel, M., *et al.* (2005) Sequence variants in SLITRK1 are associated with Tourette's syndrome. *Science*, **310**, 317–320.

123 Clop, A., Marcq, F., Takeda, H., Pirottin, D., Tordoir, X., Bibe, B., Bouix, J., Caiment, F., Elsen, J.M., Eychenne, F., *et al.* (2006) A mutation creating a potential illegitimate microRNA target site in the myostatin gene affects muscularity in sheep. *Nat. Genet.*, **38**, 813–818.

124 Delay, C., Calon, F., Mathews, P., and Hébert, S.S. (2011) Alzheimer-specific variants in the 3′UTR of amyloid precursor protein affect microRNA function. *Mol. Neurodegener.*, **6**, 70.

125 Hébert, S.S., Horré, K., Nicolaï, L., Papadopoulou, A.S., Mandemakers, W., Silahtaroglu, A.N., Kauppinen, S., Delacourte, A., and De Strooper, B. (2008) Loss of microRNA cluster miR-29a/b-1 in sporadic Alzheimer's disease correlates with increased BACE1/beta-secretase expression. *Proc. Natl. Acad. Sci. U.S.A.*, **105** (17), 6415–6420.

126 Smith, P., Al Hashimi, A., Girard, J., Delay, C., and Hébert, S.S. (2011) *In vivo* regulation of amyloid precursor protein neuronal splicing by microRNAs. *J. Neurochem.*, **116** (2), 240–247.

127 Boissonneault, V., Plante, I., Rivest, S., and Provost, P. (2009) MicroRNA-298 and microRNA-328 regulate expression of mouse beta-amyloid precursor protein-converting enzyme 1. *J. Biol. Chem.*, **284**, 1971–1981.

128 Wang, W.X., Rajeev, B.W., Stromberg, A.J., Ren, N., Tang, G., Huang, Q., *et al.* (2008) The expression of microRNA miR-107 decreases early in Alzheimer's disease and may accelerate disease progression through regulation of beta-site amyloid precursor protein-cleaving enzyme 1. *J. Neurosci.*, **28**, 1213–1223.

129 Martin, I., Dawson, V.L., and Dawson, T.M. (2011) Recent advances in the genetics of Parkinson's disease. *Annu. Rev. Genomics Hum. Genet.*, **12**, 301–325.

130 Kim, J., Inoue, K., Ishii, J., Vanti, W.B., Voronov, S.V., Murchison, E., *et al.* (2007) A microRNA feedback circuit in midbrain dopamine neurons. *Science*, **317**, 1220–1224.

131 Junn, E., Lee, K.W., Jeong, B.S., Chan, T.W., Im, J.Y., and Mouradian, M.M. (2009) Repression of alpha-synuclein expression and toxicity by microRNA-7. *Proc. Natl. Acad. Sci. U.S.A.*, **106**, 13052–13057.

132 Wang, G., van der Walt, J.M., Mayhew, G., Li, Y.J., Zuchner, S., Scott, W.K., *et al.* (2008) Variation in the miRNA-433 binding site of FGF20 confers risk for

Parkinson disease by overexpression of alpha-synuclein. *Am. J. Hum. Genet.*, **82**, 283–289.
133 Gehrke, S., Imai, Y., Sokol, N., and Lu, B. (2010) Pathogenic LRRK2 negatively regulates microRNA-mediated translational repression. *Nature*, **466**, 637–641.
134 Finkbeiner, S. (2011) Huntington's disease. *Cold Spring Harb. Perspect. Biol.*, **3** (6), a007476.
135 Eidelberg, D. and Surmeier, D.J. (2011) Brain networks in Huntington disease. *J. Clin. Invest.*, **121** (2), 484–492.
136 Johnson, R., Zuccato, C., Belyaev, N.D., Guest, D.J., Cattaneo, E., and Buckley, N.J. (2008) A microRNA-based gene dysregulation pathway in Huntington's disease. *Neurobiol. Dis.*, **29**, 438–445.
137 Packer, A.N., Xing, Y., Harper, S.Q., Jones, L., and Davidson, B.L. (2008) The bifunctional microRNA miR-9/miR-9* regulates REST and CoREST and is downregulated in Huntington's disease. *J. Neurosci.*, **28**, 14341–14346.
138 Lee, V.M.Y., Goedert, M., and Trojanowski, J.Q. (2001) Neurodegenerative tauopathies. *Annu. Rev. Neurosci.*, **24**, 1121–1159.
139 Rademakers, R., Eriksen, J.L., Baker, M., Robinson, T., Ahmed, Z., Lincoln, S.J., *et al.* (2008) Common variation in the miR-659 binding-site of GRN is a major risk factor for TDP43-positive frontotemporal dementia. *Hum. Mol. Genet.*, **17**, 3631–3642.
140 Buratti, E., De Conti, L., Stuani, C., Romano, M., Baralle, M., and Baralle, F. (2010) Nuclear factor TDP-43 can affect selected microRNA levels. *FEBS J.*, **277** (10), 2268–2281.
141 Ling, S.C., Albuquerque, C.P., Han, J.S., Lagier-Tourenne, C., Tokunaga, S., Zhou, H., and Cleveland, D.W. (2010) ALS-associated mutations in TDP-43 increase its stability and promote TDP-43 complexes with FUS/TLS. *Proc. Natl. Acad. Sci. U.S.A.*, **107** (30), 13318–13323.
142 Wang, X., Arai, S., Song, X., Reichart, D., Du, K., Pascual, G., Tempst, P., Rosenfeld, M.G., Glass, C.K., and Kurokawa, R. (2008) Induced ncRNAs allosterically modify RNA-binding proteins in cis to inhibit transcription. *Nature*, **454**, 126–130.
143 Lee, Y., Samaco, R.C., Gatchel, J.R., Thaller, C., Orr, H.T., and Zoghbi, H.Y. (2008) miR-19, miR-101 and miR-130 co-regulate ATXN1 levels to potentially modulate SCA1 pathogenesis. *Nat. Neurosci.*, **11**, 1137–1139.
144 Saba, R., Goodman, C.D., Huzarewich, R.L., Robertson, C., and Booth, S.A. (2008) A miRNA signature of prion induced neurodegeneration. *PLoS One*, **3**, e3652.
145 Hebert, S.S., Horre, K., Nicolai, L., Bergmans, B., Papadopoulou, A.S., Delacourte, A., *et al.* (2009) MicroRNA regulation of Alzheimer's amyloid precursor protein expression. *Neurobiol. Dis.*, **33**, 422–428.

Further Reading

Ambros, V. (2004) The functions of animal microRNAs. *Nature*, **431**, 350–355.
Badcock, C. and Crespi, B. (2006) Imbalanced genomic imprinting in brain development: an evolutionary basis for the aetiology of autism. *J. Evol. Biol.*, **19**, 1007–1032.
Barbato, C., Ruberti, F., and Cogoni, C. (2009) Searching for MIND: microRNAs in neurodegenerative diseases. *J. Biomed. Biotechnol.*, **2009**, 871313.
Bernard, D., Prasanth, K.V., Tripathi, V., Colasse, S., Nakamura, T., Xuan, Z., Zhang, M.Q., Sedel, F., Jourdren, L., Coulpier, F., Triller, A., Spector, D.L., and Bessis, A. (2010) A long nuclear-retained non-coding RNA regulates synaptogenesis by modulating gene expression. *EMBO J.*, **29** (18), 3082–3093.
Bond, A.M., Vangompel, M.J., Sametsky, E.A., Clark, M.F., Savage, J.C., Disterhoft, J.F., and Kohtz, J.D. (2009) Balanced gene regulation by an embryonic brain ncRNA is critical for adult hippocampal GABA circuitry. *Nat. Neurosci.*, **12** (8), 1020–1027.
Bushati, N. and Cohen, S.M. (2008) MicroRNAs in neurodegeneration. *Curr. Opin. Neurobiol.*, **18**, 292–296.
Cao, X., Pfaff, S.L., and Gage, F.H. (2007) A functional study of miR-124 in the

developing neural tube. *Genes Dev.*, **21**, 531–536.

Carthew, R.W. and Sontheimer, E.J. (2009) Origins and mechanisms of miRNAs and siRNAs. *Cell*, **136**, 642–655.

Chalfie, M., Horvitz, H.R., and Sulston, J.E. (1981) Mutations that lead to reiterations in the cell lineages of *C. elegans*. *Cell*, **24**, 59–69.

Czech, B., Malone, C.D., Zhou, R., Stark, A., Schlingeheyde, C., Dus, M., Perrimon, N., Kellis, M., Wohlschlegel, J.A., Sachidanandam, R., *et al.* (2008) An endogenous small interfering RNA pathway in *Drosophila*. *Nature*, **453**, 798–U797.

Herbert, M.R. (2006) Autism: parsing heterogeneity of causes and features. *Neurotoxicol. Teratol.*, **28**, 413–413.

Kapranov, P., Willingham, A.T., and Gingeras, T.R. (2007) Genome-wide transcription and the implications for genomic organization. *Nat. Rev. Genet.*, **8**, 413–423.

Kawamura, Y., Saito, K., Kin, T., Ono, Y., Asai, K., Sunohara, T., Okada, T.N., Siomi, M.C., and Siomi, H. (2008) *Drosophila* endogenous small RNAs bind to Argonaute 2 in somatic cells. *Nature*, **453**, 793–U795.

Kurokawa, R. (2011) Promoter-associated long noncoding RNAs repress transcription through a RNA binding protein TLS. *Adv. Exp. Med. Biol.*, **722**, 196–208.

Mattick, J.S. (2003) Challenging the dogma: the hidden layer of non-protein-coding RNAs in complex organisms. *Bioessays*, **25**, 930–939.

Mattick, J.S. (2004) RNA regulation: a new genetics? *Nat. Rev. Genet.*, **5**, 316–323.

Mercer, T.R., Dinger, M.E., Sunkin, S.M., Mehler, M.F., and Mattick, J.S. (2008) Specific expression of long noncoding RNAs in the mouse brain. *Proc. Natl. Acad. Sci. U.S.A.*, **105**, 716–721.

Mercer, T.R., Qureshi, I.A., Gokhan, S., Dinger, M.E., Li, G., Mattick, J.S., and Mehler, M.F. (2010) Long noncoding RNAs in neuronal-glial fate specification and oligodendrocyte lineage maturation. *BMC Neurosci.*, **11**, 14.

Miller, D.T., Shen, Y., Weiss, L.A., Korn, J., Anselm, I., Bridgemohan, C., Cox, G.F., Dickinson, H., Gentile, J., Harris, D.J., *et al.* (2008) Microdeletion/duplication at 15q13.2q13.3 among individuals with features of autism and other neuropsychiatric disorders. *J. Med. Genet.*, **46**, 242–248.

Nguyen, H.T. and Frasch, M. (2006) MicroRNAs in muscle differentiation: lessons from *Drosophila* and beyond. *Curr. Opin. Genet. Dev.*, **16**, 533–539.

Okamura, K., Chung, W.J., Ruby, J.G., Guo, H.L., Bartel, D.P., and Lai, E.C. (2008) The *Drosophila* hairpin RNA pathway generates endogenous short interfering RNAs. *Nature*, **453**, 803–U808.

Persico, A.M. and Bourgeron, T. (2006) Searching for ways out of the autism maze: genetic, epigenetic and environmental clues. *Trends Neurosci.*, **29**, 349–358.

Royo, H., Basyuk, E., Marty, V., Marques, M., Bertrand, E., and Cavaillé, J. (2007) Bsr, a nuclear-retained RNA with monoallelic expression. *Mol. Biol. Cell*, **18** (8), 2817–2827.

Schanen, N.C. (2006) Epigenetics of autism spectrum disorders. *Hum. Mol. Genet.*, **15**, R138–R150.

Simon, D.J., Madison, J.M., Conery, A.L., Thompson-Peer, K.L., Soskis, M., Ruvkun, G.B., Kaplan, J.M., and Kim, J.K. (2008) The microRNA miR-1 regulates a MEF-2-dependent retrograde signal at neuromuscular junctions. *Cell*, **133**, 903–915.

Siomi, H. and Siomi, M.C. (2008) Interactions between transposable elements and Argonautes have (probably) been shaping the evolution *Drosophila* genome throughout. *Curr. Opin. Genet. Dev.*, **18**, 181–187.

Smith, P.Y., Delay, C., Girard, J., Papon, M.A., Planel, E., Sergeant, N., Buée, L., and Hébert, S.S. (2011) MicroRNA-132 loss is associated with tau exon 10 inclusion in progressive supranuclear palsy. *Hum. Mol. Genet.*, **20** (20), 4016–4024.

Sridhar, J. and Rafi, Z.A. (2007) Small RNA identification in Enterobacteriaceae using synteny and genomic backbone retention. *OMICS*, **11**, 74–U72.

Underwood, J.G., Uzilov, A.V., Katzman, S., Onodera, C.S., Mainzer, J.E., Mathews, D.H., Lowe, T.M., Salama, S.R., and Haussler, D. (2010) FragSeq: transcriptome-wide RNA structure probing using high-throughput sequencing. *Nat. Methods*, **7** (12), 995–1001.

Wienholds, E., Kloosterman, W.P., Miska, E., Alvarez-Saavedra, E., Berezikov, E., de Bruijn, E., Horvitz, H.R., Kauppinen, S., and Plasterk, R.H.A. (2005) MicroRNA expression in zebrafish embryonic development. *Science*, **309**, 310–311.

Wilkinson, L.S., Davies, W., and Isles, A.R. (2007) Genomic imprinting effects on brain development and function. *Nat. Rev. Neurosci.*, **8**, 832–843.

Zhang, L.L., Wang, T., Wright, A.F., Suri, M., Schwartz, C.E., Stevenson, R.E., and Valle, D. (2006) A microdeletion in Xp11.3 accounts for co-segregation of retinitis pigmentosa and mental retardation in a large kindred. *Am. J. Med. Genet. A*, **140A**, 349–357.

10
The Evolution of the Modern RNA World

Ying Chen, Hongzheng Dai, and Manyuan Long

How RNAs evolve remains an essential question for understanding the evolution of the fundamental molecular processes that control cellular life activities. Over more than two decades since the discovery of ribozyme [1, 2], the fascinating antiquity of the RNAs has been explored for its possible role in the origin of life (e.g., [3], coined as the RNA World by Walter Gilbert [4]). The recent evolution of RNAs has just begun to catch the attention of evolutionary biologists, who find that the evolutionary dynamics of present-day RNAs, which we call the modern RNA world, is also important as a general problem of evolution. Numerous valuable reports in the evolution of ncRNA genes, evolution of transcriptional regulation, and evolution of posttranscriptional gene regulation have provided an emerging picture of how RNA machineries evolved in multicellular organism. In this chapter, we focus on this modern RNA world and summarize the progresses in these areas of RNA evolution.

10.1
Evolution of Noncoding RNA Gene

For a long time, noncoding RNA (ncRNA) genes have escaped the focus of molecular evolutionary biologist, living in the shadow of protein-coding genes because of the limited number of reported ncRNA genes and the challenge in the methodology to identify and characterize such genes. Sporadic data were often taken as anecdote, although biologically, such genes showed interesting characteristics [5–7]. However, the tide has begun to turn recently with the advance of RNA biology and the application of genomic tiling arrays, which have revealed massive expression of intronic and intergenic regions [8–12]. The large amount of ncRNA genes identified make it possible to unravel evolutionary histories, measure the rates of evolution and origination, and, in a rare circumstance, even a phenotypic effect of ncRNA genes using sophisticated technology of reverse genetics.

Posttranscriptional Gene Regulation: RNA Processing in Eukaryotes, First Edition. Edited by Jane Wu.

10.1.1
Conserved and Nonconserved Nature of ncRNA Genes

ncRNAs include a variety of different families of genes. Some of the better defined members include microRNAs (miRNA) and small interference RNA (siRNA). Functioning in a similar pathway, miRNA and siRNA guide effector complexes RISC (RNA-induced silencing complex) to target messenger RNAs (mRNAs) through base-pair complementarity, facilitating translational repression or site-specific cleavage [13, 14]. The miRNAs are derived from sequences that are antisense to known genes and are typically transcribed from intergenic regions, introns, or precursors that have been previously identified as ncRNA [13–16].

Many miRNAs are evolutionarily conserved and have very specific tissue and/or time-specific expressions presumably controlled by unique cis-acting regulatory elements that coevolved with the miRNA sequences [17]. Genes encoding miRNA are thought to originate from inverted duplication of target gene sequences [18, 19] or from transposable elements and other genome repeats [20–22]. Despite their unusually small sizes, the evolutionary history of miRNA gene families seems to be similar to their protein-coding counterparts [23].

Yet, there are reasons to believe that miRNA and small nucleolar RNAs (snoRNAs) might only be a small proportion of what we called ncRNA. Many of the poorly studied larger ncRNA elude easily classification and lack strong sequence conservation, prompting suggestions that they might be nonfunctional [24]. However, certain long functional ncRNAs such as Air and Xist are also poorly conserved [25–28]. Recent study shows that unlike miRNAs and snoRNAs, many of the longer functional ncRNAs display rapid sequence evolution [29]. It has been suggested that longer ncRNAs are under the influence of different evolutionary constraints and that the lack of conservation displayed by the thousands of candidate ncRNAs does not necessarily signify an absence of function [29].

10.1.2
Lineage-Specific ncRNA Genes

Relatively few cases of lineage-specific ncRNA gene has been characterized; yet, it has been reported that genes with chimeric origins sometimes lose their coding capability due to disruption of the original open reading frame. Under certain circumstances, this can lead to origination of ncRNA genes rather than pseudogenes. In a survey of species-specific genes, Wang *et al.* identified a new non-protein-coding gene, *sphinx*, that originated and became fixed in a single species, *Drosophila melanogaster*, within the last two to three million years (mys) [6, 7]. The *sphinx* gene was formed by the insertion of a retroposed sequence of the ATP synthase F chain gene (*ATPS-F*) from chromosome 2 into the 102F region of chromosome 4, recruiting sequences upstream to form a new exon and intron, a region we refer to as 102F-EI [6]. Two alternative transcripts in adult males were detected from this locus [6]. Figure 10.1 summarized the evolutionary process of *sphinx* [6, 7].

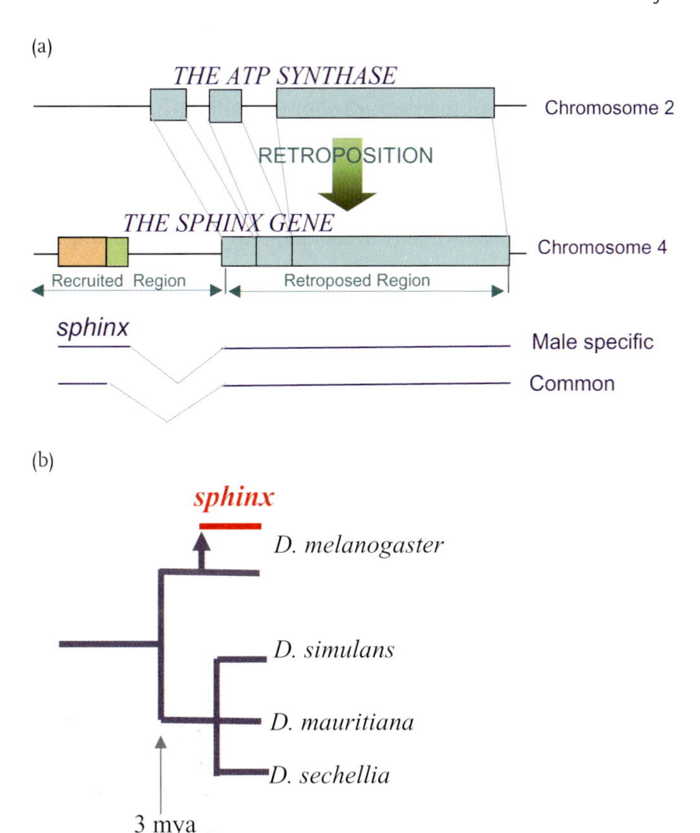

Figure 10.1 Origination process of *sphinx*. (a) Molecular evolutionary process. The arrow with the letter P indicates the promoter region identified in this study. (b) The distribution of *sphinx* (red) and its parent, the ATP-synthase-F gene (cyan) in the related *Drosophila* species.

The *sphinx* gene appears to be functional since the gene only contains indel polymorphisms in the nonexonic sequences, it has a rate of evolution significantly above neutral expectations, suggesting rapid adaptive evolution, and it has a very specific pattern of expression from the reverse transcription-polymerase chain reaction (RT-PCR) experiment [6]. However, although it is derived in part from a protein-coding gene, it is most likely an ncRNA gene because its parental-inherited coding regions are disrupted by several nonsense mutations. The further investigation of the phenotypic effects of this gene, using gene replacement [30], supported that this ncRNA gene is functional with an important role in evolution of courtship behaviors [31, 32] (see Section 10.4 for a further discussion).

The evolutionary histories of other ncRNA genes have also been explored. For example, Xist, the gene that controlled dosage compensation by inactivation of one X chromosome in mammals, was found to be originated via a process of gene duplication and pseudogenization in the common ancestor of mammals [33]. An extremely conserved ncRNA gene was also identified in vertebrates and shown to control the development of the human brain [34]. Finally, Lu *et al.* [35] identified a large number of young miRNA genes in the *Drosophila* subgroup species and found that the life and death process of an miRNA gene is within a short period less than several million years.

10.2 Evolution of Transcriptional Regulation

Changes in the patterns of gene expression are of interest because they are believed to underlie many of the phenotypic differences within and between species. Recent advances in array technology have allowed unbiased genome-wide analysis of RNA transcription. The evolutionary analyses of gene expression divergence using array technology are generally motivated by two different levels of interests: study general rules that govern expression evolution and identify the molecular mechanisms that underlie phenotypic features.

Although changes in gene expression level are ultimately caused by changes in DNA sequences, there is a fundamental difference between studying sequences and expression levels in terms of the complexity of measurement and modeling. The neutral theory is the accepted null model for the evolution at DNA sequence level. It postulated that the vast majority of nucleotide substitutions are neutral or slightly deleterious [36, 37]. The fixations of these mutations are subjected to random genetic drift. The alternative hypothesis, which might be more appealing at the phenotypic level, is that most differences between species are adaptive and driven by positive selection. Recent studies on gene expression both within and between species posed a question on whether gene expression evolution follows a neutral process, or what fraction of genes may evolve under neutrality [38, 39].

In studies within primate lineage, transcriptome divergence between species correlates positively with intraspecific expression diversity and accumulates approximately linearly with time. It was also found that the rates of transcriptome divergence between a set of expressed pseudogenes and intact genes do not differ significantly. These observations led to the suggestion that a neutral model of evolution may apply to the transcriptome, that is, that the majority of genes expressed in a certain tissue change over evolutionary time as the result of stochastic processes that are limited in their extent by negative selection rather then as the result of positive Darwinian selection [39]. However, it is worth noticing that two tissues that stands out with regard to gene expression are the testes, where positive selection has been implied in both humans and chimpanzee, and the brain, where gene expression has changed less than in other organs but acceleration might have occurred in human ancestors [40, 41].

Drosophila species, with its large effective population size, might present a different picture. Rifkin *et al.* [42] compared patterns of gene expression variation within species with the variation between species to identify genes for which a neutral model could not be rejected. Based on a series of tests, they could not reject the overall low variation for 44% of the expressed genes, could not reject the species-specific gene expression patterns for 39% of the genes, and could not reject a model consistent with neutrality for the remaining 17% of the genes [42]. They interpreted these results to indicate a dominant signature for stabilizing selection in gene expression evolution with smaller, but important, roles for directional selection and neutral evolution, respectively [42]. Another comparison of the gene-expression profiles between adults of *D. melanogaster* and *Drosophila simulans* suggest that sex-dependent selection may drive changes in expression of many of the most rapidly evolving genes in the *Drosophila* transcriptome [43].

There have been growing interest in studying differences in RNA profiling between populations and species, to identify genetic basis of reproductive isolation, adaptation, or a particular phenotypic trait of interest, an approach known as expression QTL (quantitative trait loci) mapping. Studies in *Drosophila* have revealed large numbers of pleiotropic genes that interact epistatically to regulate quantitative traits, and large numbers of QTLs with sex-, environment-, and genotype-specific effects [44]. Multiple molecular polymorphisms in regulatory regions of candidate genes are often associated with variation for complex traits [44]. These observations offer valuable lessons for understanding the genetic basis of variation for complex traits in other organisms, including humans.

10.3
Evolution of Posttranscriptional Gene Regulation

Expression of a gene can be controlled at various levels. While the majority of studies have focused on transcriptional control, the importance of posttranscriptional mechanism in regulation gene expression in eukaryotes is becoming increasingly clear. Common posttranscriptional gene regulatory mechanisms include mRNA splicing, mRNA stability, translation, and posttranslational events. In this section, we will present three aspects of evolutionary changes that alter gene expression at posttranscriptional level: alternative splicing, RNA secondary structure, and posttranscriptional gene silencing (PTGS).

10.3.1
Evolution of Alternative Splicing

Alternative premessenger RNA (pre-mRNA) splicing is an important posttranscriptional regulation mechanism that increases transcription diversity in higher organism. Numerous studies have shown that alternative splicing is prevalent in mammalian genomes. It was estimated that at least 74% of all human genes undergo alternative splicing [45]. Comparative genomic studies investigating the conservation of alternative splicing pattern in different species provide first insight into the evolution dynamic of alternative splicing. Nurtdinov *et al.* [46] compared alternative splicing isoforms of 166 pairs of orthologous human and mouse genes, and found that approximately 50% of alternatively spliced genes have nonconserved isoforms. Another investigation by Thanaraj *et al.* [47] based on exon junctions estimates an overall conservation of 61% of alternative and 74% of constitutive splice junctions. The differences between these estimations can be explained partly by the use of different data sets and partly by the degree of conservatism of the work. Overall, many alternative splicing events are conserved, with the exception of minor forms (splice forms that are present in less than 50% of the total transcripts of a particular gene), which are often associated with exon creation or loss [47, 48].

Changes of splicing patterns directly affect the pool of mRNAs. However, splicing changes may also have implications on evolutionary changes at the sequence level. Modrek and Lee [49] analyzed a set of 9434 orthologous genes in humans and mice. They found that the levels of conservation in constitutive as well as exons in major alternative transcripts were highly conserved (98%), but that alternative exons in minor transcripts were much less well conserved (25%). Based on these observations, Modrek and Lee proposed that alternative splicing might play a major role in genome evolution by allowing new exons to evolve with less constraint. When a new exon is incorporated into a gene and alternatively spliced, it would probably first be included in only a few of the transcripts and would be free to evolve as the original transcript form would still accomplish its function. Without alternative splicing, those new exons would probably be under selection constraint, lowering the capacity for evolution of an organism [50].

Like gene duplication, newly evolved splice variance can be a very important contributor to genomic complexity and gene function [51, 52]. Yet, it is not clear whether alternative splicing plays a role in species evolution. The only identified case that relates changes of splicing pattern with speciation event is the work by Terai *et al.* on a pigmentation gene *hagoromo* [53]. They found that the complexity of alternative splicing of *hagoromo* mRNAs is increased in an explosively speciated lineage in East African cichlids. Their results suggest that alternative splicing may play a role in mate choice, leading to cichlid speciation through diversification of gene function by production of multiple mRNAs from a single gene [53]. A counter example comes from a comparative study of conserved alternatively spliced (CAS) exons (orthologous exons that are alternatively spliced both in humans and mice) and non-CAS exons in humans [54]. Sorek *et al.* [54] claimed that many of the nonconserved splice variants in the human genome are nonfunctional, as these non-CAS exons have very different properties comparing to CAS exons that are presumably functional. But even if non-CAS exons are not functional, they might have an important evolutionary role according to the model proposed by Modrek and Lee [50].

Comparison between distantly related species (e.g., human vs. mouse) to study changes in alternative splicing pattern over time is often accompanied with difficulties of accurately assigning orthologous genes. Ideally, one would like to look for newly evolved splice variance by doing large-scale comparisons between closely related species where homology is fairly well preserved. We did a phylogenetic survey on 12 genes with a well-characterized tissue and/or development-specific splicing pattern in *D. melanogaster*. Splicing patterns of the 12 genes surveyed in this study are well conserved among the *melanogaster* subgroup with less than 10 million years' divergence time, while changes do happen along the *pseudoobscura* lineage. Roughly speaking, 5 out of 27 mRNA types undergo changes in 25 million years' evolutionary time between *D. melanogaster* and *Drosophila pseudoobscura*.

Significantly lower than genome average dN/dS values of alternatively spliced genes suggests this group of genes evolved under constraint. It is possible that the slow rate of evolution is due to an additional selective pressure imposed by regulating alternative splicing. Alternatively, essential genes might be more likely to associate with multiple RNA transcripts, which can also lead to lower dN/ds values of alternatively spliced genes. The later scenario to some extent explains why we saw more changes in *pseudoobscura* if one allows for genes to lose essentiality over time.

10.3.2
Compensatory Evolution of RNA Secondary and Tertiary Structure

RNAs not only serve as templates for protein translation, but also often function as a catalyst and regulatory element by forming secondary or tertiary structures, perhaps reflecting their important role in the earliest stage of evolution. Higher-order structure of modern RNA is important in various aspects such as ribozyme interaction, mRNA stability, RNA binding, and other enzymatic reactions. For example, several studies indicated that pre-mRNA secondary structure can be important in splice-site choice and splicing efficiency of both constitutively and alternatively spliced introns [55–57].

The structure of RNA comprises two levels, the secondary structure, which is dominated by Watson–Crick base pairs that form A-form double helices, and the tertiary structure. An RNA hairpin is the most common secondary structure found in almost every RNA folding prediction. It consists of single-stranded (loop) and double-stranded regions (stem), where the double-stranded stems are formed by base paring and loops are mismatches and bulges (i.e., unpaired sequences within the stem). Many of the pathways utilizing and/or responding to RNA hairpins have evolved independently and are not linked to others. Yet, it is conceivable that, generally, there are epistatic fitness interactions between nucleotides that are important in maintaining these hairpin structures. Compensatory evolution associated with RNA secondary structures occurs in ribosomal RNA (rRNA) and mRNA [58, 59], and is expected to be important also for transfer RNAs (tRNAs) and ribozymes.

The most thorough demonstration of selection maintaining nucleotide sequence variation associated with mRNA secondary structure is the study on *Drosophila* alcohol dehydrogenase (*Adh*) gene. Evidence for the evolutionary maintenance of a hairpin structure possibly involved in intron processing had been found in intron 1 of the *Adh* in diverse *Drosophila* species [59]. It was further demonstrated that mutations that disrupt the putative hairpin structure right upstream of the intron branch point cause a significant reduction in both splicing efficiency and ADH protein production [58]. In contrast, the compensatory double mutant that restores the putative hairpin structure was indistinguishable from the wild type in both splicing efficiency and ADH level [58]. However, because the fitness effect associates with sequence variation within ADH locus is rather small and the rate of evolution in stems is only five times slower than unpaired regions, it was argued that the selection against deleterious intermediate state is not very strong, comparable with weak selection governing codon usage bias or base composition (which are usually considered weak) [58].

10.3.3
Evolution of Posttranscriptional Gene Silencing

The phenomenon of posttranscriptional gene silencing (PTGS) was reported in the early 1990s, when plant transgenic engineering encountered problems of unexpected silencing of transgenes (as cosuppression) in a sequence homology-dependent manner. At the posttranscriptional level, transgenes mRNA was "accidentally" degraded through the pathway by which plants defend themselves against the invasion of plant viruses (90% of those viruses are RNA virus) [60]. Later, similar silencing components were observed in other eukaryotes in defending transposable elements: fungi, nematode, insects, and mammals, which suggests that such RNA-mediated gene silencing might be an evolutionary conserved mechanism across kingdoms [61, 62].

Two key components play important roles in the PTGS process: an RNAse III enzyme, like Dicer or Drosha, with the help of cofactors, processes double-stranded RNA or premature miRNA into siRNA; an RNAse H protein, Argonaute/P-element-induced wimpy testis (PIWI), in the assembly of RISC loads siRNA and degrades targeted mRNA or inhibits the translation process. In some eukaryotes, an RNA-dependent RNA polymerase is required for PTGS [63]. By sequence comparison, Cerutti and Casas-Mollano [64] investigated the distribution of those three components in 25 eukaryotes species. In all the examined five supergroup of eukaryotes, these key proteins of PTGS-related machinery were identified. Combined with the observations of RNA interference (RNAi)-related phenomenon in most of those species, it was suggested that the primitive machinery of PTGS was already present in the last common ancestor of eukaryotes [64].

The PTGS pathways in diverse eukaryote organisms can evolve through two aspects: the endogenous trigger loci to produce siRNA/miRNA, the components of processing machinery that could further split into biogenesis factor and effectors, which lead to the proliferation and specialization of distinct pathways [65]. On one hand, the lost of some or all three key components of PTGS in different lineages was observed, and the sequence differences between homolog proteins were analyzed with their diverse functions [64, 65]. On the other hand, frequent birth and death of miRNA genes was suggested through the analysis of nonconserved miRNA genes identified by sequencing and computational efforts in *Arabidopsis* [66, 67]. Target validation assays of miRNA in *Arabidopsis* not only reveal the complexity between the parental gene where the miRNAs come from and the target gene that the miRNAs silence: they are not always coupled [66], but also suggest the autoregulatory mechanism for the expression of silencing machinery components [67]. Indeed, it was observed that the feedback loops between miRNA and the components of silencing machinery occur multiple times in both bryophytes and angiosperms, which suggests the importance of the interaction between miRNA and the processing machinery in the evolution of silencing pathways [68].

Recent findings indicate that some components of the PTGS machinery are also important in transcriptional gene silencing (TGS) through the modification of chromatin in a similar sequence-homolog-dependent manner [61, 69, 70]. TGS is also suggested in defending genome from parasite or invading DNA elements. Together, PTGS and TGS might reflect a reminiscent picture in which primitive cellular life forms struggled for living with their fragile genome/transcriptome in the ancient DNA/RNA world. Only later in the evolution, these machineries were exploited into a broad field to regulate gene expression.

10.4 Phenotypic Evolution by the Origination of New ncRNA Genes and Perspectives between Protein-Coding and ncRNA Genes

The ncRNA genes originated with a broad range of functions, resulting in the evolution of phenotypes and conferring organisms with a new adaptation to the environments they live. The new addition of ncRNA genes most likely functions through the interaction with protein-coding genes, with some exceptions that the ncRNA genes may only play the structural role (e.g., *Xist*). In this section, we will summarize the known data and present conceivable interactions as the current data may show.

10.4.1 Evolution of Phenotypes Contributed by New ncRNA Genes

The evolutionary process in which new ncRNA genes contributed to phenotypic changes is coming into the scene in recent publications. These new findings revealed various molecular processes that control important functions and phenotypes, ranging from regulating early development in vertebrates [32] to silencing transcription of an X chromosome for dosage compensation in mammals (*Xist*) [25]. We recently investigated the phenotypic effect of *sphinx*, aforementioned lineage-specific ncRNA gene in *D. melanogaster* [31].

To understand the phenotypic effects of *sphinx*, we inactivated the male-specific expression by using gene replacement to knock out the gene in a *D. melanogaster* strain (*W1118*) [30, 31]. The loss-of-function mutant was created with the male-specific expression inactivated. Compared with a wild-type strain in the same species, the knockout male showed more than 10 times longer of courtship time for male–male courtship than the males in the wild-type strain. Because this is a loss-of-function mutation, it is likely that the detected male–male courtship phenotype was an ancient phenotype before *sphinx* originated two to three million years ago. This conjecture was supported by the evidence that in the four species related to *D. melanogaster* 3–25 million years ago, a high proportion of males showed male–male courtship behaviors.

Investigations of male–male mating behavior in other *Drosophila* species have suggested that, compared with wild-type *D. melanogaster*, male–male courtship is significantly more prevalent in three closely related species, *D. simulans*, *Drosophila mauritiana*, and *Drosophila yakuba*, and one distantly related species, *D. pseudoobscura*. While the male–male courtship phenotypes in the related species of *D. melanogaster* can be maintained by various evolutionary genetic forces [32], it seems likely that the *sphinx* gene was recruited in the *D. melanogaster* lineage to inhibit male–male courtship [31]. This example shows that the fixation of a single ncRNA gene can impact on the fitness of the entire population [31].

These analyses suggested that the origination of *sphinx* might be driven by sexual selection and reduced male–male courtship behavior in *D. melanogaster*. This case also suggests that the genetic control of courtship was an evolving system driven by newly evolved ncRNA gene that may regulate other courtship genes.

10.4.2 Perspectives between Protein-Coding and ncRNA Genes

Numerical reports have revealed that the functions of ncRNA genes are mostly regulatory. The regulatory functions can be in different levels. For example, *Xist* can inactivate an entire female X chromosome in mammals in a structural interaction between the gene and the chromosome to accomplish dosage compensation. Another ncRNA can bind the PIWI proteins to suppress the transposable elements (e.g., [70]). Pseudogenes from a protein-coding gene can be transcribed and act as interfering RNAs to regulate the oocyte gene expression in mice [71]. It was found that >90% and >70% of the intergenic regions in mammalian and *Drosophila* genomes, respectively, were transcribed [72–74]. Interestingly, similar to *sphinx* in *D. melanogaster*, numerous long intergenic noncoding RNAs (lincRNAs) with lengths of thousands of long nucleotides were recently predicted in mammalian genomes (e.g., [75]), and a few of these genes have been characterized for their molecular function. For example, the lincRNA HOTAIR that is transcribed from HOXC (>2000 nucleotides) was found to be a scaffold of histone modification complexes [76]. Are these ncRNAs and lincRNAs generally functional? How were they originated and evolved? What is the distinct evolutionary feature of ncRNA genes compared with its targets in regulation, the protein-coding genes? The ncRNA genes evolved more rapidly than protein-coding genes, showing significantly lower evolutionary constraint. This evolutionary feature revealed a great plasticity that adapts these ncRNA genes to regulate diverse protein functions. The huge amount of ncRNA genes in *D. melanogaster* [73, 74] revealed the gigantic regulatory diversity, resulting from a high level of sequence plasticity of the ncRNA genes.

References

1 Cech, T.R. (1983) RNA splicing: three themes with variations. *Cell*, **34**, 713–716.

2 McClain, W.H., Guerrier-Takada, C., and Altman, S. (1987) Model substrates for an RNA enzyme. *Science*, **238**, 527–530.

3 Gesteland, R.F., Cech, T.R., and Atkins, J.F. (eds) (2006) *The RNA World*, 3rd edn, Cold Spring Harbor Laboratory Press Cold Spring Harbor, New York.

4 Gilbert, W. (1978) Why genes in pieces? *Nature*, **271**, 501.

5 Eddy, S.R. (2001) Non-coding RNA genes and the modern RNA world. *Nat. Rev. Genet.*, **2**, 919–929.

6 Wang, W., Brunet, F.G., Nevo, E., and Long, M. (2002) Origin of *sphinx*, a young chimeric RNA gene in *Drosophila melanogaster*. *Proc. Natl. Acad. Sci. U.S.A.*, **99**, 4448–4453.

7 Wang, W., Thornton, K., Berry, A., and Long, M. (2002) Nucleotide variation along the *Drosophila melanogaster* fourth chromosome. *Science*, **295**, 134–137.

8 Cheng, J., Kapranov, P., Drenkow, J., Dike, S., Brubaker, S., Patel, S., Long, J., Stern, D., Tammana, H., Helt, G., Sementchenko, V., Piccolboni, A., Bekiranov, S., Bailey, D.K., Ganesh, M., Ghosh, S., Bell, I., Gerhard, D.S., and Gingeras, T.R. (2005) Transcriptional maps of 10 human chromosomes at 5-nucleotide resolution. *Science*, **308**, 1149–1154.

9 Kapranov, P., Drenkow, J., Cheng, J., Long, J., Helt, G., Dike, S., and Gingeras, T.R. (2005) Examples of the complex architecture of the human transcriptome revealed by RACE and high-density tiling arrays. *Genome Res.*, **15**, 987–997.

10 Singh-Gasson, S., Green, R.D., Yue, Y., Nelson, C., Blattner, F., Sussman, M.R., and Cerrina, F. (1999) Maskless fabrication of light-directed oligonucleotide microarrays using a digital micromirror array. *Nat. Biotechnol.*, **17**, 974–978.

11 Wong, G.K., Passey, D.A., and Yu, J. (2001) Most of the human genome is transcribed. *Genome Res.*, **11**, 1975–1977.

12 Yamada, K., Lim, J., Dale, J.M., Chen, H., Shinn, P., Palm, C.J., Southwick, A.M., Wu, H.C., Kim, C., Nguyen, M., Pham, P., Cheuk, R., Karlin-Newmann, G., Liu, S.X., Lam, B., Sakano, H., Wu, T., Yu, G., Miranda, M., Quach, H.L., Tripp, M., Chang, C.H., Lee, J.M., Toriumi, M., Chan, M.M., Tang, C.C., Onodera, C.S., Deng, J.M., Akiyama, K., Ansari, Y., Arakawa, T., Banh, J., Banno, F., Bowser, L., Brooks, S., Carninci, P., Chao, Q., Choy, N., Enju, A., Goldsmith, A.D., Gurjal, M., Hansen, N.F., Hayashizaki, Y., Johnson-Hopson, C., Hsuan, V.W., Iida, K., Karnes, M., Khan, S., Koesema, E., Ishida, J., Jiang, P.X., Jones, T., Kawai, J., Kamiya, A., Meyers, C., Nakajima, M., Narusaka, M., Seki, M., Sakurai, T., Satou, M., Tamse, R., Vaysberg, M., Wallender, E.K., Wong, C., Yamamura, Y., Yuan, S., Shinozaki, K., Davis, R.W., Theologis, A., and Ecker, J.R. (2003) Empirical analysis of transcriptional activity in the *Arabidopsis* genome. *Science*, **302**, 842–846.

13 Ambros, V. (2004) The functions of animal microRNAs. *Nature*, **431**, 350–355.

14 Bartel, D.P. (2004) MicroRNAs: genomics, biogenesis, mechanism, and function. *Cell*, **116**, 281–297.

15 Lagos-Quintana, M., Rauhut, R., Yalcin, A., Meyer, J., Lendeckel, W., and Tuschl, T. (2002) Identification of tissue-specific microRNAs from mouse. *Curr. Biol.*, **12**, 735–739.

16 Lau, N.C., Lim, L.P., Weinstein, E.G., and Bartel, D.P. (2001) An abundant class of tiny RNAs with probable regulatory roles in *Caenorhabditis elegans*. *Science*, **294**, 858–862.

17 Fukao, T., Fukuda, Y., Kiga, K., Sharif, J., Hino, K., Enomoto, Y., Kawamura, A., Nakamura, K., Takeuchi, T., and Tanabe, M. (2007) An evolutionarily conserved mechanism for microRNA-223 expression revealed by microRNA gene profiling. *Cell*, **129**, 617–631.

18 Allen, E., Xie, Z., Gustafson, A.M., Sung, G.H., Spatafora, J.W., and Carrington, J.C. (2004) Evolution of microRNA genes by inverted duplication of target gene sequences in *Arabidopsis thaliana*. *Nat. Genet.*, **36**, 1282–1290.

19 Tanzer, A. and Stadler, P.F. (2004) Molecular evolution of a microRNA cluster. *J. Mol. Biol.*, **339**, 327–335.

20 Piriyapongsa, J. and Jordan, I.K. (2007) A family of human microRNA genes from miniature inverted-repeat transposable elements. *PLoS One*, **2**, e203.

21 Piriyapongsa, J., Marino-Ramirez, L., and Jordan, I.K. (2007) Origin and evolution of human microRNAs from transposable elements. *Genetics*, **176**, 1323–1337.

22 Smalheiser, N.R. and Torvik, V.I. (2005) Mammalian microRNAs derived from genomic repeats. *Trends Genet.*, **21**, 322–326.

23 Li, A. and Mao, L. (2007) Evolution of plant microRNA gene families. *Cell Res.*, **17**, 212–218.

24 Wang, J., Zhang, J., Zheng, H., Li, J., Liu, D., Li, H., Samudrala, R., Yu, J., and Wong, G.K. (2004) Mouse transcriptome: neutral evolution of 'non-coding' complementary DNAs. *Nature*, **431**, 757.

25 Chureau, C., Prissette, M., Bourdet, A., Barbe, V., Cattolico, L., Jones, L., Eggen, A., Avner, P., and Duret, L. (2002) Comparative sequence analysis of the X-inactivation center region in mouse, human, and bovine. *Genome Res.*, **12**, 894–908.

26 Hong, Y.K., Ontiveros, S.D., and Strauss, W.M. (2000) A revision of the human XIST gene organization and structural comparison with mouse Xist. *Mamm. Genome*, **11**, 220–224.

27 Nesterova, T.B., Slobodyanyuk, S.Y., Elisaphenko, E.A., Shevchenko, A.I., Johnston, C., Pavlova, M.E., Rogozin, I.B., Kolesnikov, N.N., Brockdorff, N., and Zakian, S.M. (2001) Characterization of the genomic Xist locus in rodents reveals conservation of overall gene structure and tandem repeats but rapid evolution of unique sequence. *Genome Res.*, **11**, 833–849.

28 Oudejans, C.B., Westerman, B., Wouters, D., Gooyer, S., Leegwater, P.A., van Wijk, I.J., and Sleutels, F. (2001) Allelic IGF2R repression does not correlate with expression of antisense RNA in human extraembryonic tissues. *Genomics*, **73**, 331–337.

29 Pang, K.C., Frith, M.C., and Mattick, J.S. (2006) Rapid evolution of noncoding RNAs: lack of conservation does not mean lack of function. *Trends Genet.*, **22**, 1–5.

30 Rong, Y.S. and Golic, K.G. (2001) A targeted gene knockout in *Drosophila*. *Genetics*, **157**, 1307–1312.

31 Dai, H., Chen, Y., Chen, S., Mao, Q., Kennedy, D., Landback, P., Eyre-Walker, A., Du, W., and Long, M. (2008) The evolution of courtship behaviors through the origination of a new gene in *Drosophila*. *Proc. Natl. Acad. Sci. U.S.A.*, **105**, 7478–7483.

32 Gavrilets, S. and Rice, W.R. (2006) Genetic models of homosexuality: generating testable predictions. *Proc. Biol. Sci.*, **273**, 3031–3038.

33 Duret, L., Chureau, C., Samain, S., Weissenbach, J., and Avner, P. (2006) The Xist RNA gene evolved in eutherians by pseudogenization of a protein-coding gene. *Science*, **312**, 1653–1655.

34 Pollard, K.S., Salama, S.R., Lambert, N., Lambot, M.A., Coppens, S., Pedersen, J.S., Katzman, S., King, B., Onodera, C., Siepel, A., Kern, A.D., Dehay, C., Igel, H., Ares, M., Jr., Vanderhaeghen, P., and Haussler, D. (2006) An RNA gene expressed during cortical development evolved rapidly in humans. *Nature*, **443**, 167–172.

35 Lu, J., Shen, Y., Wu, Q., Kumar, S., He, B., Shi, S., Carthew, R.W., Wang, S.M., and Wu, C.I. (2008) The birth and death of microRNA genes in *Drosophila*. *Nat. Genet.*, **40**, 351–355.

36 Kimura, M. (1983) Rare variant alleles in the light of the neutral theory. *Mol. Biol. Evol.*, **1**, 84–93.

37 Ohta, T. (1995) Synonymous and nonsynonymous substitutions in mammalian genes and the nearly neutral theory. *J. Mol. Evol.*, **40**, 56–63.

38 Gilad, Y., Oshlack, A., and Rifkin, S.A. (2006) Natural selection on gene expression. *Trends Genet.*, **22**, 456–461.

39 Khaitovich, P., Paabo, S., and Weiss, G. (2005) Toward a neutral evolutionary model of gene expression. *Genetics*, **170**, 929–939.

40 Khaitovich, P., Enard, W., Lachmann, M., and Paabo, S. (2006) Evolution of primate

gene expression. *Nat. Rev. Genet.*, **7**, 693–702.

41 Khaitovich, P., Tang, K., Franz, H., Kelso, J., Hellmann, I., Enard, W., Lachmann, M., and Paabo, S. (2006) Positive selection on gene expression in the human brain. *Curr. Biol.*, **16**, R356–R358.

42 Rifkin, S.A., Kim, J., and White, K.P. (2003) Evolution of gene expression in the *Drosophila melanogaster* subgroup. *Nat. Genet.*, **33**, 138–144.

43 Ranz, J.M., Castillo-Davis, C.I., Meiklejohn, C.D., and Hartl, D.L. (2003) Sex-dependent gene expression and evolution of the *Drosophila* transcriptome. *Science*, **300**, 1742–1745.

44 Mackay, T.F. (2004) The genetic architecture of quantitative traits: lessons from *Drosophila*. *Curr. Opin. Genet. Dev.*, **14**, 253–257.

45 Johnson, J.M., Castle, J., Garrett-Engele, P., Kan, Z., Loerch, P.M., Armour, C.D., Santos, R., Schadt, E.E., Stoughton, R., and Shoemaker, D.D. (2003) Genome-wide survey of human alternative pre-mRNA splicing with exon junction microarrays. *Science*, **302**, 2141–2144.

46 Nurtdinov, R.N., Artamonova, I.I., Mironov, A.A., and Gelfand, M.S. (2003) Low conservation of alternative splicing patterns in the human and mouse genomes. *Hum. Mol. Genet.*, **12**, 1313–1320.

47 Thanaraj, T.A., Clark, F., and Muilu, J. (2003) Conservation of human alternative splice events in mouse. *Nucleic Acids Res.*, **31**, 2544–2552.

48 Boue, S., Letunic, I., and Bork, P. (2003) Alternative splicing and evolution. *Bioessays*, **25**, 1031–1034.

49 Modrek, B. and Lee, C.J. (2003) Alternative splicing in the human, mouse and rat genomes is associated with an increased frequency of exon creation and/or loss. *Nat. Genet.*, **34**, 177–180.

50 Modrek, B. and Lee, C. (2002) A genomic view of alternative splicing. *Nat. Genet.*, **30**, 13–19.

51 Modrek, B., Resch, A., Grasso, C., and Lee, C. (2001) Genome-wide detection of alternative splicing in expressed sequences of human genes. *Nucleic Acids Res.*, **29**, 2850–2859.

52 Mironov, A.A., Fickett, J.W., and Gelfand, M.S. (1999) Frequent alternative splicing of human genes. *Genome Res.*, **9**, 1288–1293.

53 Terai, Y., Morikawa, N., Kawakami, K., and Okada, N. (2003) The complexity of alternative splicing of hagoromo mRNAs is increased in an explosively speciated lineage in East African cichlids. *Proc. Natl. Acad. Sci. U.S.A.*, **100**, 12798–12803.

54 Sorek, R., Shamir, R., and Ast, G. (2004) How prevalent is functional alternative splicing in the human genome? *Trends Genet.*, **20**, 68–71.

55 Eng, F.J. and Warner, J.R. (1991) Structural basis for the regulation of splicing of a yeast messenger RNA. *Cell*, **65**, 797–804.

56 Goguel, V. and Rosbash, M. (1993) Splice site choice and splicing efficiency are positively influenced by pre-mRNA intramolecular base pairing in yeast. *Cell*, **72**, 893–901.

57 Libri, D., Piseri, A., and Fiszman, M.Y. (1991) Tissue-specific splicing *in vivo* of the beta-tropomyosin gene: dependence on an RNA secondary structure. *Science*, **252**, 1842–1845.

58 Chen, Y. and Stephan, W. (2003) Compensatory evolution of a precursor messenger RNA secondary structure in the *Drosophila melanogaster* Adh gene. *Proc. Natl. Acad. Sci. U.S.A.*, **100**, 11499–11504.

59 Kirby, D.A., Muse, S.V., and Stephan, W. (1995) Maintenance of pre-mRNA secondary structure by epistatic selection. *Proc. Natl. Acad. Sci. U.S.A.*, **92**, 9047–9051.

60 Matzke, M.A. and Matzke, A. (1995) How and why do plants inactivate homologous (trans)genes? *Plant Physiol.*, **107**, 679–685.

61 Almeida, R. and Allshire, R.C. (2005) RNA silencing and genome regulation. *Trends Cell Biol.*, **15**, 251–258.

62 Hammond, S.M. (2005) Dicing and slicing: the core machinery of the RNA interference pathway. *FEBS Lett.*, **579**, 5822–5829.

63 Wassenegger, M. and Krczal, G. (2006) Nomenclature and functions of RNA-directed RNA polymerases. *Trends Plant Sci.*, **11**, 142–151.

64 Cerutti, H. and Casas-Mollano, J.A. (2006) On the origin and functions of RNA-mediated silencing: from protists to man. *Curr. Genet.*, **50**, 81–99.

65 Chapman, E.J. and Carrington, J.C. (2007) Specialization and evolution of endogenous small RNA pathways. *Nat. Rev. Genet.*, **8**, 884–896.

66 Fahlgren, N., Howell, M.D., Kasschau, K.D., Chapman, E.J., Sullivan, C.M., Cumbie, J.S., Givan, S.A., Law, T.F., Grant, S.R., Dangl, J.L., and Carrington, J.C. (2007) High-throughput sequencing of *Arabidopsis* microRNAs: evidence for frequent birth and death of MIRNA genes. *PLoS One*, **2**, e219.

67 Rajagopalan, R., Vaucheret, H., Trejo, J., and Bartel, D.P. (2006) A diverse and evolutionarily fluid set of microRNAs in *Arabidopsis thaliana. Genes Dev.*, **20**, 3407–3425.

68 Axtell, M.J., Snyder, J.A., and Bartel, D.P. (2007) Common functions for diverse small RNAs of land plants. *Plant Cell*, **19**, 1750–1769.

69 Hutvagner, G. and Simard, M.J. (2008) Argonaute proteins: key players in RNA silencing. *Nat. Rev. Mol. Cell Biol.*, **9**, 22–32.

70 Vaucheret, H., Beclin, C., and Fagard, M. (2001) Post-transcriptional gene silencing in plants. *J. Cell Sci.*, **114**, 3083–3091.

71 Tam, O.H., Aravin, A.A., Stein, P., Girard, A., Murchison, E., Cheloufi, S., Hodges, E., Anger, M., Sachidanandam, R., Schultz, R.M., and Hannon, G.J. (2008) Pseudogene-derived small interfering RNAs regulate gene expression in mouse oocytes. *Nature*, **453**, 534–538.

72 Birney, E., Stamatoyannopoulos, J.A., Dutta, A., Guigó, R., Gingeras, T.R., Margulies, E.H., Weng, Z., Snyder, M., Dermitzakis, E.T., Thurman, R.E., *et al.* (2007) Identification and analysis of functional elements in 1% of the human genome by the ENCODE pilot project. *Nature*, **447**, 799–816.

73 Li, Z., Liu, M., Zhang, L., Zhang, W., Gao, G., Zhu, Z., Wei, L., Fan, Q., and Long, M. (2009) Detection of intergenic non-coding RNAs expressed in the main developmental stages in *Drosophila melanogaster. Nucleic Acids Res.*, **37**, 4308–4314.

74 Manak, J.R., Dike, S., Sementchenko, V., Kapranov, P., Biemar, F., Long, J., Cheng, J., Bell, I., Ghosh, S., Piccolboni, A., and Gingeras, T.R. (2006) Biological function of unannotated transcription during the early development of *Drosophila melanogaster. Nat. Genet.*, **38**, 1151–1158.

75 Wilusz, J.E., Sunwoo, H., and Spector, D.L. (2009) Long noncoding RNAs: functional surprises from the RNA world. *Genes Dev.*, **23**, 1494–1504.

76 Tsai, M.C., Manor, O., Wan, Y., Mosammaparast, N., Wang, J.K., Lan, F., Shi, Y., Segal, E., and Chang, H.Y. (2010) Long noncoding RNA as modular scaffold of histone modification complexes. *Science*, **329**, 689–693.

Index

Note: Page references in *italics* denote figures; those in **bold** refer to tables.

Posttranscriptional Gene Regulation: RNA Processing in Eukaryotes, First Edition. Edited by Jane Wu.

t